Fluid Power
with Applications

Fluid Power

with Applications

Third Edition

Anthony Esposito

Professor Emeritus
Department of Manufacturing Engineering
Miami University
Oxford, Ohio

REGENTS/PRENTICE HALL
Englewood Cliffs, New Jersey 07632

Library of Congress Cataloging-in-Publication Data

Esposito, Anthony
   Fluid power with applications / Anthony Esposito. — 3rd ed.
      p.   cm.
   Includes index.
   ISBN 0-13-772468-3
   1. Fluid power technology.   I. Title.
TJ843.E86   1993
620.1′06—dc20   93-4031
                                                       CIP

Acquisitions Editor: Rob Koehler
Cover Design: Marianne Frasco
Prepress Buyer: Ilene Levy Sanford
Manufacturing Buyer: Ed O'Dougherty
Marketing Manager: Ramona Baron

© 1994, 1988, 1980 by Regents/Prentice Hall
A Division of Simon & Schuster
Englewood Cliffs, New Jersey 07632

Printed in the United States of America

10   9   8   7   6   5   4   3   2   1

0-13-772468-3

Prentice-Hall International (UK) Limited, *London*
Prentice-Hall of Australia Pty. Limited, *Sydney*
Prentice-Hall Canada Inc., *Toronto*
Prentice-Hall Hispanoamericana, S.A., *Mexico*
Prentice-Hall of India Private Limited, *New Delhi*
Prentice-Hall of Japan, Inc., *Tokyo*
Simon & Schuster Asia Pte. Ltd., *Singapore*
Editora Prentice-Hall do Brasil, Ltda., *Rio de Janeiro*

*To my wife, Mary,
and to my grandchildren:
Carly, Chelsea, and Kevin*

Contents

3 *ENERGY AND POWER IN HYDRAULIC SYSTEMS*   **78**

4 *THE DISTRIBUTION SYSTEM*   **119**

5   BASICS OF HYDRAULIC FLOW IN PIPES   142

6   THE SOURCE OF HYDRAULIC POWER: PUMPS   175

7   FLUID POWER ACTUATORS   220

8 CONTROL COMPONENTS IN HYDRAULIC SYSTEMS   273

9 HYDRAULIC CIRCUIT DESIGN AND ANALYSIS   322

10 PNEUMATIC COMPONENTS AND CIRCUITS   369

11 FLUID LOGIC CONTROL SYSTEMS   422

INDEX

Preface

The primary purpose of the third edition of *Fluid Power with Applications* remains the same as that of the previous editions: to provide the student with a sound, basic background in the vast field of fluid power. As such, this book covers those subjects essential to understanding the design, analysis, operation, and maintenance of fluid power systems. Similarly, it is written for engineering technologists, engineering technicians, and apprentices of industrial training programs.

As in the previous editions, although theory is presented where desirable, the emphasis is placed on understanding how and why fluid power systems operate and on practical applications as well. In this way, the student learns not only the "how" but also the "why" of fluid power system operation.

Based on input from users of the second edition and from my colleagues, the following major changes have been incorporated in this book:

1. A section (12.11) on programmable logic controllers (PLCs) has been added. This reflects the current trend toward increasing use of PLCs in industrial control applications of fluid power. Unlike general-purpose computers, PLCs are designed to operate in industrial environments where high ambient temperature and humidity levels may exist, as is typically the case for fluid power applications. Unlike electromechanical relays, PLCs are not hardwired to perform specific functions. Thus, when system operation requirements change, a PLC software program is readily changed instead of having to physically rewire relays.

2. A section on cartridge valves (8.6) has been added. Market pressures and worldwide competition make the need for more efficient and economical

hydraulic systems greater than ever. The use of cartridge valves with manifolds to produce integrated hydraulic circuits offers a proven way to achieve these improvements. Advantages include reduced oil leakages and contamination due to fewer fittings, less system installation and service time, and smaller space requirements of overall systems.

3. Numerous changes have been made to the end-of-chapter exercises. Many additional exercises of varied difficulty and applications have been added to provide the student with the necessary practice to master concepts and problem-solving techniques. Specifically, the total number of exercises has been increased from 437 to 601. The instructor can adapt many of these exercises for student examination purposes.

4. A section (13.5) on the Beta ratio of filters has been added. Beta ratio is defined mathematically and is based on laboratory test results. The use of the Beta ratio parameter provides a better means for establishing how well a given filter will trap contaminants in comparison to using filter ratings in nominal and absolute values.

5. Many photographs and illustrations have been updated to reflect current fluid power technology. These include the areas of programmable logic controllers, microprocessors, electrohydraulic servo systems, and integrated systems.

As in the case of the previous editions, I am indebted to the numerous fluid power equipment manufacturing companies for permitting the inclusion of their photographs and other illustrations in this textbook.

I also wish to thank the users of the previous editions for their many constructive suggestions, which have been incorporated in this textbook.

Anthony Esposito

1

Introduction to
Fluid Power

1.1 WHAT IS FLUID POWER?

Fluid power is the technology that deals with the generation, control, and transmission of power-using pressurized fluids. It can be said that fluid power is the muscle that moves industry. This is because fluid power is used to push, pull, regulate, or drive virtually all the machines of modern industry. For example, fluid power steers and brakes automobiles, launches spacecraft, moves earth, harvests crops, mines coal, drives machine tools, controls airplanes, processes food, and even drills teeth. In fact, it is almost impossible to find a manufactured product that hasn't been ''fluid-powered'' in some way at some stage of its production or distribution.

Since a fluid can be either a liquid or a gas, fluid power is actually the general term used for hydraulics and pneumatics. Hydraulic systems use liquids such as petroleum oils, water, synthetic oils, and even molten metals. The first hydraulic fluid to be used was water because it is readily available. However, water has many deficiencies. It freezes readily, is a relatively poor lubricant, and tends to rust metal components. Hydraulic oils are far superior and hence are widely used in lieu of water. Pneumatic systems use air as the gas medium because air is very abundant and can be readily exhausted into the atmosphere after completing its assigned task.

It should be realized that there are actually two different types of fluid systems: fluid transport and fluid power.

Fluid transport systems have as their sole objective the delivery of a fluid from one location to another to accomplish some useful purpose. Examples include pumping stations for pumping water to homes, cross-country gas lines, and systems where chemical processing takes place as various fluids are brought together.

Fluid power systems are designed specifically to perform work. The work is accomplished by a pressurized fluid bearing directly on an operating cylinder or fluid motor, which, in turn, provides the muscle to do the desired work. Of course, control components are also needed to ensure that the work is done smoothly, accurately, efficiently, and safely.

Liquids provide a very rigid medium for transmitting power and thus can provide huge forces to move loads with utmost accuracy and precision. On the other hand, pneumatic systems exhibit spongy characteristics due to the compressibility of air. However, pneumatic systems are less expensive to build and operate. In addition, provisions can be made to control the operation of the pneumatic actuators that drive the loads.

Fluid power equipment ranges in size from huge hydraulic presses to miniature fluid logic components used to build reliable control systems.

How versatile is fluid power? In terms of brute power, a feather touch by an operator can control hundreds of horsepower and transmit it to any location where a hose or pipe can go. In terms of precision such as applications in the machine tool industry, tolerances of one ten-thousandth of an inch can be achieved and repeated over and over again. Fluid power is not merely a powerful muscle; it is a controlled, flexible muscle that provides power smoothly, efficiently, safely, and precisely to accomplish useful work.

Figure 1-1 shows a pneumatically controlled dextrous hand designed to study machine dexterity and human manipulation in applications such as robotics and tactile sensing. Servo-controlled pneumatic actuators give the hand human-like grasping and manipulating capability. Key operating characteristics include high speed in performing manipulation tasks, strength to easily grasp hand-sized objects that have varying densities, and force grasping control. The hand possesses three fingers and an opposing thumb, each with four degrees of freedom. Each joint is positioned by two pneumatic actuators (located in an actuator pack with the controlling servo valve) driving a high-strength tendon. Performance and configuration constraints concerning the weight, size, geometry, cleanliness, and availability of individual actuators led to the choice of pneumatic actuation.

1.2 HISTORY OF FLUID POWER

Fluid power is probably as old as civilization itself. Ancient historical accounts show that water was used for centuries to produce power by means of waterwheels, and air was used to turn windmills and propel ships. However, these early uses of fluid power required the movement of huge quantities of fluid because of the relatively low pressures provided by nature.

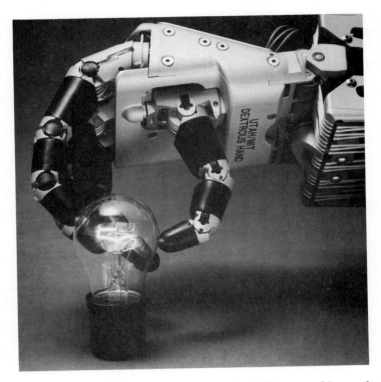

Figure 1-1.   Pneumatically controlled dextrous hand. (*Courtesy of Sarcos, Inc., Salt Lake City, Utah.*)

Fluid power technology actually began in 1650 with the discovery of Pascal's law: *Pressure is transmitted undiminished in a confined body of fluid.*

Pascal found that when he rammed a cork down into a jug completely full of wine, the bottom of the jug broke and fell out. Pascal's law indicated that the pressures were equal at the top and bottom of the jug. However, the jug has a small opening area at the top and a large area at the bottom. Thus, the bottom absorbs a greater force due to its larger area.

In 1750, Bernoulli developed his law of conservation of energy for a fluid flowing in a pipeline. Pascal's law and Bernoulli's law operate at the very heart of all fluid power applications and are used for analysis purposes. However, it was not until the Industrial Revolution of 1850 in Great Britain that these laws would actually be applied to industry. Up to this point in time, electrical energy had not been developed to power the machines of industry. Instead, it was fluid power that, by 1870, was being used to drive hydraulic equipment such as cranes, presses, winches, extruding machines, hydraulic jacks, shearing machines, and riveting machines. In these systems, steam engines drove hydraulic water pumps, which delivered water at moderate pressures through pipes to industrial plants for powering the various machines. These early hydraulic systems had a number of deficiencies such as sealing problems because the designs had evolved more as an art than a science.

Then, late in the nineteenth century, electricity emerged as a dominant technology. This resulted in a shift of development effort away from fluid power. Electrical power was soon found to be superior to hydraulics for transmitting power over great distances. There was very little development in fluid power technology during the last 10 years of the nineteenth century.

The modern era of fluid power is considered to have begun in 1906 when a hydraulic system was developed to replace electrical systems for elevating and controlling guns on the battleship USS *Virginia*. For this application, the hydrau-

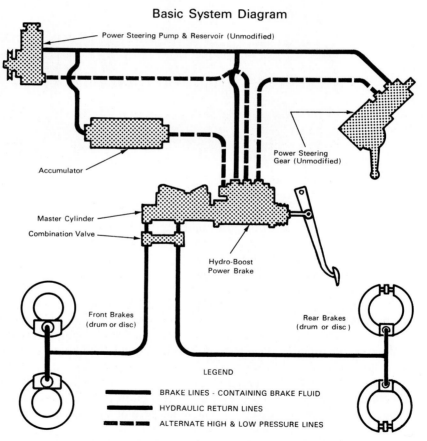

Figure 1-2. Bendix Hydro-Boost Power Brake System. (*Courtesy of Bendix Corp., South Bend, Indiana.*)

lic system developed used oil instead of water. This change in hydraulic fluid and the subsequent solution of sealing problems were significant milestones in the rebirth of fluid power.

In 1926 the United States developed the first unitized, packaged hydraulic system consisting of a pump, controls, and actuator. The military requirements leading up to World War II kept fluid power applications and developments going at a good pace. The naval industry had used fluid power for cargo handling, winches, propeller pitch control, submarine control systems, operation of shipboard aircraft elevators, and drive systems for radar and sonar.

The demands of World War II and the expanding economy that followed led to the present situation where there are virtually a limitless number of fluid power applications. Today fluid power is used extensively in practically every branch of industry. Some typical applications are in automobiles, tractors, airplanes, missiles, boats, and machine tools. In the automobile alone, fluid power is utilized in hydraulic brakes, automotive transmissions, power steering, power brakes, air conditioning, lubrication, water coolant, and gasoline pumping systems.

Relative to automotive applications, Fig. 1-2 is a diagram showing the Bendix Hydro-Boost Power Brake System. The basic system consists of an open center spool valve and hydraulic cylinder assembled in a single unit (see Fig. 1-3). Operating pressure is supplied by the power steering pump. Hydro-Boost provides a power assist to operate a dual master-cylinder braking system. Normally mounted on the engine compartment fire wall, it is designed to provide specific "brake-feel" characteristics throughout a wide range of pedal forces and travel. A spring accumulator stores energy for reverse stops. From one to three stops are available depending on the magnitude and duration of the brake application. This system was developed by Bendix Corporation as an answer to crowded engine compartments and replaces the larger vacuum units.

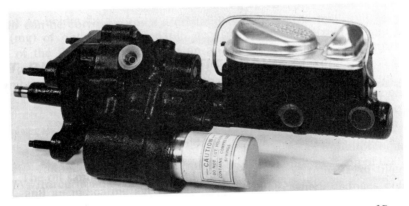

Figure 1-3.   Bendix Hydro-Boost attached to master cylinder. (*Courtesy of Bendix Corp., South Bend, Indiana.*)

1.3 ADVANTAGES OF FLUID POWER

There are only three basic methods of transmitting power: electrical, mechanical, and fluid power. Most applications actually use a combination of the three methods to obtain the most efficient overall system. To properly determine which principle method to use, it is important to know the salient features of each type. For example, fluid systems can transmit power more economically over greater distances than can mechanical types. However, fluid systems are restricted to shorter distances than are electrical systems.

The secret of fluid power's success and widespread use is its versatility and manageability. Fluid power is not hindered by the geometry of the machine as is the case in mechanical systems. Also, power can be transmitted in almost limitless quantities because fluid systems are not so limited by the physical limitations of materials as are the electrical systems. For example, the performance of an electromagnet is limited by the saturation limit of steel. On the other hand, the power limit of fluid systems is limited only by the strength capacity of the material.

Industry is going to depend more and more on automation in order to increase productivity. This includes remote and direct control of production operations, manufacturing processes, and materials handling. Fluid power is the muscle of automation because of advantages in the following four major categories.

1. Ease and accuracy of control. By the use of simple levers and push buttons, the operator of a fluid power system can readily start, stop, speed up or slow down, and position forces that provide any desired horsepower with tolerances as precise as one ten-thousandth of an inch. Figure 1-4 shows a fluid power system

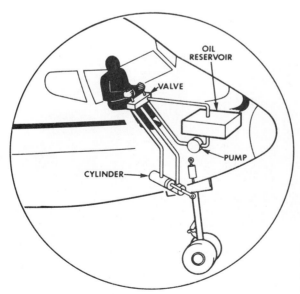

Figure 1-4. Hydraulic operation of aircraft landing gear. (*Courtesy of National Fluid Power Association, Milwaukee, Wisconsin.*)

that allows an aircraft pilot to raise and lower his landing gear. When the pilot moves a small control valve in one direction, oil under pressure flows to one end of the cylinder to lower the landing gear. To retract the landing gear, the pilot moves the valve lever in the opposite direction, allowing oil to flow into the other end of the cylinder.

2. Multiplication of force. A fluid power system (without using cumbersome gears, pulleys, and levers) can multiply forces simply and efficiently from a fraction of an ounce to several hundred tons of output. Figure 1-5 shows an application where a rugged, powerful drive is required for handling huge logs. In this case, a turntable, which is driven by a hydraulic motor, can carry a 20,000-lb load at a 10-ft radius under rough operating conditions.

3. Constant force or torque. Only fluid power systems are capable of providing constant force or torque regardless of speed changes. This is accomplished whether the work output moves a few inches per hour, several hundred inches per minute, a few revolutions per hour, or thousands of revolutions per minute. Figure 1-6 depicts an application in oceanography that involves the exploration and development of the ocean's resources for the benefit of mankind. In this instance, it is important for the operator to apply a desired constant grabbing force through the use of the grappling hooks.

Figure 1-5. Hydraulically driven turntable for handling huge logs. (*Courtesy of Eaton Corp., Fluid Power Division, Eden Prairie, Minnesota.*)

Figure 1-6. Fluid power application in oceanography. (*Courtesy of National Fluid Power Association, Milwaukee, Wisconsin.*)

4. Simplicity, safety, economy. In general, fluid power systems use fewer moving parts than comparable mechanical or electrical systems. Thus, they are simpler to maintain and operate. This, in turn, maximizes safety, compactness, and reliability. Figure 1-7 shows a power steering control designed for off-highway vehicles. The steering unit (shown attached to the steering wheel column in Fig. 1-7) consists of a manually operated directional control valve and meter in a single body. See Fig. 1-8 for a cutaway of this steering unit. Because the steering unit is fully fluid-linked, mechanical linkages, universal joints, bearings, reduction gears, etc., are eliminated. This provides a simple, compact system. In addition, very

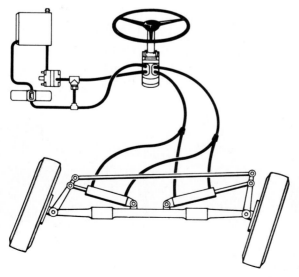

Figure 1-7. Power steering control system for off-highway vehicles. Typical Orbitrol Power Steering Systems. (*Courtesy of Eaton Corp., Fluid Power Division, Eden Prairie, Minnesota.*)

Figure 1-8.   Cutaway of steering unit used in power steering control systems. (*Courtesy of Eaton Corp., Fluid Power Division, Eden Prairie, Minnesota.*)

little input torque is required to produce the control needed for the toughest applications. This is important where limitations of control space require a small steering wheel and it becomes necessary to reduce operator fatigue. The compact design and versatility of the control system allow the unit to control many large and high-powered systems with a high degree of reliability. The steering unit shown in Fig. 1-8 contains a check valve that converts the unit to a hand-operated pump for emergency power-off steering.

Additional benefits of fluid power systems include instantly reversible motion, automatic protection against overloads, and infinitely variable speed control. Fluid power systems also have the highest horsepower per weight ratio of any known power source.

In spite of all these highly desirable features of fluid power, it is not a panacea for all power transmission problems. Hydraulic systems also have some drawbacks. Hydraulic oils are messy, and leakage is impossible to eliminate completely. Hydraulic lines can burst, possibly resulting in human injury due to flying objects, if proper design is not implemented. Also, most hydraulic oils can cause fires if an oil leak occurs in an area of hot equipment. Therefore, each application must be studied thoroughly to determine the best overall design. It is hoped that this book will not only assist the reader in developing the ability to make these types of system selection decisions but also present in a straightforward way the techniques for designing, analyzing, and troubleshooting basic fluid power systems.

1.4 APPLICATIONS OF FLUID POWER

Although a number of cases of fluid power have already been presented in this chapter, the following additional applications should give the reader a broader view of the widespread use of fluid power in today's world.

1. Fluid power drives high-wire overhead tram. Most overhead trams require a haulage or tow cable to travel up or down steep inclines. However, the 22-passenger, 12,000-lb hydraulically powered and controlled Skytram shown in Fig. 1-9 is unique. It is self-propelled and travels on a stationary cable. Because the tram moves instead of the cable, the operator can stop, start, and reverse any one car completely independently of any other car in the tram system. Integral to the design of the Skytram drive is a pump (driven by a standard eight-cylinder gasoline engine), which supplies pressurized fluid to four hydraulic motors. Each motor drives two friction drive wheels.

Eight drive wheels on top of the cables support and propel the tram car. On steep inclines, high driving torque is required for ascent and high braking torque for descent. Dual compensation of the four hydraulic motors provides efficient proportioning of available horsepower to meet the variable torque demands.

2. Fluid power is applied to harvest corn. The world's dependence on the United States for food has resulted in a great demand for agricultural equipment development. Fluid power is being applied to solve many of the problems dealing with the harvesting of food crops. Figure 1-10 shows a hydraulically driven elevator conveyor system, which is used to send harvested, husked ears of corn to a

Figure 1-9. Hydraulically powered and controlled Skytram. Hydraulic Motor Dual Compensation Feature, designed by Les Haisen, Sperry Vickers, Torrance, California. (*Courtesy of Skytram Systems, Inc., Scottsdale, Arizona.*)

Figure 1-10. Hydraulically driven elevator conveyor for use in harvesting of corn. (*Courtesy of Eaton Corp., Fluid Power Division, Eden Prairie, Minnesota.*)

wagon trailer. Mounted directly to the chain-drive conveyor, a hydraulic motor delivers full-torque rotary power from start-up to full rpm.

3. Hydraulics power brush drives. Figure 1-11 shows a fluid-power-driven brush drive used for cleaning roads, floors, etc., in various industrial locations. Mounted directly at the hub of the front and side sweep-scrub brushes, compact

Figure 1-11. Hydraulic power brush drive. (*Courtesy of Eaton Corp., Fluid Power Division, Eden Prairie, Minnesota.*)

hydraulic motors place power right where it's needed. They eliminate bulky mechanical linkages for efficient, lightweight machine design. The result is continuous, rugged industrial cleaning action at the flip of a simple valve.

4. Fluid power positions and holds parts for welding. In Fig. 1-12, we see an example of a welding operation in which a farm equipment manufacturer applied hydraulics for positioning and holding parts while welding is done. It is a typical example of how fluid power can be used in manufacturing and production operations to reduce costs and increase production. This particular application required a sequencing system for fast, positive holding. This was accomplished by placing a restrictor (sequence valve) on the flow of oil in the line leading to the second of the two cylinders (rams), as illustrated in Fig. 1-13. The first cylinder extends to the end of its stroke. Oil pressure then builds up, overcoming the restrictor setting, and the second cylinder extends to complete the "hold" cycle. This unique welding application of hydraulics was initiated to increase productivity by making more parts per hour. In addition, the use of hydraulics reduced scrap rates and operator fatigue as well as increasing productivity from 5 pieces per shift to more than 20—a 400% increase.

5. Fluid power performs bridge maintenance. A municipality had used fluid power for years as a means for removing stress from structural members of bridges, making repairs, and replacing beams. As many as four or five bulky, low-pressure hand pumps and jacking ram setups were used to remove stress from beams needing replacement. Labor costs were high, and no accurate methods

Figure 1-12. Welding equipment using hydraulic controls. (*Courtesy of Owatonna Tool Co., Owatonna, Minnesota.*)

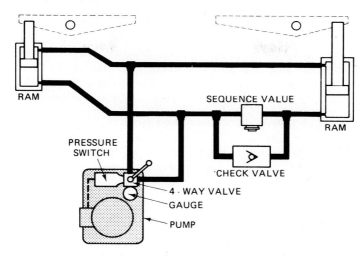

Figure 1-13. Hydraulic circuit for welding application. (*Courtesy of Owatonna Tool Co., Owatonna, Minnesota.*)

existed for recording pressures. An excessive downtime problem dictated that a new system be designed for the job. A modern fluid power system was designed that located several 100-ton rams on the bridge structure, as illustrated in Fig. 1-14. One portable pump was used to actuate all of the rams by the use of a special manifold. This made it easy to remove stress from members needing repair or replacement. This new fluid power system cut the setup time and labor costs for each repair job to one-third that required with the hand pump and jacking ram setups previously used.

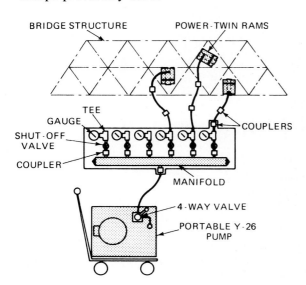

Figure 1-14. Fluid power system for relieving stresses in structural members of bridge. (*Courtesy of Owatonna Tool Co., Owatonna, Minnesota.*)

6. Fluid power is the muscle in industrial lift trucks. Figure 1-15 shows an industrial hydraulic lift truck having a 5000-lb capacity. The hydraulic system includes dual-action tilt cylinders and a hoist cylinder. Tilting action is smooth and sure for better load stability and easier load placement. A lowering valve in the hoist cylinder controls the speed of descent even if the hydraulic circuit is broken. Hydrostatic power steering is available as an optional feature.

7. Fluid power drives front-end loaders. Figure 1-16 shows a front-end loader filling a dump truck with soil scooped up by a large hydraulic-powered bucket. Excellent load control is made possible with a specially designed flow control valve. The result is low effort and precise control; this keeps the operator

Figure 1-15.   Industrial hydraulic lift truck. (*Courtesy of Eaton Corp., Industrial Truck Division, Philadelphia, Pennsylvania.*)

Figure 1-16.   Hydraulic-powered front-end loader. (*Courtesy of Koehring Husco Division, Waukesha, Wisconsin.*)

working on the job longer and more efficiently. Thus, reduced operator fatigue is accompanied by increased production.

   8. Fluid power preserves the heartbeat of life. Dr. Robert Jarvik made medical history with the design of an artificial, pneumatically actuated heart, which sustained the life of Dr. Barney Clark for over 100 days. Figure 1-17 shows this air-driven artificial heart that preserves the very heartbeat of life. Other health applications include artificial kidneys and valve-assisted bladders, which employ fluid power principles of pressure and flow. Miniature, oxygen-tight pumps are implanted in patients to provide continuous medication. These microdelivery pumps can either be permanent for internal use or disposable for external infusion of medicine.

   9. Hydraulics power robotic dextrous arm. Figure 1-18 shows a hydraulically powered robotic arm that has the strength and dexterity to torque down bolts with its fingers and yet can gingerly pick up an eggshell. This robotic arm is adept at using human tools such as hammers, electric drills, and tweezers and can even bat a baseball. The arm has a hand with a thumb and two fingers, as well as a wrist, elbow, and shoulder. It has ten degrees of freedom including a three-degree-of-freedom end effector (hand) designed to handle human tools and other objects

Figure 1-17. Dr. Robert Jarvik's air-driven artificial heart. (*Courtesy of Symbion, Inc., Salt Lake City, Utah.*)

with humanlike dexterity. The servo control system is capable of accepting computer or human operator control inputs. The system can be designed for carrying out hazardous applications in the subsea, utilities, or nuclear environments, and it is also available in a range of sizes from human proportions to six feet long.

1.5 COMPONENTS OF A FLUID POWER SYSTEM

Virtually all hydraulic circuits are essentially the same regardless of the application. There are six basic components required in a hydraulic circuit (refer to Fig. 1-19):

1. A tank (reservoir) to hold the liquid, which is usually hydraulic oil.
2. A pump to force the liquid through the system.
3. An electric motor or other power source to drive the pump.
4. Valves to control liquid direction, pressure, and flow rate.
5. An actuator to convert the energy of the liquid into mechanical force or torque to do useful work. Actuators can either be cylinders to provide linear

Figure 1-18. Hydraulically Powered Dextrous Arm. (*Courtesy of Sarcos, Inc., Salt Lake City, Utah.*)

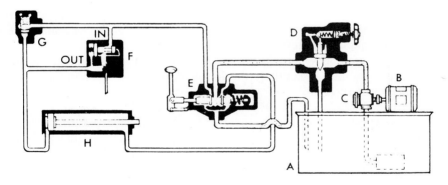

List of Components

A—Reservoir
B—Electric Motor
C—Pump
D—Maximum Pressure
    (Relief) Valve

E—Directional Value
F—Flow Control Valve
G—Right-Angle
       Check Valve
H—Cylinder

Figure 1-19. Basic hydraulic system with linear hydraulic actuator (cylinder). (*Courtesy of Sperry Vickers, Sperry Rand Corp., Troy, Michigan.*)

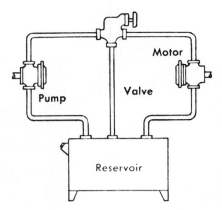

Figure 1-20. Basic hydraulic system with rotary hydraulic actuator (motor). (*Courtesy of Sperry Vickers, Sperry Rand Corp., Troy, Michigan.*)

motion such as shown in Fig. 1-19 or motors (hydraulic) to provide rotary motion as shown in Fig. 1-20.

6. Piping, which carries the liquid from one location to another.

Of course, the sophistication and complexity of hydraulic systems will vary depending on the specific application. This is also true of the individual components that comprise the hydraulic system. As an example, refer to Fig. 1-21, which shows two different-sized, complete, hydraulic power units designed for two uniquely different applications. Each unit is a complete, packaged power system containing its own electric motor, pump, shaft coupling, reservoir and miscellaneous piping, pressure gages, valves, and other components as required for proper operation. These hydraulic components and systems are studied in detail in subsequent chapters.

Pneumatic systems have components that are similar to those used in hydraulic circuits. Essentially the following six basic components are required for pneumatic circuits:

1. An air tank to store a given volume of compressed air

2. A compressor to compress the air that comes directly from the atmosphere

3. An electric motor or other prime mover to drive the compressor

4. Valves to control air direction, pressure, and flow rate

5. Actuators, which are similar in operation to hydraulic actuators

6. Piping to carry the pressurized air from one location to another

Figure 1-22 shows a compact, self-contained pneumatic power unit complete with tank, compressor, electric motor, and miscellaneous components such as valves, piping, and pressure gages.

It should be noted in pneumatic systems that after the pressurized air is spent driving actuators, it is then exhausted back into the atmosphere. On the

Figure 1-21.   Two different-sized, complete, hydraulic power units. (*Courtesy of Continental Hydraulics, Division of Continental Machines, Inc., Savage, Minnesota.*)

other hand, in hydraulic systems the spent oil drains back to the reservoir and is repeatedly reused after being repressurized by the pump as needed by the system.

1.6 TYPES OF FLUID POWER SYSTEMS

We have already established that fluid power includes both hydraulic and pneumatic systems. The field of fluid power can be further subdivided into two additional major categories: open-loop and closed-loop systems.

A closed-loop system is one that uses feedback. This means that the state of the output from the system is automatically sampled and compared (fed back) to the input or command signal by means of a device called a *feedback transducer*. If there is a difference between the command and feedback signals, action is taken to correct the system output until it matches the requirement imposed on the system. Closed-loop systems are frequently called *servo systems*, and the valves used to direct fluid to the actuators are typically called *servo valves*.

An open-loop system does not use feedback. The output performance of the system, therefore, depends solely on the characteristics of the individual components and how they interact in the circuit. Most hydraulic circuits are of the open-

Figure 1-22. Compact, self-contained pneumatic power unit with electric-driven compressor. (*Courtesy of Ingersoll-Rand, Phillipsburg, New Jersey.*)

loop type, which are generally not so complex or so precise as closed-loop systems. This is because any errors such as slippage (oil leakage past seals, the magnitude of which depends on system pressure and temperature) are not compensated for in open-loop systems. For example, the viscosity of a hydraulic fluid decreases (fluid becomes thinner) as its temperature rises. This increases oil leakage past seals inside pumps, which, in turn, causes the speed of an actuator, such as a hydraulic motor, to drop. In a closed-loop system, a feedback transducer (for example, a tachometer, which generates a signal proportional to the speed at which it is rotated) would sense this speed reduction and feed a proportional signal back to the command signal. The difference between the two signals is used to control a servo valve, which would then increase the fluid flow rate to the hydraulic motor until its speed is at the required level.

A further classification of fluid power systems is whether or not they are electrically controlled. For example, electrical components such as pressure switches, limit switches, and relays can be used to operate electrical solenoids to control the operation of valves that direct fluid to the hydraulic actuators. An electric solenoid control system permits the design of a very versatile fluid power circuit. Automatic machines such as those used in the machine-tool industry rely principally on electrical components to control the hydraulic muscles for doing the required work. The aircraft and mobile equipment industries have also found that fluid power and electricity work very well together, especially where remote control is needed. By merely pressing a simple push-button switch, an operator can control a very complex machine to perform hundreds of machinery operations to manufacture a complete product. An electrically controlled fluid power system can be either of the open-loop or closed-loop type, depending on the precision required.

Another major classification of fluid power systems is whether or not they interact with fluid logic devices for control purposes instead of with electrical devices. Two such fluid logic systems are called moving-part logic (MPL) and fluidics, which perform a wide variety of sensory and control functions. Among these control functions are AND/NAND, OR/NOR, and FLIPFLOP, logic capability. Fluid logic devices switch a fluid, usually air, from one outlet of the device to another outlet. Hence an output of a fluid logic device is either ON or OFF as it rapidly switches from one state to the other by the application of a control signal.

MPL devices are miniature valve-type devices that, by the action of internal moving parts, perform switching operations in fluid logic systems. These MPL devices are typically available as spool, poppet, and diaphragm valves, which can be actuated by means of mechanical displacement, electric voltage, or fluid pressure. Figure 1-23 shows an electronic driven MPL valve that readily interfaces with electric and electronic circuits. This valve converts low voltage (12 to 24 volts) signals into high pressure (100 psi) pneumatic outputs. There are no sliding parts and the total travel of the poppet (the only moving part) is a mere 0.007 inches. As a result, low power consumption (0.67 watts) and long life are major benefits of this design. Additionally, the very fast response time (5–10 milliseconds) and small size make this MPL valve well suited for a wide range of applications in biomedical, environmental test equipment, textile machines, packaging machinery, computerized industrial automation, and portable systems.

Figure 1-24 shows how either 8 or 12 of the electronic valves of Fig. 1-23 can be mounted on a manifold card to provide added convenience in interfacing electronics with pneumatics. The self-contained card includes a manifold mount for single air supply, a fully wired circuit board, and instant plug-in with a 25 pin RS-232 connector. This system allows low voltage signals from controllers, computers, or other sources to operate powerful pneumatic valves with a minimum of piping and hookup.

Figure 1-25 shows an industrial application using MPL controls: a micro-orifice drilling machine, which is capable of drilling accurate holes down to 0.006 inches in diameter. The operator inserts the part to be drilled on the tip of the lower probe, then presses two buttons to initiate the cycle. The bottom probe raises and inserts the part into the drill jig. This maintains the clamping force and automatically actuates the drill quill. The machine completes one cycle, the probe retracts and the part is removed.

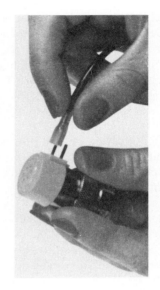

Figure 1-23. Miniature electronic MPL valve. (a) Assembled valve, (b) disassembled valve, (c) wire being connected to spade lug of valve. (*Courtesy of Clippard Instrument Laboratory, Inc., Cincinnati, Ohio.*)

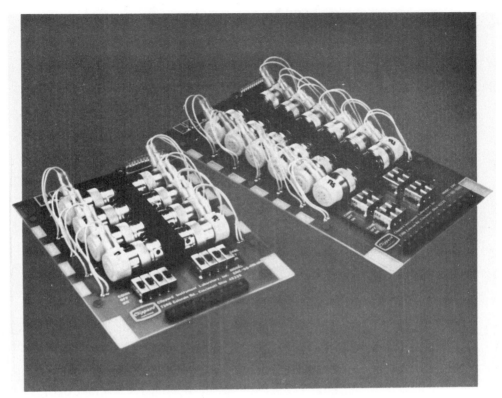

Figure 1-24.   Electronic manifold cards using MPL valves. (*Courtesy of Clippard Instrument Laboratory, Inc., Cincinnati, Ohio.*)

The second fluid logic system, *fluidics*, utilizes fluid flow phenomena in components and circuits also to perform a wide variety of sensory and control functions. Fluidic components (when kept free of contaminants, which can obstruct critical air passageways) are reliable because they contain no moving parts.

Fluidics is an offshoot of fluid power technology and is equivalent to electronics as an offshoot of the technology of electrical power. Just as electronic devices use tiny currents as opposed to the huge currents flowing in electrical power lines, fluidic devices use small flow rates at low pressures in contrast to the high pressure and large flow rates required to operate a huge hydraulic press. Fluidic systems (unlike electrical controls) cannot cause fire hazards due to sparks in a potentially explosive environment. Because they operate with fluids, fluidic components also interface readily within fluid power systems.

A final major classification of fluid power systems is where programmable logic controllers (PLCs) are used to control system operation. In recent years, PLCs have increasingly been used in lieu of electromechanical relays to control fluid power systems. A PLC is a user-friendly electronic computer designed to

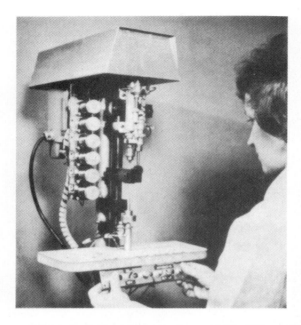

Figure 1-25. Micro-orifice drilling machine using MPL controls. (*Courtesy of Clippard Instrument Laboratory, Inc., Cincinnati, Ohio.*)

perform logic functions such as AND, OR, and NOT for controlling the operation of industrial equipment and processes. A PLC consists of solid-state digital logic elements for making logic decisions and providing corresponding outputs. Unlike general-purpose computers, a PLC is designed to operate in industrial environments where high ambient temperature and humidity levels may exist. PLCs offer a number of advantages over electromechanical relay control systems. Unlike electromechanical relays, PLCs are not hard-wired to perform specific functions. Thus, when system operation requirements change, a software program is readily changed instead of having to physically rewire relays. In addition, PLCs are more reliable, faster in operation, smaller in size, and can be readily expanded.

Figure 1-26 shows a PLC-based synchronous lift system used for precise lifting and lowering of high-tonnage objects on construction jobs. Unlike complex and costly electronic lift systems, this hydraulic system has a minimum number of parts and can be run effectively and efficiently by one person. The PLC enables the operator to quickly and easily set the number of lift points, stroke limit, system accuracy, and other operating parameters from a single location. The PLC receives input signals from electronic sensors located at each lift point, and in turn sends output signals to the solenoid valve that controls fluid flow to each hydraulic cylinder to maintain the relative position and accuracy selected by the operator. Because the sensors are attached directly to the load, they assure more exact measurement of the load movement. The system accommodates loads of any size and is accurate to ±0.040 in. (1 mm).

The PLC unit of this system (see Fig. 1-27) contains a LCD display that shows the position of the load at each lift point and the status of all system

Figure 1-26.   PLC synchronous lift system. (*Courtesy of Enerpac, Applied Power Inc., Butler, Wisconsin.*)

operations so the operator can stay on top of every detail throughout the lift. The PLC unit, which weighs only 37 pounds and has dimensions of only 16 in. by 16 in. by 5 in., can control up to eight lifting points. The system diagram is shown in Fig. 1-28, in which components are identified using letters as follows:

A: Programmable logic controller

B: Solenoid directional control valve

C: Electronic load displacement sensors

D: Sensor cables

E: Flow control valves for regulating movement of hydraulic cylinders

Figure 1-29 shows a PLC-based system that contains electronic/pneumatic interface valves mounted on a manifold card, a microprocessor on a board below the manifold card, and a microcomputer. This system provides machine control

Figure 1-27. PLC unit of synchronous lift system. (*Courtesy of Enerpac, Applied Power Inc., Butler, Wisconsin.*)

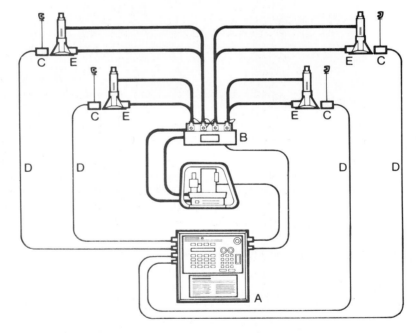

Figure 1-28. Diagram of PLC synchronous lift system. (*Courtesy of Enerpac, Applied Power Inc., Butler, Wisconsin.*)

Figure 1-29.   PLC system with electronic/pneumatic interface valves. (*Courtesy of Clippard Instrument Laboratory, Inc., Cincinnati, Ohio.*)

utilizing pneumatic outputs and actuators through a programmable series of sequentially executed steps. The system is easy to program through user-friendly software and allows wide choices of programming devices including personal computers and hand-held data-entry terminals. Figure 1-30 shows the microprocessor, a computer on a board, that is arranged for ease of connection and use with the interface valves mounted on the manifold card.

The various types of fluid power systems are studied in more detail throughout this book. They are introduced here so that the reader can better appreciate the total technology of fluid power at this time.

1.7 THE FLUID POWER INDUSTRY

The fluid power industry is huge as evidenced by its present annual sales figure for system components of $9.6 billion. It is also a fast-growing industry with a 65% increase in terms of equipment sales during the period 1981–1990. This is reflected in the fact that nearly all U.S. manufacturing plants rely on fluid power in the production of goods. Over half of all industrial products have fluid power systems or components as part of their basic design. As shown in Fig. 1-31, the fluid power industry is larger than many better-known industries such as mining, machinery, construction equipment, and machine tools.

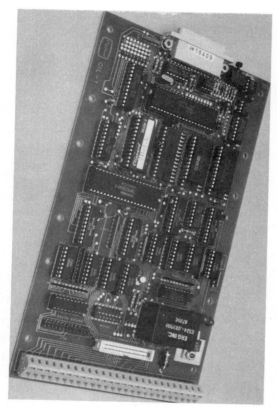

Figure 1-30. Microprocessor of PLC system. (*Courtesy of Clippard Instrument Laboratory, Inc., Cincinnati, Ohio.*)

Personnel in industry who work in the fluid power field can generally be placed into three categories:

1. Fluid power mechanics. Workers in this category are responsible for repair and maintenance of fluid power equipment. They generally are high school graduates who have undertaken an apprenticeship training program. Such a program usually consists of three or four years of paid, on-the-job training plus corresponding classroom instruction.

2. Fluid power technicians. These people usually assist engineers in areas such as design, troubleshooting, testing, maintenance, and installation of fluid power systems. They generally are graduates of two-year technical colleges, which award an associate degree in fluid power. The technician can advance into supervisory positions in sales, manufacturing, or service management.

3. Fluid power engineers. This category consists of people who perform advanced design, development, and testing of new, sophisticated fluid power components or systems. The fluid power engineer typically is a graduate of a four-

THE EXPANDING SIZE AND SCOPE OF FLUID POWER

Nearly all U.S. manufacturing plants rely on fluid power in the production of goods. Over half of all industrial products have fluid power systems or components as part of their basic design.

Fluid power is a $5 billion industry. This level of shipments makes fluid power manufacturing larger than many better-known industries, such as mining machinery, construction equipment and machine tools.

There are over 400 U.S. companies manufacturing fluid power products. These companies operate an estimated 1,200 plants. In addition, an estimated 1,200 U.S. companies are involved in the distribution, service and repair of fluid power products.

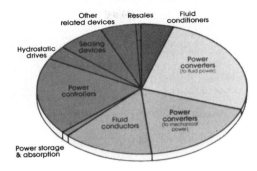

KEY FLUID POWER PRODUCTS AS A PERCENTAGE OF TOTAL SHIPMENTS* (See Key)

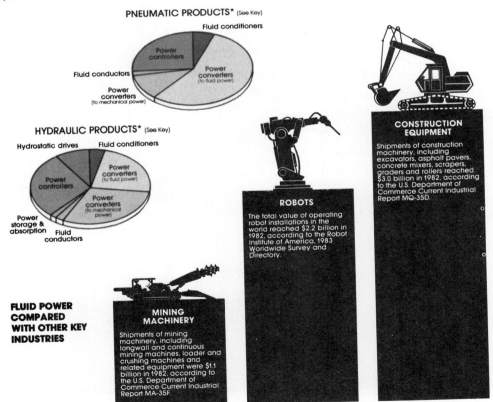

FLUID POWER COMPARED WITH OTHER KEY INDUSTRIES

Figure 1-31. The expanding size and scope of the fluid power industry. (*Courtesy of National Fluid Power Association, Milwaukee, Wisconsin.*)

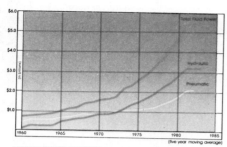

**TOTAL FLUID POWER
PRODUCT SHIPMENTS
HISTORICAL TREND** (moving average)

FLUID POWER

Hydraulic and pneumatic
product shipments reached
$5.1 billion in 1982, according
to two U.S. Department of
Commerce Current Industrial
Reports, MA-35N and MA-35P.
This includes fluid power
pumps and motors, cylinders
and rotary actuators, valves,
hose and tube fittings,
accumulators, shock
absorbers and air
compressors.

**METALWORKING
MACHINERY**

Metal-cutting and metal-
forming machine tool
shipments were $3.7 billion in
1982, according to the U.S.
Department of Commerce
Current Industrial Report MA-
35F. Boring, drilling, grinding,
turning, and milling
machinery are included in
metal-cutting type. Punching,
shearing and forging
machines and presses are
among metal-forming type.

TRACTORS

Shipments for agricultural
and other types of tractors
were $3.1 billion in 1983,
according to the U.S.
Department of Commerce
Current Industrial Report
M35S.

*KEY

Fluid conditioners:
Hydraulic and air filters,
FRLs, mufflers, air dryers,
heat exchangers.

**Power converters (to
fluid power):** Hydraulic
power units, intensifiers,
pumps, air
compressors, vacuum
pumps.

**Power converters (to
mechanical power):**
Hydraulic and air
cylinders, motors, rotary
actuators, positioners
and hand tools.

Fluid conductors:
hydraulic and air hose
and fittings, manifolds
and tubes.

**Power storage and
absorption:**
Accumulators, shock
absorbers, cushions.

Power controllers:
Hydraulic and air
valves.

Sealing devices: Seals,
packings, o-rings.

Others: Instrumentation,
accessories, reservoirs,
test bench parts.

Figure 1-31.   *(Continued)*

Figure 1-32. Fluid power engineer using microcomputer. (*Courtesy of Festo Corp., Hauppage, NY.*)

year college program. Most engineers who work on fluid power systems are manufacturing, sales, or mechanical design oriented. They can advance into management positions in design, manufacturing, or sales.

The future of the fluid power industry looks very promising, especially when one considers that the vast majority of all manufactured products have been processed in some way by fluid power systems. As a result, career opportunities are very bright. The fantastic growth of the fluid power industry has opened many new opportunities in all areas, including supervisors, engineers, technicians, mechanics, sales personnel, servicemen, and operators.

Figure 1-32 shows a fluid power engineer designing a fluid power circuit using CAD software, a microcomputer, and a plotter.

In addition, a critical shortage of trained, qualified instructors presently exists. This shortage exists at universities and colleges as well as at two-year technical colleges and high schools. It is hoped that this book will help in some way in the education of these fluid-power-inspired people.

EXERCISES

1-1. List five fields of application where fluid power can be used more effectively than other power sources.

1-2. Why is hydraulic power especially useful when performing heavy work?

1-3. Compare the use of fluid power to a mechanical system by listing the advantages and disadvantages of each.

1-4. What is the basic law that is important in applying fluid power, and what is its significance?

1-5. Comment on the difference between using pneumatic fluid power and hydraulic fluid power.

1-6. Define fluid power.

1-7. What hydraulic device creates a force that can push or pull a load?

1-8. What hydraulic device creates a force that can rotate a shaft?

1-9. What two factors are responsible for the high responsiveness of hydraulic devices?

1-10. Why can't air be used for all fluid power applications?

1-11. What is the prime mover?

1-12. Name the six basic components required in a hydraulic circuit.

1-13. Name the six basic components required in a pneumatic circuit.

1-14. Obtain from the Fluid Power Society or Fluid Power Foundation the requirements to become a certified fluid power technician.

1-15. From publications such as *Hydraulics and Pneumatics* (The Magazine of Fluid Power and Control Systems), trace the economic growth of the fluid power industry since World War II.

1-16. From publications such as the *Hydraulics and Pneumatics*, trace the economic growth of pneumatics in the fluid power industry since World War II.

1-17. From publications such as *Hydraulics and Pneumatics*, trace the history of moving-part logic (MPL) applications in the fluid power industry.

1-18. Take a plant tour of a company that manufactures fluid power components such as pumps, cylinders, valves, or motors. Write a report stating how at least one component is manufactured. List the specifications and include potential customer applications.

1-19. Why are some fluid power circuits controlled by electricity?

1-20. List five applications of fluid power in the automotive industry.

1-21. What is the difference between a closed-loop and an open-loop fluid power system?

1-22. What three types of personnel work in the fluid power field in industry?

1-23. What is the significance of the phrase "fluid power preserves the heartbeat of life"? Cite two examples that support the subject phrase.

1-24. Discuss the phrase "the expanding size and scope of fluid power." Cite two facts that show the size and two additional facts that show the growth of the fluid power industry.

1-25. What are moving-part logic devices?

1-26. Name three ways moving-part logic devices can be actuated.

1-27. What are fluidic devices?

1-28. What must be done to ensure that fluidic devices will operate reliably?

1-29. Give one reason why automotive hydraulic brakes might exhibit a spongy feeling when a driver pushes on the brake pedal.

1-30. Relative to the automobile, is cruise control an open-loop or closed-loop system? Explain your answer.

1-31. What is the difference between the terms *fluid power* and *hydraulics and pneumatics*?

1-32. Differentiate between "fluid transport" and "fluid power" systems.

1-33. Name one source of error that is compensated for in a closed-loop system.

1-34. Name five hydraulic applications and five pneumatic applications.

1-35. Name the components on one hydraulic application and give their functions.

1-36. Name the components on one pneumatic application and give their functions.

1-37. Contact the National Fluid Power Association to determine the requirements for becoming a fluid power engineer.

1-38. From publications such as *Hydraulics and Pneumatics*, trace the history of programmable logic control (PLC) applications in the fluid power industry.

1-39. What is a programmable logic controller?

1-40. How does a PLC differ from a general-purpose computer?

1-41. What is the difference between a PLC and an electromechanical relay control?

1-42. Name three advantages that PLCs provide over electromechanical relay control systems.

2

Properties of

Hydraulic Fluids

2.1 INTRODUCTION

The single most important material in a hydraulic system is the working fluid itself. Hydraulic fluid characteristics have a crucial effect on equipment performance and life. It is important to use a clean, high-quality fluid in order to achieve efficient hydraulic system operation.

Most modern hydraulic fluids are complex compounds that have been carefully prepared to meet their demanding tasks. In addition to having a base fluid, hydraulic fluids contain special additives to provide desired characteristics.

Essentially, a hydraulic fluid has four primary functions:

1. To transmit power
2. To lubricate moving parts
3. To seal clearances between mating parts
4. To dissipate heat

To accomplish properly these primary functions and be practical from a safety and cost point of view, a hydraulic fluid should have the following properties:

1. Good lubricity
2. Ideal viscosity
3. Chemical and environmental stability
4. Compatibility with system materials

   5. Large bulk modulus

   6. Fire resistance

   7. Good heat-transfer capability

   8. Low density

   9. Foam resistance

   10. Nontoxic

   11. Low volatility

   12. Inexpensive

   13. Readily available

This is a challenging list, and no single hydraulic fluid posseses all of these desirable characteristics. The fluid power designer must select the fluid that comes the closest to being ideal overall for a particular application.

Hydraulic fluids must also be changed periodically, the frequency depending not only on the fluid but also on the operating conditions. Laboratory analysis is the best method for determining when a fluid should be changed. Generally speaking, a fluid should be changed when its viscosity and acidity increase due to fluid breakdown or contamination. Preferably, the fluid should be changed while the system is at operating temperature. In this way, most of the impurities are in suspension and will be drained off.

Much hydraulic fluid has been discarded in the past due to the possibility that contamination existed—it costs more to test the fluid than to replace it. This situation has changed as the need to conserve hydraulic fluids has developed. Figure 2-1 shows a hydraulic fluid test kit that provides a quick, easy method to

Figure 2-1.   Hydraulic fluid test kit. (*Courtesy of Gulf Oil Corp., Houston, Texas.*)

test hydraulic system contamination. Even small hydraulic systems may be checked. The test kit may be used on the spot to determine whether fluid quality permits continued use. Three key quality indicators can be evaluated: viscosity, water content, and foreign particle contamination level.

In this chapter we discuss the important characteristics of hydraulic fluids and their effect on fluid power system operation. We also study the methods used to measure various fluid properties. The proper understanding of the total function of hydraulic fluids and of the effect each fluid property has on fluid power component and system operation is an absolute must.

2.2 FLUIDS: LIQUIDS AND GASES

The term *fluid* refers to both gases and liquids. A liquid is a fluid that, for a given mass, will have a definite volume independent of the shape of its container. This means that even though a liquid will assume the shape of the container, it will fill only that part of the container whose volume equals the volume of the quantity of the liquid. For example, if water is poured into a vessel and the volume of water is not sufficient to fill the vessel, a free surface will be formed, as shown in Fig. 2-2(a). A free surface is also formed in the case of a body of water, such as a lake exposed to the atmosphere [see Fig. 2-2(b)].

Liquids are considered to be incompressible so that their volume does not change with pressure changes. This is not exactly true, but the change in volume due to pressure changes is so small that it is ignored for most engineering applications. Variations from this assumption of incompressibility are discussed in Section 2.7 where the parameter bulk modulus is defined.

Gases, on the other hand, are fluids that are readily compressible. In addition, their volume will vary to fill the vessel containing them. This is illustrated in Fig. 2-3 where a gas is allowed to enter an empty vessel. As shown, the gas molecules always fill the entire vessel. Therefore, unlike liquids, which have a

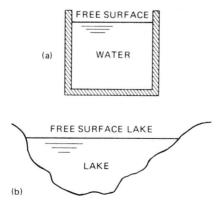

Figure 2-2.   Free surface of a liquid.

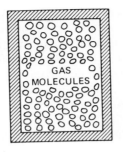

Figure 2-3.   A gas always fills its entire vessel.

definite volume for a given mass, the volume of a given mass of a gas will increase to fill the vessel that contains the gas. Gases are greatly influenced by the pressure to which they are subjected. An increase in pressure causes the volume of the gas to decrease, and vice versa.

Air is the only gas commonly used in fluid power systems because it is inexpensive and readily available. Air also has the following advantages as a power fluid:

1. It is fire resistant.
2. It is not messy.
3. It can be exhausted back into the atmosphere.

Its disadvantages include:

1. Due to its compressibility, it cannot be used in an application where accurate positioning or rigid holding is required.
2. Because it is compressible, it tends to be sluggish.
3. Air can be corrosive since it contains oxygen and water.

2.3 *WEIGHT, DENSITY, AND SPECIFIC GRAVITY*

All objects, whether solids or fluids, are pulled toward the center of the earth by a force of attraction. This force is called the weight of the object and is proportional to the object's mass, as defined by

$$F = W = mg \qquad\qquad (2\text{-}1)$$

where F = force in units of lb,
$\quad W$ = weight in units of lb,
$\quad m$ = mass of object in units of slugs,
$\quad g$ = proportionality constant called the acceleration of gravity, which equals 32.2 ft/sec/sec or ft/sec^2.

From Eq. (2-1), W equals 32.2 lb if m is 1 slug. Therefore, 1 slug is the amount of mass that weighs 32.2 lb on the surface of the earth.

Figure 2-4 shows a cubic container full of water as an example for discussion purposes. Since each side of the container has a dimension of 1 ft, the volume of the container is 1 ft³, as calculated from

$$\text{volume} = (\text{area of base}) \times (\text{height})$$

$$\text{volume} = (1 \text{ ft} \times 1 \text{ ft}) \times (1 \text{ ft}) = 1 \text{ ft}^3 \tag{2-2}$$

It has been found by measurement that 1 ft³ of water weighs 62.4 lb. This brings us to the concept of density and specific gravity. Weight density (or specific weight, as it is sometimes called) is defined as weight per unit volume. Stated mathematically, we have

$$\text{weight density} = \frac{\text{weight}}{\text{volume}}$$

or

$$\gamma = \frac{W}{V} \tag{2-3}$$

where γ = weight density (lb/ft³),
  W = weight (lb),
  V = volume (ft³).

We can now calculate the weight density of water using Eq. (2-3):

$$\gamma_{\text{water}} = \frac{W}{V} = \frac{62.4 \text{ lb}}{1 \text{ ft}^3} = 62.4 \text{ lb/ft}^3$$

If we want to calculate the weight density in units of lb/in.³, we can perform the following unit manipulation:

$$\underset{\substack{\uparrow \\ \text{units wanted}}}{\gamma \,(\text{lb/in}^3)} = \underset{\substack{\uparrow \\ \text{units have}}}{\gamma \,(\text{lb/ft}^3)} \times \underset{\substack{\uparrow \\ \text{conversion factor}}}{\frac{1}{1728} \,(\text{ft}^3/\text{in.}^3)}$$

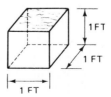

Figure 2-4. Cubic container full of water.

The conversion factor of $\frac{1}{1728}$ is valid since 1 ft³ = 1728 in.³ and provides a consistent set of units of lb/in.³ on both sides of the equal sign.

Substituting the known value for the weight density of water in units of lb/ft³, we have

$$\gamma_{\text{water}} \ (\text{lb/in.}^3) = 62.4 \times \frac{1}{1728} = 0.0361 \ \text{lb/in.}^3$$

Most oils have a weight density of about 56 lb/ft³ or 0.0325 lb/in.³ However, depending on the type of oil, the density can vary from a low of 55 lb/ft³ to a high of 58 lb/ft³.

The specific gravity (S_g) of a given fluid is defined as the weight density of the fluid divided by the weight density of water. Therefore, the specific gravity of water is unity by definition. The specific gravity of oil can be found using

$$S_{g \ \text{oil}} = \frac{\gamma_{\text{oil}}}{\gamma_{\text{water}}} \qquad\qquad (2\text{-}4)$$

Substituting known values, we have

$$S_{g \ \text{oil}} = \frac{56 \ \text{lb/ft}^3}{62.4 \ \text{lb/ft}^3} = 0.899$$

Note that specific gravity is a dimensionless parameter.

EXAMPLE 2-1

Air at 68°F and under atmospheric pressure has a specific weight of 0.0752 lb/ft³. Find its specific gravity.

Solution

$$S_{g \ \text{air}} = \frac{\gamma_{\text{air}}}{\gamma_{\text{water}}} = \frac{0.0752 \ \text{lb/ft}^3}{62.4 \ \text{lb/ft}^3} = 0.00121$$

Thus water is 1/0.00121 times or about 830 times as heavy as air at 68°F and under atmospheric pressure. It should be noted that since air is highly compressible, the value of 0.00121 for S_g is valid only at 68°F and under atmospheric pressure.

The Greek symbol gamma (γ) is used to denote weight density. We can also talk about mass density, which is defined as mass per unit volume:

$$\rho = \frac{m}{V} \qquad\qquad (2\text{-}5)$$

where ρ = Greek symbol rho = mass density (slugs/ft³),
     m = mass (slugs),
     V = volume (ft³).

Since mass is proportional to weight, it follows that specific gravity can also be defined as the mass density of the given fluid divided by the mass density of water.

Mass density is related to weight density as follows:

$$\rho = \frac{\gamma}{g} \tag{2-6}$$

where γ has units of lb/ft³,
     g has units of ft/sec²,
     ρ has units of slugs/ft³.

Thus the mass density of water can be found:

$$\rho_{\text{water}} = \frac{62.4}{32.2} = 1.94 \text{ slug/ft}^3$$

Knowing that the weight density (specific weight) of air at 68°F and under atmospheric pressure is 0.0752 lb/ft³, we can find the corresponding value of mass density:

$$\rho_{\text{air at 68°F}} = \frac{0.0752}{32.2} = 0.00234 \text{ slug/ft}^3$$

2.4 PRESSURE, HEAD, AND FORCE

The concept of pressure must be thoroughly mastered in order to clearly understand fluid power. In fluid power systems it is pressure rather than force that is transmitted equally in all directions. Pressure is defined as force per unit area. Hence, pressure is the amount of force acting over a unit area, as indicated by

$$P = \frac{F}{A} \tag{2-7}$$

where P = pressure,
    F = force,
    A = area.

P will have units of lb/ft² if F and A have units of lb and ft², respectively. Similarly, by changing the units of A from ft² to in.², the units for P will become lb/in.^{2} Let's go back to our 1 ft^3 container of Fig. 2-4. The pressure acting at the bottom of the container can be calculated using Eq. (2-7), knowing that the total force acting at the bottom equals the 62.4-lb weight of the water:

$$P = \frac{62.4 \text{ lb}}{1 \text{ ft}^2} = 62.4 \text{ lb/ft}^2 = 62.4 \text{ psf}$$

Note that units of lb/ft^2 are commonly written as psf. Also, since 1 ft^2 = 144 in.^{2}, we have

$$P = \frac{62.4 \text{ lb}}{144 \text{ in.}^2} = 0.433 \text{ lb/in.}^2 = 0.433 \text{ psi}$$

Units of lb/in.^{2} are commonly written as psi.

We can now conclude that a 1-ft column of water develops at its base a pressure of 0.433 psi. The 1-ft height of water is commonly called a pressure head.

Let's now refer to Fig. 2-5, which shows a 1-ft^2 section of water 10-ft high. Since there are 10 ft^3 of water and each cubic foot weighs 62.4 lb, the total weight of water is 624 lb. The pressure at the base is

$$P = \frac{F}{A} = \frac{624 \text{ lb}}{144 \text{ in.}^2} = 4.33 \text{ psi}$$

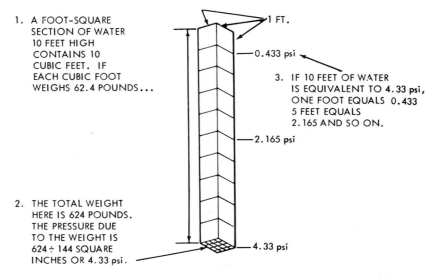

1. A FOOT-SQUARE
    SECTION OF WATER
    10 FEET HIGH
    CONTAINS 10
    CUBIC FEET. IF
    EACH CUBIC FOOT
    WEIGHS 62.4 POUNDS...

3. IF 10 FEET OF WATER
    IS EQUIVALENT TO 4.33 psi,
    ONE FOOT EQUALS 0.433
    5 FEET EQUALS
    2.165 AND SO ON.

2. THE TOTAL WEIGHT
    HERE IS 624 POUNDS.
    THE PRESSURE DUE
    TO THE WEIGHT IS
    624 ÷ 144 SQUARE
    INCHES OR 4.33 psi.

Figure 2-5. Pressure developed by a 10-ft column of water. (*Courtesy of Sperry Vickers, Sperry Rand Corp., Troy, Michigan.*)

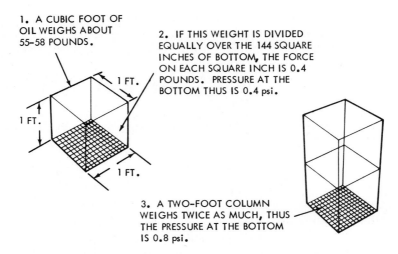

1. A CUBIC FOOT OF OIL WEIGHS ABOUT 55–58 POUNDS.

2. IF THIS WEIGHT IS DIVIDED EQUALLY OVER THE 144 SQUARE INCHES OF BOTTOM, THE FORCE ON EACH SQUARE INCH IS 0.4 POUNDS. PRESSURE AT THE BOTTOM THUS IS 0.4 psi.

1 FT.

1 FT.

1 FT.

3. A TWO-FOOT COLUMN WEIGHS TWICE AS MUCH, THUS THE PRESSURE AT THE BOTTOM IS 0.8 psi.

Figure 2-6. Pressures developed by 1- and 2-ft columns of oil. (*Courtesy of Sperry Vickers, Sperry Rand Corp., Troy, Michigan.*)

Thus each foot of the 10-ft head produces a pressure of 0.433 psi.

What happens to the pressure if the fluid is not water? Figure 2-6 shows a 1-ft³ volume of oil. Assuming a density of 57 lb/ft³, the pressure at the base is

$$P = \frac{F}{A} = \frac{57 \text{ lb}}{144 \text{ in.}^2} = 0.4 \text{ psi}$$

Therefore, as depicted in Fig. 2-6, a 2-ft column of oil produces a pressure at its bottom of 0.8 psi. These values for oil are slightly less than for water because the density of oil is somewhat less than that for water.

Equation (2-8) is a handy mathematical tool for calculating the pressure produced at the bottom of any column of any liquid. Observe, per the equation, that the pressure does not depend on the area of the bottom of the column but only on the column height and the specific gravity of the fluid. The reason is simple: Changing the cross-sectional area of the column changes the force a proportional amount, and thus F/A (which equals pressure) remains constant.

$$P = 0.433 \, HS_g \qquad\qquad (2\text{-}8)$$

where P = pressure (psi),
$\quad H$ = head (ft),
$\quad S_g$ = specific gravity.

$$S_g = \frac{\gamma}{\gamma_{water}}$$

EXAMPLE 2-2

Find the pressure on a skin diver who has descended to a depth of 60 ft in fresh water.

Solution

$$P = (0.433)(60)(1) = 26.0 \text{ psi}$$

What about the pressure developed on the surface of the earth due to the force of attraction between the atmosphere and the earth? For all practical purposes we live at the bottom of a huge sea of air, which extends hundreds of miles above us. Equation (2-8) cannot be used to find this pressure because of the compressibility of air. This means that the specific gravity (S_g) of the air in the atmosphere is not a constant. The density is greatest at the earth's surface and diminishes as the distance from earth increases.

Let's refer to Fig. 2-7, which shows a column of air 1 in.^{2} in cross section and as high as the atmosphere. This column weighs 14.7 lb at sea level and thus produces a pressure of 14.7 psi, which is called atmospheric pressure.

Why, then, does a deflated automobile tire read zero pressure instead of 14.7 psi when using a pressure gage? The answer lies in the fact that a pressure gage reads gage pressure and not absolute pressure. Gage pressures are measured relative to the atmosphere, whereas absolute pressures are measured relative to a

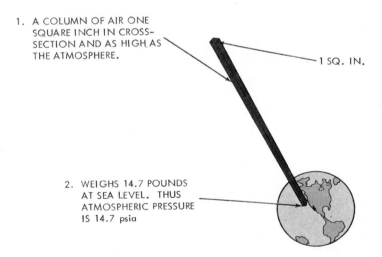

1. A COLUMN OF AIR ONE SQUARE INCH IN CROSS-SECTION AND AS HIGH AS THE ATMOSPHERE.

1 SQ. IN.

2. WEIGHS 14.7 POUNDS AT SEA LEVEL. THUS ATMOSPHERIC PRESSURE IS 14.7 psia

Figure 2-7.   The atmosphere as a pressure head. (*Courtesy of Sperry Vickers, Sperry Rand Corp., Troy, Michigan.*)

perfect vacuum such as that which exists in outer space. To distinguish between them, gage pressures are labeled psig or simply psi, whereas absolute pressures are labeled psia.

This means that atmospheric pressure is really 14.7 psia instead of 14.7 psi. Thus, atmospheric pressure is always an absolute pressure and must be measured with special devices called barometers. Figure 2-8 shows how a mercury barometer works. The atmospheric pressure to be measured can support a column of mercury equal to 29.92 in. because this head produces a pressure of 14.7 psi.

This can be checked by using Eq. (2-8) and noting that the specific gravity of mercury is 13.6:

$$P = 0.433 HS_g$$

$$14.7 = 0.433 H(13.6)$$

or

$$H = 2.49 \text{ ft} = 29.9 \text{ in. of mercury}$$

EXAMPLE 2-3

How high would the tube of a barometer have to be if water were used instead of mercury?

Solution

$$14.7 = 0.433 H(1)$$

or

$$H = 34 \text{ ft}$$

Thus a water barometer would be impractical because it takes a 34-ft column of water to produce a pressure of 14.7 psi at its base.

Of course, the atmospheric pressure changes a small amount depending on the elevation of the particular surface on the earth and the weather conditions, which affect the density of the air.

Figure 2-9 has a chart showing the difference between gage and absolute pressures. Let's examine two pressure levels: P_1 and P_2. Relative to a perfect vacuum they are

$P_1 = 4.7$ psia   (a pressure less than atmospheric pressure)

$P_2 = 24.7$ psia   (a pressure greater than atmospheric pressure)

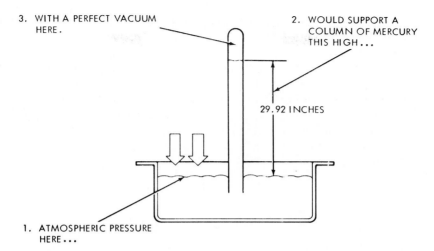

Figure 2-8. Operation of a mercury barometer. (*Courtesy of Sperry Vickers, Sperry Rand Corp., Troy, Michigan.*)

Relative to the atmosphere, they are

$$P_1 = 10 \text{ psig suction (or vacuum)} = -10 \text{ psig}$$

$$P_2 = 10 \text{ psig}$$

The use of the terms suction or vacuum and the use of the minus sign mean that pressure P_1 is 10 psi below atmospheric pressure. Also note that the terms psi and psig are used interchangeably. Hence, P_1 also equals -10 psi and P_2 equals 10 psi.

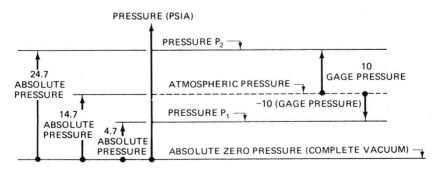

Figure 2-9. Difference between absolute and gage pressures.

As can be seen from Fig. 2-9, the following rule can be used in pressure conversion calculations:

absolute pressure = gage pressure + atmospheric pressure

EXAMPLE 2-4

Convert a −5-psi pressure to an absolute pressure.

Solution

$$\text{absolute pressure} = -5 + 14.7 = 9.7 \text{ psia}$$

EXAMPLE 2-5

Find the absolute pressure on the skin diver of Example 2-2.

Solution

$$\text{absolute pressure} = 26.0 + 14.7 = 40.7 \text{ psia}$$

It should be noted that vacuum or suction pressures exist in certain locations of fluid power systems (for example, in suction lines of pumps). Therefore, it is important to understand the meaning of pressures below atmospheric pressure. One way to generate a suction pressure is to remove some of the fluid from a closed vessel initially containing fluid at atmospheric pressure.

2.5 THE METRIC-SI SYSTEM

The metric-SI system was standardized in June 1966 when the International Organization for Standardization approved a metric system called *Le Systeme International d'Unites*. This system, which is abbreviated SI, has supplanted the old MKS metric system, and U.S. adoption of the metric SI system is considered to be imminent.

In the metric-SI system, the units of measurement are as follows:

Length is the meter (m).

Mass is the kilogram (kg).

Force is the newton (N).

Time is the second (s).

One meter equals 39.4 in. or 3.28 ft.

A newton is the force required to give a mass of 1 kg an acceleration of 1 m/s². Stated mathematically, we have

$$1 \text{ newton} = 1 \text{ kilogram} \times 1 \text{ meter/s}^2$$

or, in abbreviated form,

$$1 \text{ N} = 1 \text{ kg} \times 1 \text{ m/s}^2$$

The conversion between newtons and pounds shows that a newton is only about one-fourth of a pound. The conversion is as follows:

$$1 \text{ N} = 0.225 \text{ lb}$$

Since the acceleration of gravity at sea level equals 9.80 m/s², a mass of 1 kg weighs 9.80 N.

The metric-SI system uses units of pascals to represent pressure. A pressure of 1 Pa is equal to a force of 1 N applied over an area of 1 m²:

$$1 \text{ pascal} = 1 \text{ newton/meter}^2$$

or, in abbreviated form,

$$1 \text{ Pa} = 1 \text{ N/m}^2$$

The conversion between pascals and psi is as follows:

$$1 \text{ Pa} = 0.000145 \text{ psi}$$

Since the pascal is a very small unit, the bar is commonly used:

$$1 \text{ bar} = 10^5 \text{ N/m}^2 = 14.5 \text{ psi}$$

Thus, atmospheric pressure equals 14.7/14.5 bars or 1.014 bars.

The following equations permit density conversions between the English and metric-SI systems:

$$\rho \text{ (kg/m}^3) = 515\rho \text{ (slugs/ft}^3)$$

$$\gamma \text{(N/m}^3) = 157\gamma \text{(lb/ft}^3)$$

Thus, the weight density of water in the metric-SI system becomes

$$\gamma_{H_2O} \text{ (N/m}^3) = 157\gamma_{H_2O} \text{ (lb/ft}^3)$$

$$= (157)(62.4) = 9797 \text{N/m}^3$$

Similarly, the mass density of water becomes

$$\rho_{H_2O} \ (kg/m^3) = 515 \ \rho_{H_2O} \ (slugs/ft^3)$$

$$= (515)(1.94) = 999.1 \ kg/m^3$$

A value of 999.1 kg/m^3 happens to be the mass density of fresh water at 60°F and atmospheric pressure.

It should be noted that temperature in the metric system is measured in units of degrees Celsius (°C). The mathematical relationship between the Fahrenheit and Celsius scales is

$$°F = 1.8 \ (°C) + 32$$

Thus, to find the equivalent Celsius temperature corresponding to room temperature (68°F), we have:

$$°C = (°F - 32)/1.8 = (68 - 32)/1.8 = 20°C$$

The following prefixes are used in the metric system to represent powers of 10:

Name	SI Symbol	Multiplication Factor
giga	G	10^9
mega	M	10^6
kilo	k	10^3
centi	c	10^{-2}
milli	m	10^{-3}
micro	μ	10^{-6}
nano	n	10^{-9}

Thus, for example:

$$1 \ kPa = 10^3 \ Pa = 1000 \ Pa$$

2.6 PASCAL'S LAW

Pascal's law reveals the underlying principle of how fluids transmit power. It states that pressure applied to a confined fluid is transmitted undiminished in all directions. This reveals why a full glass bottle can break if a stopper is forced into

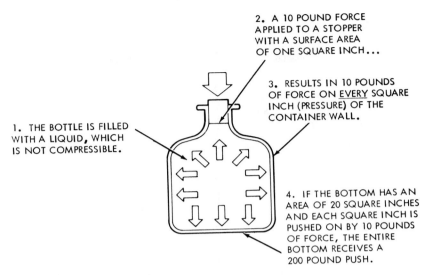

2. A 10 POUND FORCE APPLIED TO A STOPPER WITH A SURFACE AREA OF ONE SQUARE INCH...

3. RESULTS IN 10 POUNDS OF FORCE ON <u>EVERY</u> SQUARE INCH (PRESSURE) OF THE CONTAINER WALL.

1. THE BOTTLE IS FILLED WITH A LIQUID, WHICH IS NOT COMPRESSIBLE.

4. IF THE BOTTOM HAS AN AREA OF 20 SQUARE INCHES AND EACH SQUARE INCH IS PUSHED ON BY 10 POUNDS OF FORCE, THE ENTIRE BOTTOM RECEIVES A 200 POUND PUSH.

Figure 2-10.   Demonstration of Pascal's law. (*Courtesy of Sperry Vickers, Sperry Rand Corp., Troy, Michigan.*)

the already full chamber. Since the liquid is incompressible, it transmits the pressure applied at the stopper throughout the container, as illustrated in Fig. 2-10. Let's assume that the area of the stopper is 1 in.^{2} and that the area of the bottom is 20 in.^{2} Then a 10-lb force applied to the stopper produces a pressure of 10 psi. This pressure is transmitted undiminished to the bottom of the bottle, acting on the full 20-in.^{2} area and producing a 200-lb force. Therefore, it is possible to break out the bottom by pushing on the stopper with a moderate force.

The example of Fig. 2-10 shows how a small force exerted on a small area can create a proportionally larger force on a larger area. Hence, the only limit to the force a machine can exert is the area to which the pressure is applied. Figure 2-11, view A, shows how Pascal's law can be applied to produce a useful amplified output force such as in a hydraulic press. An input force of 10 lb is applied to a 1-in.^{2} piston. This develops a 10-psi pressure throughout the container. This 10-psi pressure acts on a 10-in.^{2} piston, producing a 100-lb output force.

It is interesting to note that there is a similarity between the single hydraulic press of view A in Fig. 2-11 and the mechanical lever of view B. Notice that the 10-lb input force has an arm (distance from force to pivot) that is 10 times as long as the output force arm. As a result, the output force is 10 times as large as the input force and thus equals 100 lb. This is the force multiplication aspect of Pascal's law, which is applied to fluid power systems throughout this book.

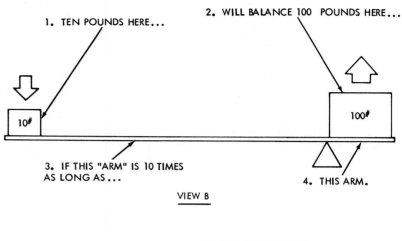

1. TEN POUNDS HERE...

2. WILL BALANCE 100 POUNDS HERE...

10#

100#

3. IF THIS "ARM" IS 10 TIMES AS LONG AS ...

4. THIS ARM.

VIEW B

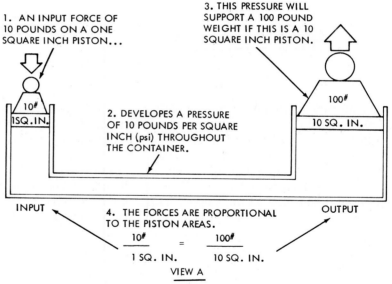

1. AN INPUT FORCE OF 10 POUNDS ON A ONE SQUARE INCH PISTON...

3. THIS PRESSURE WILL SUPPORT A 100 POUND WEIGHT IF THIS IS A 10 SQUARE INCH PISTON.

10#
1SQ.IN.

100#
10 SQ. IN.

2. DEVELOPES A PRESSURE OF 10 POUNDS PER SQUARE INCH (psi) THROUGHOUT THE CONTAINER.

INPUT

OUTPUT

4. THE FORCES ARE PROPORTIONAL TO THE PISTON AREAS.

$$\frac{10\#}{1\ \text{SQ. IN.}} = \frac{100\#}{10\ \text{SQ. IN.}}$$

VIEW A

Figure 2-11.   Force multiplication aspects of Pascal's law. (*Courtesy of Sperry Vickers, Sperry Rand Corp., Troy, Michigan.*)

2.7 BULK MODULUS

The highly favorable horsepower-to-weight ratio and the stiffness of hydraulic systems make them the frequent choice for most high-power applications. The stiffness of a hydraulic system is directly related to the incompressibility of the oil. Bulk modulus is a measure of this incompressibility. The higher the bulk modulus, the less compressible or stiffer the fluid. Mathematically the bulk modulus is defined by Eq. (2-9), where the minus sign accounts for the fact that as the pressure increases, the volume decreases:

$$\beta = -V\left(\frac{\triangle P}{\triangle V}\right) \tag{2-9}$$

where β = bulk modulus(psi),
$\quad\quad V$ = original volume (in.^{3}),
$\quad\quad \triangle P$ = change in pressure (psi),
$\quad\quad \triangle V$ = change in volume (in.^{3}).

The bulk modulus of an oil changes somewhat with pressure and temperature, but within the operating ranges in most fluid power systems, this factor can be neglected. A typical value for oil is 250,000 psi.

EXAMPLE 2-6

A 10-in.^{3} sample of oil is compressed in a cylinder until its pressure is increased from 100 to 2000 psi. If the bulk modulus equals 250,000 psi, find the change in volume of the oil.

Solution   Rewriting Eq. (2-9) to solve for $\triangle V$, we have

$$\triangle V = -V\left(\frac{\triangle P}{\beta}\right) = -10\left(\frac{1900}{250,000}\right) = -0.076 \text{ in.}^3$$

This represents only a 0.76% decrease in volume, which shows that the oil is highly incompressible.

2.8 VISCOSITY AND VISCOSITY INDEX

Viscosity is probably the single most important property of a hydraulic fluid. It is a measure of the sluggishness with which a fluid moves. When the viscosity is low, the fluid flows easily because it is thin and has a low body. A fluid that flows with difficulty has a high viscosity and is thick in appearance with a high body.

In reality, the ideal viscosity for a given hydraulic system is a compromise. Too high a viscosity results in

1. High resistance to flow, which causes sluggish operation.
2. Increased power consumption due to frictional losses.
3. Increased pressure drop through valves and lines.
4. High temperatures caused by friction.

On the other hand, if the viscosity is too low, the result is

1. Increased leakage losses past seals.
2. Excessive wear due to breakdown of the oil film between moving parts. These parts may be internal components of a pump or even a sliding valve spool inside its valve body, as shown in Fig. 2-12.

The concept of viscosity can be understood by examining two parallel plates separated by an oil film of thickness y, as illustrated in Fig. 2-13. The lower plate is stationary, whereas the upper plate moves with a velocity v as it is being pushed by a force F as shown. Due to viscosity, the oil adheres to both surfaces. Thus, the velocity of the layer of fluid in contact with the lower plate is zero, and the velocity of the layer in contact with the top plate is v. The consequence is a linearly varying velocity profile whose slope is v/y. The absolute viscosity of the oil can be represented mathematically:

$$\mu = \frac{\tau}{v/y} = \frac{\text{shear stress in oil}}{\text{slope of velocity profile}}$$ (2-10)

where τ = Greek symbol tau = the shear stress in the fluid in units of force per unit area (lb/ft^2); the shear stress is caused by the adjacent sliding layers of oil film;
v = velocity of the moving plate (ft/sec);
y = oil film thickness (ft);
μ = Greek symbol mu = the absolute viscosity of the oil.

Checking units for μ in the English system, we have

$$\mu = \frac{lb/ft^2}{(ft/sec)/ft} = \frac{lb \cdot sec}{ft^2}$$

If the moving plate has unit surface area in contact with the fluid and the upper plate velocity and oil film thickness are given unit values, Eq. (2-10) becomes

$$\mu = \frac{\tau}{v/y} = \frac{\tau}{1/1} = \tau = \frac{\text{shear force}}{\text{shear area}} = \frac{\text{shear force}}{1} = F$$

We can, therefore, define absolute viscosity as the force required to move a flat plate of unit area at unit distance from a fixed plate with unit relative velocity when the space between the plates is filled with a fluid whose viscosity is to be measured.

If inches are used for the dimension of length, the units of μ become $lb \cdot sec/in.^2$ A viscosity of 1 $lb \cdot sec/in.^2$ is called a Reyn.

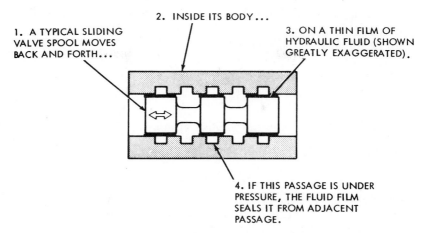

1. A TYPICAL SLIDING VALVE SPOOL MOVES BACK AND FORTH...

2. INSIDE ITS BODY...

3. ON A THIN FILM OF HYDRAULIC FLUID (SHOWN GREATLY EXAGGERATED).

4. IF THIS PASSAGE IS UNDER PRESSURE, THE FLUID FILM SEALS IT FROM ADJACENT PASSAGE.

Figure 2-12.   Fluid film lubricates and seals moving parts. (*Courtesy of Sperry Vickers, Sperry Rand Corp., Troy, Michigan.*)

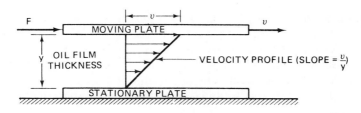

Figure 2-13.   Fluid velocity profile between parallel plates due to viscosity.

In the metric system, the units are

$$\mu = \frac{\tau}{v/y} = \frac{\text{dyne/cm}^2}{(\text{cm/s})/\text{cm}} = \text{dyne·s/cm}^2 = \text{Poise}$$

where a dyne is the force that will accelerate a 1-gram mass at a rate of 1 cm per sec². The conversions between dynes and newtons and between centimeters (cm) and meters (m) are as follows:

$$1 \text{ N} = 10^5 \text{ dynes}$$

$$1 \text{ m} = 100 \text{ cm}$$

A viscosity of 1 dyne · s/cm² is called a Poise.

Both the Reyn and the Poise are very large units. More convenient units are the microreyn (one-millionth of a Reyn) and the centipoise (one-hundredth of a Poise). A centipoise is commonly abbreviated cP.

Calculations in hydraulic systems often involve the use of kinematic viscosity rather than absolute viscosity. Kinematic viscosity (ν) equals absolute viscosity (μ) divided by mass density (ρ):

$$\nu = \frac{\mu}{\rho} \qquad\qquad (2\text{-}11)$$

Checking units for ν in the English system, we have

$$\nu = \frac{\text{lb·sec/ft}^2}{\text{slugs/ft}^3} = \frac{\text{lb·sec/ft}^2}{(\text{lb/ft}^3)/(\text{ft/sec}^2)} = \text{ft}^2/\text{sec}$$

A viscosity of 1 in.^{2}/sec is called a Newt.
In the metric system the units for ν are

$$\nu = \frac{\text{dyne·s/cm}^2}{\text{g/cm}^3} = \frac{\text{dyne·s/cm}^2}{(\text{dyne/cm}^3)/(\text{cm/s}^2)} = \text{cm}^2/\text{s}$$

where 1 kilogram (kg) equals 1000 grams (g).

A viscosity of 1 cm^2/sec is called a stoke.
Because the stoke is a large unit, viscosity is often reported in centistokes (cS). A centistoke equals one-hundredth of a stoke.
The conversion formula changing absolute viscosity to microreyns from Centipoise is

$$\mu(\text{microreyns}) = 0.145 \times \mu(\text{cP}) \qquad\qquad (2\text{-}12)$$

The conversion formula changing kinematic viscosity to Newts from centistokes is

$$\nu(\text{Newts}) = 0.00155 \times \nu(\text{cS}) \qquad\qquad (2\text{-}13)$$

The viscosity of a fluid is usually measured by a Saybolt viscosimeter, which is shown schematically in Fig. 2-14. Basically, this device consists of an inner chamber containing the sample of oil to be tested. A separate outer compartment, which completely surrounds the inner chamber, contains a quantity of oil whose temperature is controlled by an electrical thermostat and heater. A standard orifice is located at the bottom of the center oil chamber. When the oil sample is at the desired temperature, the time it takes to fill a 60-cm^3 container through the metering orifice is then recorded. The time (t), measured in seconds, is the viscosity of the oil in official units called Saybolt Universal Seconds (SUS). Since a thick liquid flows slowly, the SUS viscosity will be higher than for a thin liquid.

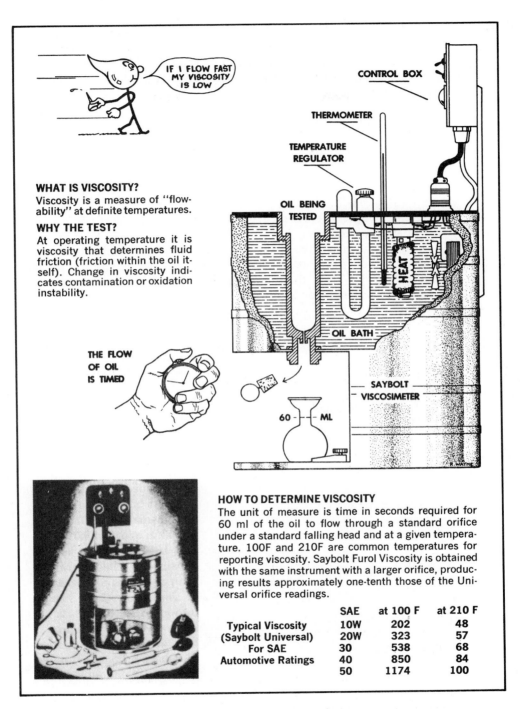

WHAT IS VISCOSITY?
Viscosity is a measure of "flow-ability" at definite temperatures.

WHY THE TEST?
At operating temperature it is viscosity that determines fluid friction (friction within the oil itself). Change in viscosity indicates contamination or oxidation instability.

THE FLOW OF OIL IS TIMED

IF I FLOW FAST MY VISCOSITY IS LOW

CONTROL BOX

THERMOMETER

TEMPERATURE REGULATOR

OIL BEING TESTED

HEAT

OIL BATH

SAYBOLT VISCOSIMETER

60 — ML

R. WAYNE

HOW TO DETERMINE VISCOSITY
The unit of measure is time in seconds required for 60 ml of the oil to flow through a standard orifice under a standard falling head and at a given temperature. 100F and 210F are common temperatures for reporting viscosity. Saybolt Furol Viscosity is obtained with the same instrument with a larger orifice, producing results approximately one-tenth those of the Universal orifice readings.

	SAE	at 100 F	at 210 F
Typical Viscosity	10W	202	48
(Saybolt Universal)	20W	323	57
For SAE	30	538	68
Automotive Ratings	40	850	84
	50	1174	100

Figure 2-14. Saybolt viscosimeter. (*Courtesy of USX Corp., Pittsburgh, Pennsylvania.*)

A relationship exists between the viscosity in SUS and the corresponding metric system units of centistokes (cS). This relationship is provided by the following empirical equations:

$$\nu(cS) = 0.226t - \frac{195}{t}, \; t \leqslant 100 \; SUS \tag{2-14}$$

$$\nu(cS) = 0.220t - \frac{135}{t}, \; t \geqslant 100 \; SUS \tag{2-15}$$

where the symbol ν represents the viscosity in centistokes and t is measured in SUS or simply seconds.

Kinematic viscosity is defined as absolute viscosity divided by mass density of the oil. Since, in the metric system, mass density equals specific gravity (because $\rho_{H_2O} = 1 \; g/cm^3$), the following equation can be used to find the absolute viscosity if the kinematic viscosity is known:

$$\nu(cS) = \frac{\mu(cP)}{S_g} \tag{2-16}$$

Since the specific gravity of most hydraulic fluids equals about 0.9, Eq. (2-16) reduces to

less exact.

$$\nu(cS) = \frac{\mu(cP)}{0.9} \tag{2-17}$$

EXAMPLE 2-7

An oil has a viscosity of 230 SUS at 150°F. Find the corresponding viscosity in units of centistokes and centipoise.

Solution

$$\nu(cS) = 0.220t - \frac{135}{t} = (0.220)(230) - \frac{135}{230} = 50 \; cS$$

$$\mu(cP) = 0.9\nu \; (cS) = (0.9)(50) = 45 \; cP$$

A quick method to determine the kinematic viscosity of fluids in cS and absolute viscosity in cP is shown in Fig. 2-15. This test measures flow through a capillary instrument using gravity flow at constant temperature. The time in seconds is then multiplied by the calibration constant for the viscosimeter to obtain

WHAT IS KINEMATIC VISCOSITY?

Kinematic viscosity is the time required for a fixed amount of an oil to flow through a capillary tube under the force of gravity. The unit of kinematic viscosity is the stoke or centistoke (1/100 of a stoke). Kinematic viscosity may be defined as the quotient of the absolute viscosity in centipoises divided by the specific gravity of a fluid, both at the same temperature—

$$\frac{\text{Centipoises}}{\text{Specific Gravity}} = \text{Centistokes}$$

WHY THE TEST?

The test is a precise viscosity measurement of fluids.

WHAT IS ABSOLUTE VISCOSITY?

Absolute viscosity is the tangential force on unit area of either one of two parallel planes at unit distance apart when the space is filled with liquid and one of the planes moves relative to the other with unit velocity in its own plane. The egs unit of absolute viscosity is the poise (which has the dimensions, grams per centimeter per second). The centipoise is 1/100 of a poise and is the unit of absolute viscosity most commonly used.

TYPICAL RESULTS

225 x 0.1200 = 27.0 Centistokes

Where 225 seconds = Flow Time

0.1200 cs/sec = Viscometer Constant

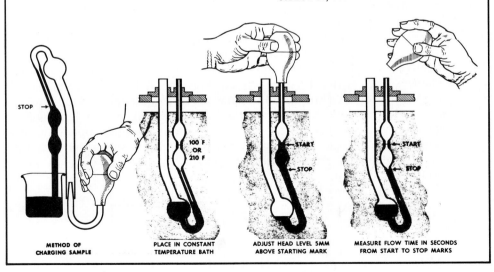

| METHOD OF CHARGING SAMPLE | PLACE IN CONSTANT TEMPERATURE BATH | ADJUST HEAD LEVEL 5MM ABOVE STARTING MARK | MEASURE FLOW TIME IN SECONDS FROM START TO STOP MARKS |

Figure 2-15. Kinematic viscosimeter. (*Courtesy of USX Corp., Pittsburgh, Pennsylvania.*)

the kinematic viscosity of the sample oil in centistokes. The absolute viscosity in centipoise is then calculated using Eq. (2-16).

EXAMPLE 2-8

An oil having a density of 0.89 g/cm^3 is tested using a kinematic viscosimeter. The given amount of oil flowed through the capillary tube in 250 s. The calibration constant is 0.100. Find the kinematic and absolute viscosities in units of cS and cP, respectively.

Solution   Kinematic viscosity equals the time in seconds multiplied by the calibration constant:

$$\nu(cS) = (250)(0.100) = 25 \text{ cS}$$

In the metric system mass density equals specific gravity. Therefore, $S_g = 0.89$. Solving Eq. (2-16) for μ, we have

$$\mu(cP) = S_g \times \nu(cS) = 0.89 \times 25 = 22.3 \text{ cP}$$

Oil becomes thicker as the temperature decreases and thins when heated. Hence, the viscosity of a given oil must be expressed at a specified temperature. For most hydraulic applications, the viscosity normally equals about 150 SUS at 100°F. It is a general rule of thumb that the viscosity should never fall below 50 SUS or rise above 4000 SUS regardless of the temperature. Where extreme temperatures are encountered, the fluid should have a high viscosity index.

Viscosity index (*VI*) is a relative measure of an oil's viscosity change with respect to temperature change. An oil having a low *VI* is one that exhibits a large change in viscosity with temperature change. A high *VI* oil is one that has a relatively stable viscosity, which does not change appreciably with temperature change. The original *VI* scale ranged from 0 to 100, representing the poorest to best characteristics known at that time. Today, with improved refining techniques and chemical additives, oils exist with *VI* values well above 100. A high *VI* oil is a good all-weather-type oil for use with outdoor machines operating in extreme temperature swings. This is where viscosity index is significant. For a hydraulic system where the temperature is relatively constant, the viscosity index of the fluid is not critical.

The *VI* of any hydraulic fluid can be found by using

$$VI = \frac{L - U}{L - H} \times 100 \tag{2-18}$$

where L = viscosity of 0-*VI* oil at 100°F,
   U = viscosity of unknown oil at 100°F,
   H = viscosity of 100-*VI* oil at 100°F.

WHAT IS VISCOSITY INDEX?

The Viscosity Index is an empirical number indicating the rate of change in viscosity of an oil within a given temperature range. A low viscosity index signifies a relatively large change in viscosity with temperature, whereas a high viscosity index shows a relatively small change in viscosity with temperature. Viscosity Index cannot be used to measure any other quality of an oil.

Viscosity Index is calculated as follows:

$$VI = \frac{L - U}{L - H} \times 100$$

U = viscosity in SUS at 100F of the oil whose viscosity index is to be calculated.

L = viscosity in SUS at 100F of an oil of 0 viscosity index having the same viscosity at 210F as the oil whose viscosity index is to be calculated.

H = viscosity in SUS at 100F of an oil of 100 viscosity index having the same viscosity at 210F as the oil whose viscosity index is to be calculated.

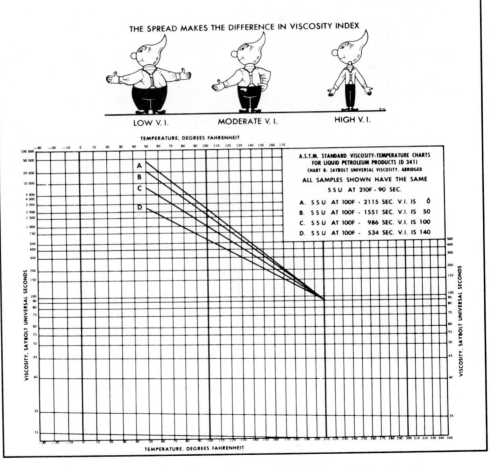

Figure 2-16. Viscosity index. (*Courtesy of USX Corp., Pittsburgh, Pennsylvania.*)

The *VI* of an unknown oil is determined from tests. A reference oil of 0 *VI* and a reference oil of 100 *VI* are selected, each of which has uniquely the same viscosity at 210°F as the unknown oil. The viscosities of the three oils are then measured at 100°F to give values for *L, U,* and *H*.

Basically, the *VI* number of an oil represents the percent the unknown oil's viscosity sensitivity is from a 0-*VI* oil toward a 100-*VI* oil. Thus, a *VI* of 50 indicates an oil having characteristics halfway between a 0-*VI* oil and a 100-*VI* oil relative to sensitivity with temperature.

Figure 2-16 shows the results of a viscosity index test where oil A is the 0-*VI* oil and oil C is the 100-*VI* oil. The two unknown oils B and D were found to have *VI* values of 50 and 140, respectively.

EXAMPLE 2-9

A sample of oil with a *VI* of 80 is tested with a 0-*VI* oil and a 100-*VI* oil whose viscosity values at 100°F are 400 and 150 SUS, respectively. What is the viscosity of the sample oil at 100°F in units of SUS?

Solution   Substitute directly into Eq. (2-18):

$$VI = \frac{L - U}{L - H} \times 100$$

$$80 = \frac{400 - U}{400 - 150} \times 100$$

$$U = 200 \text{ SUS}$$

Another characteristic relating to viscosity is called the pour point, which is the lowest temperature at which a fluid will flow. It is a very important parameter to specify for hydraulic systems that will be exposed to extremely low temperatures. As a rule of thumb, the pour point should be at least 20°F below the lowest temperature to be experienced by the hydraulic system.

2.9 OXIDATION AND CORROSION PREVENTION

Oxidation, which is caused by the chemical reaction of oxygen from the air with particles of oil, can seriously reduce the service life of a hydraulic fluid. Petroleum oils are especially susceptible to oxidation because oxygen readily unites with both carbon and hydrogen molecules. Most products of oxidation are soluble in oil and are acidic in nature, which can cause corrosion of parts throughout the system. The products of oxidation include insoluble gums, sludge, and varnish, which tend to increase the viscosity of the oil.

There are a number of parameters that hasten the rate of oxidation once it begins. Included among these are heat, pressure, contaminants, water, and metal surfaces. However, oxidation is most dramatically affected by temperature. The rate of oxidation is very slow below 140°F but doubles for every 20°F temperature rise. Additives are incorporated in many hydraulic oils to inhibit oxidation. Since this increases the costs, they should be specified only if necessary, depending on temperature and other environmental conditions.

Rust and corrosion are two different phenomenons, although they both contaminate the system and promote wear. Rust is the chemical reaction between iron or steel and oxygen. The presence of moisture in the hydraulic system provides the necessary oxygen. One primary source of moisture is from atmospheric air, which enters the reservoir through the breather cap. Figure 2-17 shows a steel part that has experienced rusting due to moisture in the oil.

Corrosion, on the other hand, is the chemical reaction between a metal and acid. The result of rusting or corrosion is the "eating away" of the metal surfaces of hydraulic components. This can cause excessive leakage past the sealing surfaces of the affected parts. Figure 2-18 shows a new valve spool and a used one, which has areas of corrosion caused by acid formation in the hydraulic oil. Rust and corrosion can be resisted by incorporating additives that plate on the metal surfaces to prevent chemical reaction.

Figure 2-19 illustrates the technique and operation for performing the inhibited oil oxidation test. The purpose of the test is to measure the resistance to oxidation by measuring the change in acidity in the oil due to absorbed oxygen. The test procedure is as follows: 300 ml of sample oil are placed in a tube and immersed in an oil bath at 95°C. Three liters per hour of oxygen are allowed to pass continuously through the sample for a period of 1000–4000 hr. The acidity of the oil is then measured by determining the neutralization number, as discussed in Sec. 2.13.

Figure 2-17. Rust caused by moisture in the oil. (*Courtesy of Sperry Vickers, Sperry Rand Corp., Troy, Michigan.*)

Figure 2-18. Corrosion caused by acid formation in the hydraulic oil. (*Courtesy of Sperry Vickers, Sperry Rand Corp., Troy, Michigan.*)

In Fig. 2-20 we see a test setup for determining the ability of oil to prevent rusting during the lubrication of ferrous parts in the presence of water. The test procedure is as follows: A steel test rod is placed in a beaker of oil and water at 140°F for 24 hr, and the amount of rusting is reported as light, moderate, or severe.

2.10 FIRE-RESISTANT FLUIDS

It is imperative that a hydraulic fluid not initiate or support a fire. Most hydraulic fluids will, however, burn under certain conditions. There are many hazardous applications where human safety requires the use of a fire-resistant fluid. Examples include coal mines, hot metal processing equipment, aircraft, and marine fluid power systems.

A fire-resistant fluid is one that can be ignited but will not support combustion when the ignition source is removed. Flammability is defined as the ease of ignition and ability to propagate a flame. The following are the usual characteristics tested for in order to determine the flammability of a hydraulic fluid:

1. *Flash point:* the temperature at which the oil surface gives off sufficient vapors to ignite when a flame is passed over the surface
2. *Fire point:* the temperature at which the oil will release sufficient vapor to support combustion continuously for five seconds when a flame is passed over the surface
3. *Autogenous ignition temperature* (AIT): the temperature at which ignition occurs spontaneously

WHAT IS THE NEUTRALIZATION NUMBER?

A number expressed in milligrams (mg) of potassium hydroxide needed to neutralize the acid in one gram of oil.

WHY THE TEST?

To show relative changes in an oil under oxidizing conditions. It measures development of injurious products in oils.

TEST PROCEDURE

A weighed amount of sample in titration solvent is titrated with a standard alcoholic potassium hydroxide solution to a definite end point. The indicator is para-naphtholbenzein solution. The color change is from orange to green or blue-green.

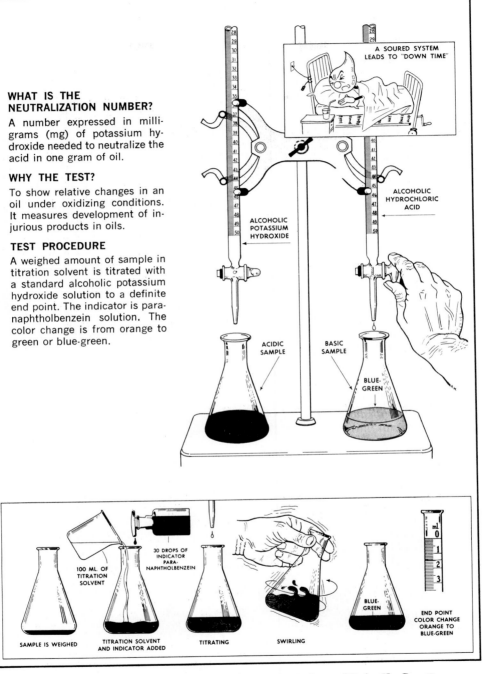

Figure 2-19. Oil oxidation test. (*Courtesy of USX Corp., Pittsburgh, Pennsylvania.*)

PASS

FAIL

A GOOD LUBRICATING OIL MUST NOT ONLY LUBRICATE BUT PREVENT RUSTING IN THE PRESENCE OF WATER

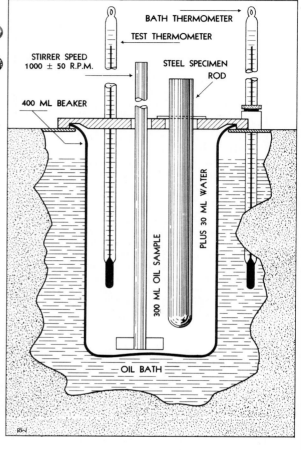

BATH THERMOMETER

TEST THERMOMETER

STIRRER SPEED 1000 ± 50 R.P.M.

STEEL SPECIMEN ROD

400 ML BEAKER

300 ML OIL SAMPLE

PLUS 30 ML WATER

OIL BATH

WHAT IS THE RUST PREVENTION TEST?

The rust prevention test indicates the ability of lubricating oil to prevent corrosion during the lubrication of ferrous parts in the presence of water.

WHY THE TEST?

Corrosion of mechanical parts, lines, and tanks in connection with lubrication contributes to high maintenance costs.

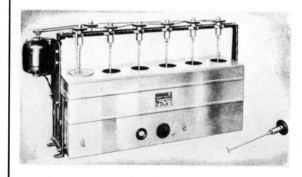

TEST PROCEDURE

The steel test rod is placed in the beaker of oil and water at 140F for 24 hours and the results reported as pass or fail. The degree of failure may be light, moderate, or severe.

Figure 2-20. Rust prevention test. (*Courtesy of USX Corp., Pittsburgh, Pennsylvania.*)

Fire-resistant fluids have been developed to reduce fire hazards. There are basically four different types of fire-resistant hydraulic fluids in common use:

1. Water-glycol solutions. This type consists of an actual solution of about 40% water and 60% glycol. These solutions have high viscosity index values, but the viscosity rises as the water evaporates. The operating temperature range runs from −10°F to about 180°F. Most of the newer synthetic seal materials are compatible with water-glycol solutions. However, metals such as zinc, cadmium, and magnesium react with water-glycol solutions and therefore should not be used. In addition, special paints must be used.

2. Water-in-oil emulsions. This type consists of about 40% water completely dispersed in a special oil base. It is characterized by the small droplets of water completely surrounded by oil. The water provides a good coolant property but tends to make the fluid more corrosive. Thus, greater amounts of corrosion inhibitor additives are necessary. The operating temperature range runs from −20°F to about 175°F. As is the case with water-glycol solutions, it is necessary to replenish evaporated water to maintain proper viscosity. Water-in-oil emulsions are compatible with most rubber seal materials found in petroleum-base hydraulic systems.

3. Straight synthetics. This type is chemically formulated to inhibit combustion and in general has the highest fire-resistant temperature. Typical fluids in this category are the phosphate esters or clorinated hydrocarbons. Disadvantages of straight synthetics include low viscosity index, incompatibility with most natural or synthetic rubber seals, and high costs. In particular, the phosphate esters readily dissolve paints, pipe thread compounds, and electrical insulation.

4. High-water-content fluids. This type consists of about 90% water and 10% concentrate. The concentrate consists of fluid additives that improve viscosity, lubricity, rust protection, and protection against bacteria growth. Advantages of high-water-content fluids include high fire resistance, outstanding cooling characteristics, and low cost, which is about 20% of the cost of petroleum-based hydraulic fluids. Maximum operating temperatures should be held to 120°F to minimize evaporation. Due to a somewhat higher density and lower viscosity compared to petroleum-based fluids, pump inlet conductors should be sized to keep fluid velocities low enough to prevent the formation of vapor bubbles, which causes cavitation. High-water-content fluids are compatible with most rubber seal materials but leather, paper, or cork materials should not be used since they tend to deteriorate in water.

2.11 FOAM-RESISTANT FLUIDS

Air can become dissolved or entrained in hydraulic fluids. For example, if the return line to the reservoir is not submerged, the jet of oil entering the liquid surface will carry air with it. This causes air bubbles to form in the oil. If these bubbles rise to the surface too slowly, they will be drawn into the pump intake. This can cause pump damage due to cavitation, which is discussed in Chapter 6.

In a similar fashion, a small leak in the suction line can cause the entrainment of large quantities of air from the atmosphere. This type of leak is difficult to find since air leaks in rather than oil leaking out. Another adverse effect of entrained and dissolved air is the great reduction in the bulk modulus of the hydraulic fluid. This can have serious consequences in terms of stiffness and accuracy of hydraulic actuators.

The amount of dissolved air can be greatly reduced by properly designing the reservoir since this is where the vast majority of the air is picked up. This subject is covered in Chapter 13.

Another method is to use premium-grade hydraulic fluids that contain foam-resistant additives. These additives are chemical compounds, which break out entrained air to separate quickly the air from the oil while it is in the reservoir.

2.12 LUBRICATING ABILITY

Hydraulic fluids must have good lubricity to prevent wear between the closely fitted working parts. Direct metal-to-metal contact is avoided by the film strength of fluids having adequate viscosity, as shown in Fig. 2-21. Hydraulic parts that are affected include pump vanes, valve spools, rings, and rod bearings.

Wear is the actual removal of surface material due to the frictional force between two mating surfaces. This can result in a change in component dimension, which can lead to looseness and subsequent improper operation.

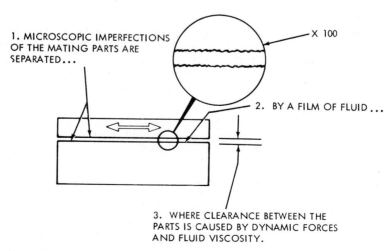

1. MICROSCOPIC IMPERFECTIONS OF THE MATING PARTS ARE SEPARATED...

X 100

2. BY A FILM OF FLUID...

3. WHERE CLEARANCE BETWEEN THE PARTS IS CAUSED BY DYNAMIC FORCES AND FLUID VISCOSITY.

Figure 2-21.   Lubricating film prevents metal-to-metal contact. (*Courtesy Sperry Vickers, Sperry Rand Corp., Troy, Michigan.*)

The friction force (F) is the force parallel to the two mating surfaces that are sliding relative to each other. This friction force actually opposes the sliding movement between the two surfaces. The greater the frictional force, the greater the wear and heat generated. This, in turn, results in power losses and reduced life, which, in turn, increase maintenance costs.

It has been determined that the friction force (F) is proportional to the normal force (N) that forces the two surfaces together. The proportionality constant (CF) is called the coefficient of friction:

$$F = (CF) \times N \qquad \left(F = \mu_i N \right) \qquad \text{(2-19)}$$

Thus, the greater the value of coefficient of friction and normal force, the greater the frictional force and hence wear. The magnitude of the normal force depends on the amount of power and forces being transmitted and thus is independent of the hydraulic fluid properties. However, the coefficient of friction depends on the ability of the fluid to prevent metal-to-metal contact of the closely fitting mating parts.

Equation (2-19) can be rewritten to solve for the coefficient of friction, which is a dimensionless parameter:

$$CF = \frac{F}{N} \qquad \text{(2-20)}$$

It can be seen now that CF can be experimentally determined to give an indication of the antiwear properties of a fluid if F and N can be measured. One such test is performed by the four-ball wear tester, which is illustrated in Fig. 2-22. This device exerts a vertical force through a rotating steel ball and measures the coefficient of friction and material displaced from three stationary balls. The device rotates the fourth ball (under specified load, speed, and temperature) against the three stationary steel balls using a lubricant. After the test is completed, the coefficient of friction is calculated using the following equation:

$$CF = 2.83 \frac{F}{L} \times \frac{r}{s} \qquad \text{(2-21)}$$

where s = diameter of each steel ball (cm),
L = total applied load on steel balls (g),
F = force acting on torque arm (g),
r = length of torque arm (cm),
CF = coefficient of friction (dimensionless).

Typical values of coefficient of friction for hydraulic fluids will range from 0.003 to 0.06. Antiwear characteristics of the fluid can be further determined by measuring the wear areas formed on the three stationary balls. This is normally done after running the test for a 1-hr period.

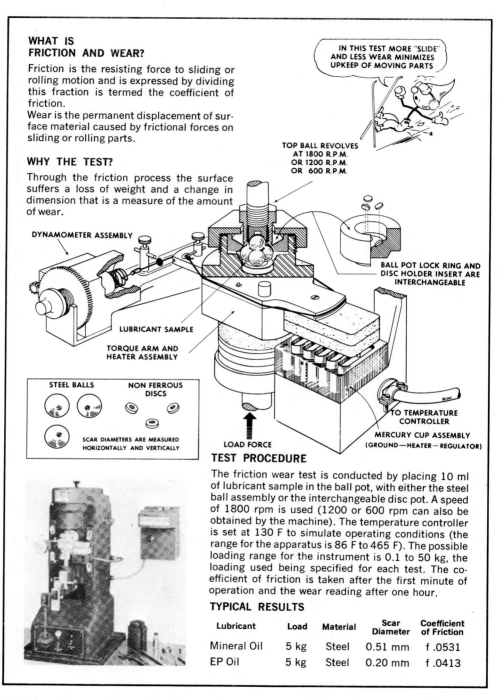

WHAT IS FRICTION AND WEAR?

Friction is the resisting force to sliding or rolling motion and is expressed by dividing this fraction is termed the coefficient of friction.

Wear is the permanent displacement of surface material caused by frictional forces on sliding or rolling parts.

WHY THE TEST?

Through the friction process the surface suffers a loss of weight and a change in dimension that is a measure of the amount of wear.

IN THIS TEST MORE "SLIDE" AND LESS WEAR MINIMIZES UPKEEP OF MOVING PARTS

TOP BALL REVOLVES
AT 1800 R.P.M.
OR 1200 R.P.M.
OR 600 R.P.M.

DYNAMOMETER ASSEMBLY

BALL POT LOCK RING AND DISC HOLDER INSERT ARE INTERCHANGEABLE

LUBRICANT SAMPLE

TORQUE ARM AND HEATER ASSEMBLY

STEEL BALLS   NON FERROUS DISCS

SCAR DIAMETERS ARE MEASURED HORIZONTALLY AND VERTICALLY

LOAD FORCE

TO TEMPERATURE CONTROLLER

MERCURY CUP ASSEMBLY
(GROUND—HEATER—REGULATOR)

TEST PROCEDURE

The friction wear test is conducted by placing 10 ml of lubricant sample in the ball pot, with either the steel ball assembly or the interchangeable disc pot. A speed of 1800 rpm is used (1200 or 600 rpm can also be obtained by the machine). The temperature controller is set at 130 F to simulate operating conditions (the range for the apparatus is 86 F to 465 F). The possible loading range for the instrument is 0.1 to 50 kg, the loading used being specified for each test. The coefficient of friction is taken after the first minute of operation and the wear reading after one hour.

TYPICAL RESULTS

Lubricant	Load	Material	Scar Diameter	Coefficient of Friction
Mineral Oil	5 kg	Steel	0.51 mm	f .0531
EP Oil	5 kg	Steel	0.20 mm	f .0413

Figure 2-22. Four-ball wear testers. (*Courtesy of USX Corp., Pittsburgh, Pennsylvania.*)

EXAMPLE 2-10

A lubrication test was run on a hydraulic fluid using a four-ball wear tester having a ball diameter (s) of 1.30 cm and a torque arm radius (r) of 8.00 cm. During the test a force (F) of 47 g was measured on the torque arm when a constant vertical load (L) of 20 kg was applied to the rotating ball. Calculate the coefficient of friction.

Solution   Use Eq. (2-21):

$$CF = (2.83)\left(\frac{47}{20,000}\right)\left(\frac{8.00}{1.30}\right) = 0.041$$

Another antiwear test is conducted using a hydraulic vane-type pump, as shown in Fig. 2-23. The tank is filled with $3\frac{1}{2}$ gal of the sample hydraulic fluid. After starting, the pressure is slowly built up to 1000 psi by adjusting the relief valve. The test is then run at a specified temperature and for a given amount of time. During the test run, periodic checks are made for viscosity, color, and neutralization number. At the conclusion of the test, the fluid is removed for final analysis. The pump parts and filter are weighed and inspected. This test measures the conditions encountered by vane-type pumps in actual operation.

2.13 NEUTRALIZATION NUMBER

The neutralization number is a measure of the relative acidity or alkalinity of a hydraulic fluid and is specified by a pH factor. A fluid having a small neutralization number is recommended because high acidity or alkalinity causes corrosion of metal parts as well as deterioration of seals and packing glands.

For an acidic fluid, the neutralization number equals the number of milligrams (mg) of potassium hydroxide necessary to neutralize the acid in a 1-g sample of the fluid. In the case of an alkaline (basic) fluid, the neutralization number equals the amount of alcoholic hydrochloric acid (expressed as an equivalent number of milligrams of potassium) that is necessary to neutralize the alkaline in a 1-g sample of the hydraulic fluid. Hydraulic fluids normally become acidic rather than basic with use.

Figure 2-24 shows a test apparatus for determining the neutralization number of a hydraulic fluid. For an acidic fluid, the test procedure is as follows:

1. The oil sample is placed in a titrating solution of toluene, distilled water, alcohol, and an indicating agent, naphthol benzene, which changes color from orange to green when neutralization occurs.

2. Alcoholic potassium hydroxide solution is added from a graduated burette a drop at a time until the solution changes color from orange to green.

Hydraulic Pump Test
ASTM D-2271

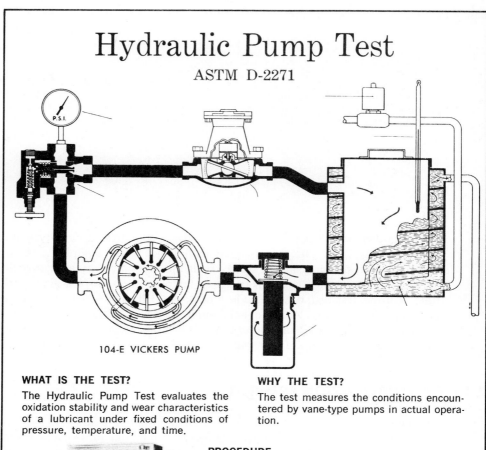

104-E VICKERS PUMP

WHAT IS THE TEST?

The Hydraulic Pump Test evaluates the oxidation stability and wear characteristics of a lubricant under fixed conditions of pressure, temperature, and time.

WHY THE TEST?

The test measures the conditions encountered by vane-type pumps in actual operation.

PROCEDURE

The machine is thoroughly flushed with a solvent. This is followed by a 24-hour preliminary run of the test oil, which is then discarded.

The machine is assembled using new pump parts that have been carefully adjusted and the weights are recorded of all parts including the filter screen. The tank is filled with 3½ gallons of sample with the Simply-trol adjusted to the desired temperature. After starting, pressure is slowly built up to 1000 psi over a two to three-hour period by adjusting the relief valve. Machine readings are recorded periodically as indicated on the report tabulation.

During the test run periodic checks are made for viscosity, color, neutralization, and precipitation numbers. At the conclusion of the test the oil is removed for final analysis, weighing, and inspection of pump parts and filter.

Figure 2-23. Hydraulic pump test. (*Courtesy of USX Corp., Pittsburgh, Pennsylvania.*)

WHAT IS INHIBITED OIL OXIDATION?

Inhibited oil oxidation is a chemical action between oil and oxygen, producing reaction products of acid and sludge. It is intended to measure the oxidation inhibitor life of inhibited turbine oils.

WHY THE TEST?

To measure resistance to oxidation by measuring the change in acidity in the oil by the oxygen absorbed.

TEST PROCEDURE

A 300-ml sample is placed in the tube and immersed in the bath at 95 C. Three liters per hour of oxygen are allowed to pass continuously through the sample for a period of 1000 or more hours. The acid number determined for neutralization value is the determining factor in the test. Limits may be established between 0.25 and 2.0 mg of KOH per gram of oil.

TYPICAL RESULTS

1000 hours to 4000 hours.

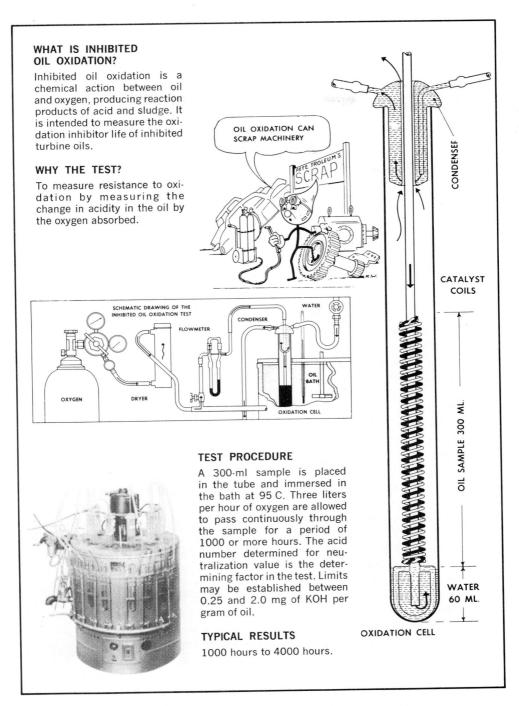

Figure 2-24. Neutralization number. (*Courtesy of USX Corp., Pittsburgh, Pennsylvania.*)

3. After each drop of the alcoholic potassium hydroxide is added, the solution is swirled.

4. Neutralization occurs when the swirling causes the color to change for at least 15 sec.

5. The neutralization number is then calculated using the following formula:

$$\text{neutralization number} = \frac{\text{total ml of titrating solution} \times 5.61}{\text{weight of sample used}}$$

Hydraulic fluids that have been treated with additives to inhibit the formation of acids are usually able to keep this number at a low value between 0 and 0.1.

2.14 GENERAL TYPES OF FLUIDS

The first major category of hydraulic fluids is the petroleum-based fluid, which is the most widely used type. If the crude oil is quality-refined, it is generally satisfactory for light services. However, additives must be included to meet with the requirements of good lubricity, high viscosity index, and oxidation resistance. Most of the desirable properties, if not already present in the petroleum oil, can be obtained by the addition of additives.

The primary disadvantage of a petroleum-based fluid is that it will burn. As a result, the second category of fluids has been developed: the fire-resistant fluid. This greatly reduces the danger of a fire. However, fire-resistant fluids generally have a higher specific gravity than do petroleum-based fluids. This may cause cavitation problems in the pump due to excessive vacuum pressure in the pump inlet line unless proper design steps are implemented. Also most fire-resistant fluids are more expensive and have more compatibility problems with seal materials. Therefore, fire-resistant fluids should be used only if hazardous operating conditions exist. Manufacturer's recommendations should be followed very carefully when changing from a petroleum-based fluid to a fire-resistant fluid and vice versa. Normally, thorough draining, cleaning, and flushing are required. It may even be necessary to change seals and gaskets on the various hydraulic components.

A third category is the conventional MS (most severe) engine-type oil, which provides increased hydraulic system life due to better lubricity. This is due to the antiwear additives used to prevent engine wear on cams and valves. This improved lubricity also provides wear resistance for the heavily loaded hydraulic components such as pumps and valves.

The fourth and final category of fluid is air itself. Air is the only gas commonly used in pneumatic fluid power systems. The reason is that air is inexpensive and readily available. One of the significant advantages of air is that it will not burn. Air can easily be made clean by the use of a filter, and any leaks are not messy since they simply dissipate into the atmosphere. Air can also be made a good lubricant by the introduction of a fine oil mist using a lubricator. Also, the use of air eliminates return lines since the spent air is exhausted into the atmosphere. Disadvantages of air include its compressibility and subsequent sluggishness and lack of rigidity. Finally, air can be corrosive since it contains oxygen and water. However, most of the water can be removed by the use of air dryers.

In summary, the single most important material in a fluid power system is the working fluid. No single fluid possesses all the ideal characteristics desired. The fluid power designer must select the fluid that comes the closest to being ideal overall for a particular application. Only if a fire hazard is present should a fire-resistant fluid be used. The most expensive fluid is not necessarily the best one to use.

Appendixes A and B give properties of common liquids and gases, respectively. The properties are applicable for a temperature of 60°F and include specific gravity, density, and viscosity. Appendix C gives density values for air at temperatures ranging from 30°F up to 300°F and pressures ranging from 0 psi up to 140 psi.

2.15 ILLUSTRATIVE EXAMPLES USING THE METRIC-SI SYSTEM

In this section we use the metric system of units to solve several example problems that were previously solved in this chapter using the English system of units. This will provide a direct comparison between the two systems of units.

EXAMPLE 2-11

Find the pressure on a skin diver who has descended to a depth of 18.3 m in fresh water (compare to example 2-2).

Solution An alternative relationship to equation (2-8) for calculating the pressure produced at the bottom of any column of liquid is:

$$P = \gamma h \qquad (2\text{-}22)$$

Substituting a consistent set of units (English units as well as metric units can be used in equation 2-22) yields:

$$P = (9797 \ N/m^3)(18.3 \ m)$$

$$= 179{,}300 \ Pa = 179.3 \ kPa = 1.793 \ bars = 26 \ psi$$

EXAMPLE 2-12

Convert a $-34,000$ Pa pressure to an absolute pressure (compare to example 2-4).

Solution

$$\text{absolute pressure} = \text{gage pressure} + \text{atmospheric pressure}$$

$$\text{where atmospheric pressure} = \frac{14.7 \text{ psi}}{0.000145 \text{ psi/Pa}} = 101,000 \text{ Pa}$$

$$\text{absolute pressure} = -34,000 + 101,000 = 67,000 \text{ Pa abs} = 9.7 \text{ psia}$$

It should be noted that the "abs" is used to indicate that the 67,000 Pa value is an absolute pressure (measured relative to a perfect vacuum).

EXAMPLE 2-13

A 164 cm^3 sample of oil is compressed in a cylinder until its pressure is increased from 687 kPa to 13,740 kPa or 13.74 MPa. If the bulk modulus equals 1718 MPa, find the percent change in volume of the oil (compare to Example 2-6).

Solution

$$\frac{\triangle V}{V} = \frac{-\triangle P}{\beta} = \frac{-(13,740 - 687)}{1,718,000} = -0.0076 = -0.76\%$$

EXERCISES

2-1. What are the four primary functions of a hydraulic fluid?

2-2. Name 10 properties that a hydraulic fluid should possess.

2-3. Generally speaking, when should a hydraulic fluid be changed?

2-4. What are the differences between a liquid and a gas?

2-5. Name two advantages and two disadvantages that air has when used in a fluid power system.

2-6. Define the terms *weight density, mass density,* and *specific gravity.*

2-7 A hydraulic fluid has a weight density of 55 lb/ft^3. What is its specific gravity? What is its mass density?

2-8. What is the pressure at the bottom of a 30-ft column of the hydraulic fluid of Exercise 2.7?

2-9. What is the difference between pressure and force?

2-10. Differentiate between gage and absolute pressures.

2-11. What is meant by the term *bulk modulus?*

2-12. A 20-in³ sample of oil is compressed in a cylinder until its pressure is increased from 50 to 1000 psi. If the bulk modulus equals 300,000 psi, find the change in volume of the oil.

2-13. Differentiate between the terms *viscosity* and *viscosity index*.

2-14. Name two undesirable results when using an oil with a viscosity that is too high.

2-15. Name two undesirable results when using an oil with a viscosity that is too low.

2-16. Relative to viscosity measurement, what is a Saybolt Universal Second (SUS)?

2-17. An oil has a viscosity of 200 SUS at 120°F. Find the corresponding viscosity in units of centistokes and centipoise.

2-18. A sample of oil with a *VI* of 70 is tested with a 0-*VI* oil and a 100-*VI* oil whose viscosity values at 100°F are 375 and 125 SUS, respectively. What is the viscosity of the sample oil at 100°F in units of SUS?

2-19. Define the term *pour point*.

2-20. What is the difference between oxidation and corrosion?

2-21. Under what conditions should a fire-resistant fluid be used?

2-22. Define the following terms: (1) *flash point*, (b) *fire point*, and (c) *autogenous ignition temperature*

2-23. Name the four different types of fire-resistant fluids.

2-24. Name three disadvantages of fire-resistant fluids.

2-25. What is a foam-resistant fluid, and why would it be used?

2-26. Why must a hydraulic fluid have good lubricating ability?

2-27. Define the term *coefficient of friction*.

2-28. A lubrication test was run on a hydraulic fluid using a four-ball wear tester having the following specifications: (1) the ball diameter equals 1.40 cm and (b) the torque arm radius equals 7.75 cm. During the test a force of 50 g was measured on the torque arm when a constant vertical load of 20 kg was applied to the rotating ball. Calculate the coefficient of friction.

2-29. What is the significance of the neutralization number?

2-30. Why should normal operating temperatures be controlled below 140°F in most hydraulic systems?

2-31. What effect does a higher specific gravity have on the inlet of a pump?

2-32. Why is it necessary to use precautions when changing from a petroleum-base fluid to a fire-resistant fluid and vice-versa?

2-33. Derive the conversion factor between viscosity in units of lb-sec/ft² and N · s/m².

2-34. Derive the conversion factor between viscosity in units of ft²/sec and m²/s.

2-35. Calculate the mass density of a hydraulic oil in units of kg/m³ knowing that the mass density is 1.74 slugs/ft³.

2-36. Find the absolute pressure on the skin diver of Example 2-2 in units of Pa.

2-37. Convert a −2kPa pressure to an absolute pressure in the same units (kPa).

2-38. A 500 cm³ sample of oil is compressed in a cylinder until its pressure is increased from one atmosphere to 50 atmospheres. If the bulk modulus equals 1750 MPa, find the percent change in volume of the oil.

2-39. Fifty gallons of hydraulic oil weighs 372 lb. What is its weight density in units of lb/ft³?

2.40. A container weighs 3 lb when empty, 53 lb when filled with water, and 66 lb when filled with glycerin. Find the specific gravity of the glycerin.

2-41. Derive Eq. (2-8) which is rewritten as follows:

$$P(\text{psi}) = 0.433\,H(\text{ft})\,S_g$$

2-42. What is meant by the term *pressure head*?

2-43. How is the mercury column supported in a barometer?

2-44. How is the viscosity of hydraulic oil affected by temperature change?

2-45. Under what condition is viscosity index important?

2-46. A 100-gallon reservoir is to be mounted within a 2ft × 2ft space. What is the minimum height of the reservoir?

2-47. What are two serious changes that can occur to a hydraulic fluid relative to the operation of a hydraulic system?

2-48. As a fluid becomes more compressible, does the bulk modulus decrease or increase?

2-49. The retracting load on a 2-inch-diameter hydraulic cylinder increases from 10,000 pounds to 15,000 pounds. Due to the compressibility of the oil, the piston retracts 0.01 in. If the volume of oil under compression is 10 in.³, what is the bulk modulus of the oil in units of psi?

2-50. Convert the data of Exercise 2-49 to metric units and determine the bulk modulus of the oil in units of MPa.

2-51. In the hydraulic press shown in Figure 2-25, a force of 100N is exerted on the small piston. Determine the upward force on the large piston. The area of the small piston is 50 cm² and the area of the large piston is 500 cm².

2-52. For the system of Exercise 2-51 shown in Figure 2-25, if the small piston moves 10 cm, how far will the large piston move? Assume the oil to be incompressible.

2-53. A 100-ft-long pipe is inclined at a 30° angle with the horizontal. It is filled with oil of specific gravity 0.90. What is the pressure at the base of the pipe if the top of the pipe is vented to the atmosphere?

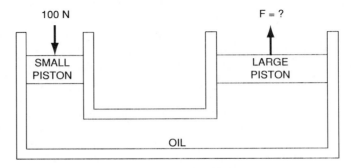

Figure 2-25. System for Exercise 2-51.

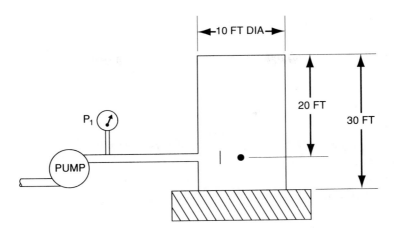

Figure 2-26.   System for Exercise 2-54.

2-54. A pump delivers oil to a cylindrical storage tank as shown in Figure 2-26. A faulty electric pressure switch, which controls the electric motor driving the pump, allows the pump to fill the tank completely. This causes the pressure P_1 near the base of the tank, to build up to 15 psig. What force is exerted on the top of the tank? What does the pressure difference between the tank top and point 1, say about Pascal's Law? What must be true about the magnitude of system pressure if the change in pressure due to elevation changes can be ignored in a fluid power system?

2-55. It is desired to select an oil of a viscosity index suitable for a hydraulic powered front end loader that will operate year round. What *VI* characteristic should be specified? Why?

3

Energy and Power in Hydraulic Systems

3.1 INTRODUCTION

In hydraulic systems, fluid enters the pump at below atmospheric pressure, called the suction pressure. As the fluid passes through the pump, its potential energy increases as evidenced by an increase in fluid pressure. Some of this energy is lost due to friction as the fluid flows through pipes, valves, and fittings. These frictional energy losses show up as heat energy and therefore are accounted for. An analysis of these energy losses due to friction is presented in Chapter 5. At the output device (hydraulic actuator) the remaining energy is transferred to the load to perform useful work. This is essentially the cycle of energy transfer in a fluid power system. Energy is added to the system by the pump and removed from the system via the actuator as it drives the output load.

A hydraulic system is not a source of energy. The energy source is the prime mover (such as an electric motor or an internal combustion engine), which drives the pump. Thus, in reality, a hydraulic system is merely an energy transfer system. Why not, then, eliminate hydraulics and simply couple the mechanical equipment directly to the prime mover? The answer is that a hydraulic system is much more versatile in its ability to transmit power. This versatility includes advantages of variable speed, reversibility, overload protection, high horsepower per weight ratio, and immunity to damage under stalled conditions.

The law of conservation of energy states that energy can neither be created nor destroyed. This means that the total energy at any location of the system remains constant. The total energy includes potential energy due to elevation and pressure and also kinetic energy due to velocity. If all the energy changes are properly accounted for, the hydraulic system will always have an energy balance. This will be accomplished by using Bernoulli's theorem, which keeps track of the changes that occur to the potential and kinetic energy of the fluid as it passes through the hydraulic system. Also included are energy losses due to friction, which transfers into heat, mechanical energy added by the pump and mechanical energy removed by the load actuators.

3.2 REVIEW OF MECHANICS

Since fluid power deals with the generation of forces to produce energy, it is essential that the basic laws of mechanics be clearly understood. Let's, therefore, have a brief review of mechanics as it relates to fluid power systems.

Forces are essential to the production of work. No motion can be generated and hence no power can be transmitted without the application of some force. It was in the late seventeenth century when Sir Isaac Newton formulated the three laws of motion dealing with the effect a force has on a body:

1. A force is required to change the motion of a body.
2. If a body is acted upon by a force, the body will have an acceleration proportional to the magnitude of the force and inversely to the mass of the body.
3. If one body exerts force on a second body, the second body must exert an equal but opposite force on the first body.

The motion of a body can be either linear or angular depending on whether the path of motion is along a straight line or circle.

Let's examine linear motion first. If a body moves, it has velocity, which is defined as the distance traveled divided by the corresponding time:

$$v = \frac{s}{t} \tag{3-1}$$

where s = distance (in. or ft),
    t = time (sec or min),
    v = velocity (in./sec, in./min, ft/sec, or ft/min).

If the velocity changes, we have a change in motion. This means we have an acceleration, which is defined as the change in velocity divided by the corresponding change in time. In accordance with Newton's first law of motion, a force is required to produce this change in motion. Per Newton's second law, we have

$$F = ma \tag{3-2}$$

where F = force (lb),
$\quad a$ = acceleration (in./sec^2 or ft/sec^2),
$\quad m$ = mass (slugs).

This now brings us to the concept of energy, which is defined as the ability to perform work. Let's assume that a force acts on a body and moves the body through a specified distance in the direction of the applied force. Then, by definition, work has been done on the body. The amount of this work equals the product of the force and distance where both the force and distance are measured in the same direction:

$$W = FS \tag{3-3}$$

where F = force (lb),
$\quad S$ = distance (in. or ft),
$\quad W$ = work (in. $\cdot$ lb or ft $\cdot$ lb).

This leads us to a discussion of power, which is defined as the rate of doing work or expending energy. Thus, power equals work divided by time:

$$P = \frac{FS}{t} = \frac{W}{t}$$

But since S/t equals v, we can rewrite the power equation as follows:

$$P = Fv \tag{3-4}$$

where F = force (lb),
$\quad v$ = velocity (in./sec, in./min, ft/sec, or ft/min),
$\quad P$ = power (in. $\cdot$ lb/sec, in. $\cdot$ lb/min, ft $\cdot$ lb/sec, or ft $\cdot$ lb/min).

Essentially, power is a measure of how fast work can be done and is usually measured in units of horsepower (hp). By definition, 1 hp equals 550 ft $\cdot$ lb/sec or 33,000 ft $\cdot$ lb/min. Thus, we have

$$HP = \frac{F(\text{lb}) \cdot v(\text{ft/sec})}{550} = \frac{F(\text{lb}) \cdot v(\text{ft/min})}{33,000} \tag{3-5}$$

EXAMPLE 3-1

A person exerts a 30-lb force to move a hand truck 100 ft in 60 sec.

   a. How much work is done?
   b. What is the power delivered by the person?

Solution

a. $W = FS = (30\ \text{lb})(100\ \text{ft}) = 3000\ \text{ft} \cdot \text{lb}$

b. $P = \dfrac{FS}{t} = \dfrac{(30\ \text{lb})(100\ \text{ft})}{60\ \text{sec}} = 50\ \text{ft} \cdot \text{lb/sec}$

$HP = \dfrac{50\ \text{ft} \cdot \text{lb/sec}}{(550\ \text{ft} \cdot \text{lb/sec})/\text{HP}} = 0.091\ \text{hp}$

Let's now examine angular-type motion. Just as in the linear case, angular motion is caused by the application of a force. Consider, for example, the force F (shown in Fig. 3-1) being applied to a disk by means of a string wrapped around its periphery. The disk rotates about its center and has a radius R. The force F creates a torque T whose magnitude equals the product of the force F and its moment arm R:

$$T = FR \tag{3-6}$$

where F = force (lb),
   R = moment arm (in. or ft),
   T = torque (in. $\cdot$ lb or ft $\cdot$ lb).

Notice that the moment arm is measured from the center of the shaft (center of rotation) perpendicularly to the line of action of the force. It is the torque T that causes the disk to rotate at some angular speed measured in units of revolutions per minute (rpm). Since the torque is clockwise as shown in Figure 3-1, the rotation is similarly in the clockwise direction. As the disk rotates and overcomes a load resistance, work is done and power is transmitted. A hydraulic motor delivers power in this fashion, and the amount of horsepower transmitted can be found from

$$HP = \frac{Tn}{63{,}000} \tag{3-7}$$

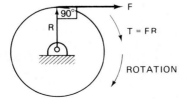

Figure 3-1. Definition of torque T using a rotating disk.

where T = torque (in. · lb),
  n = rotational speed (rpm),
 HP = horsepower.

EXAMPLE 3-2

How much torque is delivered by a 2-hp, 1800-rpm hydraulic motor?

Solution

$$HP = \frac{Tn}{63,000}$$

Substituting known values, we have

$$2 = \frac{T(1800)}{63,000}$$

$$T = 70 \text{ in. · lb}$$

Efficiency, another significant parameter when dealing with work and power, is defined as output power divided by input power:

$$\eta = \frac{P_o}{P_i} \tag{3-8}$$

where P_o = output power,
  P_i = input power,
  η = Greek letter eta = efficiency.

The efficiency of any system or component is always less than 100% and is calculated to determine power losses. In hydraulic systems, these losses are due to fluid leakage past close-fitting parts and mechanical friction due to fluid movement, rubbing parts and seals, and the operation of mechanical couplings. Efficiency measures the amount of power that is actually delivered in comparison to the power received. The power difference $(P_o - P_i)$ is loss since it is transformed into heat due to frictional effects. The output power is usually computed from force, distance, and time data, which describe the speed at which the load is moved. The input power is normally computed on parameters associated with the prime mover.

EXAMPLE 3-3

An elevator raises a 3000-lb load through a distance of 50 ft in 10 sec. If the efficiency of the entire system is 80%, how much input horsepower is required by the elevator hoist motor?

Solution

$$P = \frac{FS}{t} = \frac{(3000 \text{ lb})(50 \text{ ft})}{(10 \text{ sec})} = 15,000 \text{ ft} \cdot \text{lb/sec}$$

$$HP = \frac{15,000}{550} = 27.3$$

$$\eta = \frac{P_o}{P_i} \times 100\%$$

$$80\% = \frac{27.3}{P_i} \times 100$$

$$P_i = 34.1 \text{ hp}$$

3.3 APPLICATIONS OF PASCAL'S LAW

In this section we examine two basic applications of Pascal's law: the hydraulic jack and the air-to-hydraulic booster.

Hydraulic Jack

This system uses a piston-type hand pump to power a single-acting hydraulic cylinder, as illustrated in Fig. 3-2. The operation is as follows.

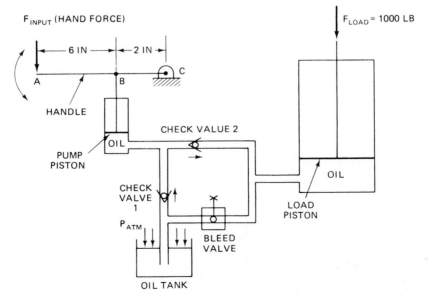

Figure 3-2.   Hand-operated hydraulic jack system.

A hand force is applied at point *A* of handle *ABC,* which pivots about point *C.* The piston rod of the hand pump is pinned to the input handle at point *B.* The hand pump essentially consists of a cylinder containing a piston, which is free to move up and down. The piston and rod are rigidly connected together. When the handle is pulled up, the piston rises and creates a vacuum in the space below it. As a result, atmospheric pressure forces oil to leave the oil tank and flow through check valve 1. This is the suction process. A check valve is a component that allows flow to pass in only one direction, as indicated by the arrow.

When the handle is then pushed down, oil is ejected from the hand pump and flows through check valve 2 and enters the bottom end of the load cylinder. The load cylinder is similar in construction to the pump cylinder and contains a piston and rod. Pressure builds up below the load piston as oil is ejected from the pump and meets resistance in finding a place to go. From Pascal's law we know that the pressure acting on the load piston equals the pressure generated by the pump below its piston. Thus, each time the input handle is cycled up and down, a specified volume of oil is ejected from the pump to raise the load cylinder a given distance against its load resistance. The bleed valve is a hand-operated valve, which, when opened, allows the load to be lowered by bleeding oil from the load cylinder back to the oil tank. It should be noted that oil enters and exits from each cylinder at only one end. Such a cylinder is called *single acting* because it is hydraulically powered in only one direction. Figure 3-3 is a photograph showing the construction of a single-acting air cylinder. The purpose of the internal spring is to return the piston to its fully retracted position. In the hydraulic jack application of Fig. 3-2 the retraction of the load piston is accomplished by the load itself when the bleed valve is opened.

Figure 3-4 is a photograph of a double-acting air cylinder, which can be powered in either direction. The construction and operation details of hydraulic and air cylinders such as these are examined in detail in Chapter 7.

Example 3-4 illustrates further how the hydraulic jack functions in terms of power and force requirements.

Figure 3-3.   Single-acting air cylinder. (*Courtesy of Sheffer Corporation, Cincinnati, Ohio.*)

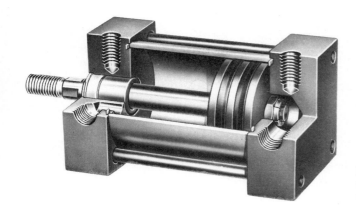

Figure 3-4. Double-acting air cylinder. (*Courtesy of Sheffer Corporation, Cincinnati, Ohio.*)

EXAMPLE 3-4

An operator makes one complete cycle per second interval using the hand pump of Fig. 3-2. Each complete cycle consists of two pump strokes (intake and power). The pump has a 1-in.-diameter piston and the load cylinder has a 3.25 in.-diameter piston. If the average hand force is 25 lb during the power stroke,

    a. How much load can be lifted?

    b. How many cycles are required to lift the load 10 in. assuming no oil leakage? The pump piston has a 2-in. stroke.

    c. What is the output *HP* assuming 100% efficiency?

    d. What is the output *HP* assuming 80% efficiency?

Solution

    a. First determine the force acting on the rod of the pump due to the mechanical advantage of the input handle:

$$F_{rod} = \frac{8}{2} \times F_{input} = \frac{8}{2}(25) = 100 \text{ lb}$$

Next, calculate the pump discharge pressure P:

$$P = \frac{\text{rod force}}{\text{piston area}} = \frac{F_{rod}}{A_{pump\ piston}} = \frac{100 \text{ lb}}{(\pi/4)(1)^2 \text{ in.}^2} = 127 \text{ psi}$$

Per Pascal's law this is also the same pressure acting on the load piston. We can now calculate the load-carrying capacity:

$$F_{load} = PA_{load\ piston} = (127) \text{ lb/in.}^2 \left[\frac{\pi}{4}(3.25)^2\right] \text{ in.}^2 = 1055 \text{ lb}$$

b. To find the load displacement, assume the oil to be incompressible. Therefore, the total volume of oil ejected from the pump equals the volume of oil displacing the load cylinder:

$$(A \times S)_{\text{pump piston}} \times (\text{no. of cycles}) = (A \times S)_{\text{load piston}}$$

Substituting, we have

$$\frac{\pi}{4} (1)^2 \text{ in.}^2 \times 2 \text{ in.} \times (\text{no. of cycles}) = \frac{\pi}{4} (3.25)^2 \text{ in.}^2 \times 10 \text{ in.}$$

$$1.57 \text{ in.}^3 \times (\text{no. of cycles}) = 82.7 \text{ in.}^3$$

$$\text{no. of cycles} = 52.7$$

Note that the 1.57-in.^{3} value is called the displacement volume of the piston pump.

c.
$$P = \frac{FS}{t} = \frac{(1055 \text{ lb}) \left(\frac{10}{12} \text{ ft}\right)}{52.7 \text{ sec}} = 16.7 \text{ ft} \cdot \text{lb/sec}$$

$$HP = \frac{16.7}{550} = 0.030 \text{ hp}$$

The *HP* output value is small, as expected, since the power comes from a human being.

d.
$$HP = (0.80)(0.03) = 0.024 \text{ hp}$$

Air-to-Hydraulic Pressure Booster

This device is used for converting shop air into the higher hydraulic pressure needed for operating cylinders requiring small to medium volumes of higher pressure oil. It consists of a large-diameter air cylinder driving a small-diameter hydraulic cylinder (see Fig. 3-5). Any shop equipped with an air line can obtain smooth, efficient hydraulic power from an air-to-hydraulic booster hooked into the air line. The alternative would be a complete hydraulic system including expensive pumps and high-pressure valves. Other benefits include a space savings and a reduction in operating costs and maintenance.

Figure 3-6 shows an application where an air-to-hydraulic booster is supplying high-pressure oil to a hydraulic cylinder used to clamp a workpiece to a machine tool table. Since shop air pressure normally operates at 100 psi, a pneumatically operated clamp would require an excessively large cylinder to rigidly hold the workpiece while it is being machined.

The air-to-hydraulic booster operates as follows (see Fig. 3-6): Let's assume that the air piston has a 10-in.^{2} area and is subjected to 100-psi air pressure. This produces a 1000-lb force on the hydraulic cylinder piston. Thus, if the area of the

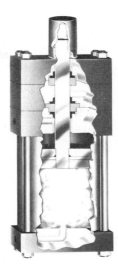

Figure 3-5. Air-to-hydraulic pressure booster. (*Courtesy of Miller Fluid Power, Bensenville, Illinois.*)

hydraulic piston is 1 in.^{2}, the hydraulic discharge oil pressure will be 1000 psi. Per Pascal's law, this produces 1000-psi oil at the small hydraulic clamping cylinder mounted on the machine tool table.

The pressure ratio of an air-to-hydraulic booster can be found by using

$$\text{pressure ratio} = \frac{\text{output oil pressure}}{\text{input air pressure}} = \frac{\text{area of air piston}}{\text{area of hydraulic piston}} \tag{3-9}$$

The pressure booster shown in Figure 3-5 is designed to provide either a single-output pressure or dual-output pressures. When connected in a single-pressure circuit, the booster produces high-pressure oil during the entire booster cylinder stroke. A single-pressure circuit is recommended when the booster cylinder approach stroke is small in comparison to the stroke of the high-pressure load cylinder being powered by the booster. In a dual-output pressure circuit, the pressure booster delivers low-pressure oil during the initial portion of the approach stroke and high pressure oil during the final portion of the stroke. The pressure booster of Figure 3-5, when connected in a dual-pressure circuit, uses as little as 10% of the amount of air required for direct air cylinder operation.

Figure 3-7 shows a booster power panel unit that contains the booster of Figure 3-5 connected in a dual-pressure circuit. This panel unit is a complete system equipped with the following components:

1. One air-hydraulic booster
2. Two air-oil tanks
3. One air line filter

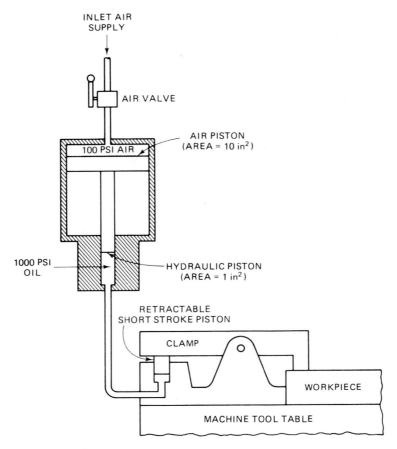

Figure 3-6. Manufacturing application of air-to-hydraulic booster.

4. One air line lubricator
5. Two air pressure regulators
6. One control valve operator
7. All necessary piping and power air valves

This panel unit permits booster reciprocation rates up to 60 strokes (30 cycles) per minute.

A riveting press, which is booster-powered, is depicted in Fig. 3-8. It is an extremely compact, space-saving unit that rivets six rivets simultaneously on aircraft parts. In Fig. 3-9 we see a milling machine using booster-powered clamps, which result in savings in clamping time and labor, simplicity, and economy of operation. This specific application uses two boosters for powering dual clamping

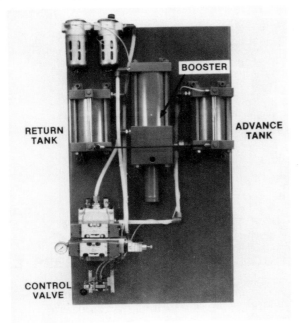

Figure 3-7. Booster power panel unit. (*Courtesy of Miller Fluid Power, Bensenville, Illinois.*)

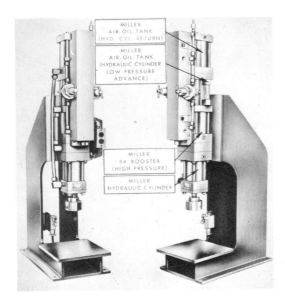

Figure 3-8. Booster-powered riveting press. (*Courtesy of Miller Fluid Power, Bensenville, Illinois.*)

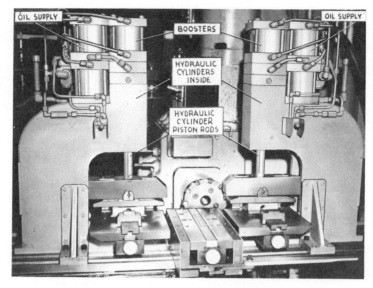

Figure 3-9. Booster-powered clamps on milling machine. (*Courtesy of Miller Fluid Power, Bensenville, Illinois.*)

fixtures on a single milling machine for reciprocal milling of small parts. The input air operates at 100 psi, and the booster produces 2000-psi oil to produce 10,000-lb forces at each clamp.

Air-to-hydraulic boosters can be purchased in a wide range of ratios and can provide hydraulic pressures up to 30,000 psi using 100-psi shop air.

EXAMPLE 3-5

Figure 3-10 shows a pressure booster system used to drive a load F via a hydraulic cylinder. The following data are given:

Inlet air pressure (P_1) = 100 psi.

Air piston area (A_1) = 20 in.^{2}

Oil piston area (A_2) = 1 in.^{2}

Load piston area (A_3) = 25 in.^{2}

Find the load-carrying capacity F of the system.

Solution   First, find the booster discharge pressure P_2:

booster input force = booster output force

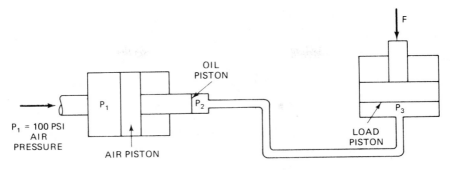

Figure 3-10. An air-to-hydraulic booster system.

$$P_1 A_1 = P_2 A_2$$

$$P_2 = \frac{P_1 A_1}{A_2} = (100)\left(\frac{20}{1}\right) = 2000 \text{ psi}$$

Per Pascal's law, $P_3 = P_2 = 2000$ psi:

$$F = P_3 A_3 = (2000)(25) = 50,000 \text{ lb}$$

To produce this force without the booster would require a 500-in.²-area load piston assuming 100-psi air pressure.

3.4 CONSERVATION OF ENERGY

The conservation of energy law states that energy can neither be created nor destroyed. This means that the total energy of the system at any location remains constant. The total energy includes potential energy due to elevation and pressure and also kinetic energy due to velocity. Let's examine each of the three types of energy.

1. Potential energy due to elevation (EPE): Figure 3-11 shows a chunk of fluid of weight W lb at an elevation Z with respect to a reference plane. The weight has potential energy (EPE) relative to the reference plane because work would have to be done on the fluid to lift it through a distance Z:

$$\text{EPE} = WZ \tag{3-10}$$

Notice that the units of EPE are either ft · lb or in. · lb.

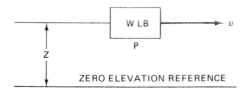

Figure 3-11. The three forms of energy as established by elevation (Z), pressure (P), and velocity (v).

2. Potential energy due to pressure (PPE): If the W lb of fluid in Fig. 3-11 possesses a pressure P, it contains pressure energy as represented by

$$\text{PPE} = W \frac{P}{\gamma} \qquad (3\text{-}11)$$

where γ is the weight density of the fluid. PPE will have units of either ft $\cdot$ lb or in. $\cdot$ lb.

3. Kinetic energy (KE): If the W lb of fluid in Fig. 3-11 is moving with a velocity v, it contains kinetic energy, which can be found using

$$\text{KE} = \frac{1}{2} \frac{W}{g} v^2 \qquad (3\text{-}12)$$

where g = acceleration of gravity
$\quad\quad\ = 32.2 \text{ ft/sec}^2 \text{ or } 386 \text{ in./sec}^2$

KE will have units of either ft $\cdot$ lb or in. $\cdot$ lb.

The total energy possessed by the W-lb chunk of fluid can neither be created nor destroyed. Mathematically the total energy E_T remains constant:

$$E_T = WZ + W \frac{P}{\gamma} + \frac{1}{2} \frac{W}{g} v^2 = \text{constant} \qquad (3\text{-}13)$$

Of course, energy can change from one form to another. For example, the chunk of fluid may lose elevation and thus have less potential energy. This, however, would result in an equal increase in either pressure energy or kinetic energy.

3.5 THE CONTINUITY EQUATION

The continuity equation states that for steady flow in a pipeline the weight flow rate is the same for all cross sections of the pipe.

To illustrate the significance of the continuity equation, refer to Fig. 3-12, which shows a pipe in which fluid is flowing at a weight flow rate w that has units

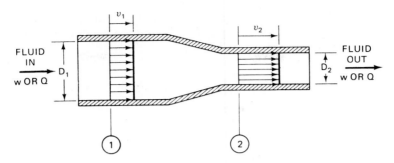

Figure 3-12.   Continuity of flow.

of weight per unit time. The pipe has two different-sized cross-sectional areas identified by stations 1 and 2. If no fluid is added or withdrawn at any location, then the weight flow rate at stations 1 and 2 must be equal:

$$w_1 = w_2$$

or

$$\gamma_1 A_1 v_1 = \gamma_2 A_2 v_2$$

where γ = weight density of fluid (lb/ft³),
$\quad\ A$ = cross-sectional area of pipe (ft²),
$\quad\ v$ = velocity of fluid (ft/sec).

Checking units, we have

$$(\text{lb/ft}^3)(\text{ft}^2)(\text{ft/sec}) = (\text{lb/ft}^3)(\text{ft}^2)(\text{ft/sec})$$

or

$$\text{lb/sec} = \text{lb/sec}$$

Assuming an incompressible fluid, we can cancel out the weight density terms from the preceding equation since $\gamma_1 = \gamma_2$. This produces the continuity equation for hydraulic fluids that are nearly incompressible:

$$A_1 v_1 = A_2 v_2 \tag{3-14}$$

Checking units, we have

$$(\text{ft}^2)(\text{ft/sec}) = (\text{ft}^2)(\text{ft/sec})$$

or

$$\text{ft}^3/\text{sec} = \text{ft}^3/\text{sec}$$

Hence, for an incompressible fluid, the volumetric flow rate (volume per unit time) is also constant in a pipeline. Using the symbol Q for volumetric flow rate, we have

$$Q = Av = A_1v_1 = Q_1 = A_2v_2 = Q_2$$

The continuity equation can be rewritten as follows:

$$\frac{v_1}{v_2} = \frac{A_2}{A_1} = \frac{(\pi/4)D_2^2}{(\pi/4)D_1^2}$$

where D_1 and D_2 are the pipe diameters at stations 1 and 2, respectively. The final result is

$$\frac{v_1}{v_2} = \frac{A_2}{A_1} = \frac{D_2^2}{D_1^2} \tag{3-15}$$

Equation (3-15) brings out the fact that the smaller the pipe size, the greater the velocity, and vice versa. It should be noted that the pipe diameters and areas are inside values and do not include the pipe wall thickness.

EXAMPLE 3-6

For the pipe in Fig. 3-12, the following data are given:

$$D_1 = 4 \text{ in.}, \ D_2 = 2 \text{ in.}$$

$$v_1 = 4 \text{ ft/sec}$$

Find

   a. The volumetric flow rate Q
   b. The fluid velocity at station 2

Solution

   a. $Q = Q_1 = A_1v_1$

$$A_1 = \frac{\pi}{4} D_1^2 = \frac{\pi}{4} \left(\frac{4}{12} \text{ ft}\right)^2 = 0.0873 \text{ ft}^2$$

$$Q = (0.0873 \text{ ft}^2)(4 \text{ ft/sec}) = 0.349 \text{ ft}^3/\text{sec}$$

   b. Solving Eq. (3-15) for v_2, we have

$$v_2 = (v_1) \frac{(D_1)^2}{(D_2)^2} = \frac{(4)(4)^2}{(2)^2} = 16 \text{ ft/sec}$$

3.6 HYDRAULIC HORSEPOWER

Now that we have established the concepts of flow rate and pressure, we can find the work done in pumping a fluid and thus the horsepower produced by a hydraulic actuator. Let's analyze the hydraulic cylinder of Fig. 3-13 by developing equations that will allow us to answer the following three questions:

1. How do we determine how big a piston diameter is required?
2. What is the required pump flow rate necessary to drive the cylinder through its stroke in the required time?
3. How much horsepower does the hydraulic cylinder deliver?

Note that the cylinder horsepower plus any horsepower losses due to friction between the pump and cylinder must be hydraulically produced by the pump.

   Answer to Question 1: A pump receives fluid on its inlet side at a pressure approximating atmospheric pressure (0 psig) and discharges the fluid on the outlet side at some elevated pressure P sufficiently high to overcome the load. This pressure P acts on the area of the piston A to produce the necessary force to overcome the load:

$$PA = F_{\text{load}}$$

Solving for the piston area A, we obtain

$$A = \frac{F_{\text{load}}}{P} \tag{3-16}$$

   Since the load is known from the application and the maximum allowable pressure is established based on the pump design, Eq. (3-16) allows us to calculate the required piston area.

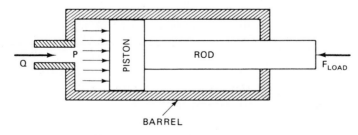

Figure 3-13.   Hydraulic horsepower of a cylinder.

Answer to Question 2: The volumetric displacement V_D of the hydraulic cylinder equals the fluid volume swept out by the piston traveling through its stroke S:

$$V_D(\text{ft}^3) = A(\text{ft}^2) \cdot S \ (\text{ft})$$

The required pump flow rate equals the volumetric displacement of the cylinder divided by the time t required for the piston to travel through its stroke:

$$Q(\text{ft}^3/\text{sec}) = \frac{V_D(\text{ft}^3)}{t(\text{sec})}$$

but since $V_D = AS$, we have

$$Q(\text{ft}^3/\text{sec}) = \frac{A(\text{ft}^2) \cdot S(\text{ft})}{t(\text{sec})} \tag{3-17}$$

Since stroke S and time t are basically known from the particular application, Eq. (3-17) permits the calculation of the required pump flow.

Recall that for a pipe we determined that $Q = Av$. Shouldn't we obtain the same equation for a hydraulic cylinder since it is essentially a pipe containing a piston? The answer is yes, as can be verified by noting that S/t can be replaced by v in Eq. (3-17) to obtain the expected result:

$$Q(\text{ft}^3/\text{sec}) = A(\text{ft}^2) \cdot v(\text{ft}/\text{sec}) \tag{3-18}$$

where v = piston velocity.

Note that the larger the piston area and velocity, the greater must be the pump flow rate.

Answer to Question 3: It has been established that energy equals force times distance:

$$\text{energy} = (F)(S) = (PA)(S)$$

Since power is the rate of doing work, we have

$$\text{power} = \frac{\text{energy}}{\text{time}} = \frac{(PA)(S)}{t} = P(Av)$$

Since $Q = Av$, the final result is

$$\text{power (ft} \cdot \text{lb/sec)} = P(\text{lb/ft}^2) \cdot Q(\text{ft}^3/\text{sec}) \tag{3-19}$$

Recalling that 1 hp = 550 ft · lb/sec, we obtain

$$HP = \frac{P(\text{lb/ft}^2) \cdot Q(\text{ft}^3/\text{sec})}{550} \qquad (3\text{-}20)$$

Since pressure in units of psi (lb/in²) and flow rate in units of gallons per minute (gpm) are in more widespread use in the English system of units, Eq. (3-20) will be modified to incorporate these units.

First, let's find the relationship between psi and lb/ft²:

$$P(\text{lb/ft}^2) = P(\text{lb/in.}^2) \times \left(\frac{144 \text{ in.}^2}{1 \text{ ft}^2}\right)$$

Therefore,

$$P(\text{lb/ft}^2) = 144 P(\text{psi})$$

Second, let's find the relationship between gpm and ft³/sec:

$$Q(\text{ft}^3/\text{sec}) = Q(\text{gpm}) \times \left(\frac{231 \text{ in.}^3}{1 \text{ gal}}\right) \times \left(\frac{1 \text{ ft}^3}{1728 \text{ in.}^3}\right) \times \left(\frac{1 \text{ min}}{60 \text{ sec}}\right)$$

Therefore,

$$Q(\text{ft}^3/\text{sec}) = \frac{Q(\text{gpm})}{448}$$

Substituting the determined relationships between psi and lb/ft² and gpm and ft³/sec into Eq. (3-20) yields the final result:

$$HP = \frac{[144 P(\text{psi})][Q(\text{gpm})/448]}{550}$$

$$HP = \frac{P(\text{psi}) \cdot Q(\text{gpm})}{1714} \qquad (3\text{-}21)$$

Observe the following power analogy between mechanical, electrical, and hydraulic systems:

Mechanical power = force × velocity.

Electrical power = volts × amps.

Hydraulic power = pressure × flow rate.

Example 3-7 illustrates the use of the hydraulic horsepower analysis technique.

EXAMPLE 3-7

A hydraulic cylinder is to compress a car body down to bale size in 10 sec. The operation requires a 10-ft stroke and a 8000-lb force. If a 1000-psi pump has been selected, find

  a. The required piston area
  b. The necessary pump flow rate
  c. The hydraulic horsepower delivered by the cylinder

Solution

a.
$$A = \frac{F_{load}}{P} = \frac{8000 \text{ lb}}{1000 \text{ lb/in.}^2} = 8 \text{ in.}^2$$

b.
$$Q(\text{ft}^3/\text{sec}) = \frac{A(\text{ft}^2) \cdot S \text{ (ft)}}{t(\text{sec})} = \frac{\left(\frac{8}{144}\right)(10)}{10} = 0.056 \text{ ft}^3/\text{sec}$$

$$Q(\text{gpm}) = 448\, Q(\text{ft}^3/\text{sec}) = (448)(0.056) = 25.1 \text{ gpm}$$

c.
$$HP = \frac{(1000)(25.1)}{1714} = 14.6 \text{ hp}$$

It should be noted that this is the theoretical horsepower delivered by the cylinder assuming it is 100% efficient.

3.7 BERNOULLI'S EQUATION

Bernoulli's equation is one of the most useful relationships for performing hydraulic circuit analysis. Its application allows us to size components such as pumps, valves, and piping for proper system operation. Bernoulli's equation can be derived by applying the conservation of energy system to a hydraulic pipeline, as shown in Fig. 3-14. At station 1 we have W lb of fluid possessing an elevation Z_1, a pressure P_1, and a velocity v_1. When this W lb of fluid arrives at station 2, its elevation, pressure, and velocity have become Z_2, P_2, and v_2, respectively.

With respect to a common zero elevation reference plane, we can identify the various energy terms as follows:

Type of Energy	Station 1	Station 2
Elevation	WZ_1	WZ_2
Pressure	$W\dfrac{P_1}{\gamma}$	$W\dfrac{P_2}{\gamma}$
Kinetic	$\dfrac{Wv_1^2}{2g}$	$\dfrac{Wv_2^2}{2g}$

Daniel Bernoulli, the eighteenth-century Swiss scientist, formulated his equation by noting that the total energy possessed by the W lb of fluid at station 1 equals the total energy possessed by the same W lb of fluid at station 2:

$$WZ_1 + W\frac{P_1}{\gamma} + \frac{Wv_1^2}{2g} = WZ_2 + \frac{WP_2}{\gamma} + \frac{Wv_2^2}{2g} \qquad (3\text{-}22)$$

If we divide both sides of Eq. (3-22) by W, we are examining the energy possessed by 1 lb of fluid rather than W lb. This yields Bernoulli's equation for an ideal frictionless system: *The total energy per pound of fluid at station 1 equals the total energy per pound of fluid at station 2:*

$$Z_1 + \frac{P_1}{\gamma} + \frac{v_1^2}{2g} = Z_2 + \frac{P_2}{\gamma} + \frac{v_2^2}{2g} \qquad (3\text{-}23)$$

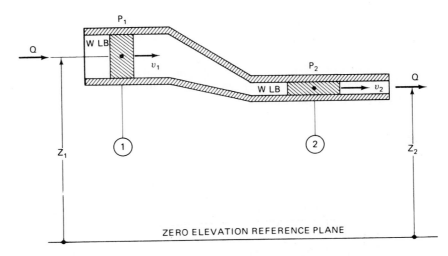

Figure 3-14. Pipeline for deriving Bernoulli's equation.

Checking units, we find that each term has units of length (i.e., ft · lb/lb = ft or in. · lb/lb = in.). This is as expected since each term represents energy per pound of fluid:

all in units
of length.

$$Z = \text{ft} \qquad\qquad \text{or} \qquad\qquad Z = \text{in.}$$

$$\frac{P}{\gamma} = \frac{\text{lb/ft}^2}{\text{lb/ft}^3} = \text{ft} \qquad \text{or} \qquad \frac{P}{\gamma} = \frac{\text{lb/in.}^2}{\text{lb/in.}^3} = \text{in.}$$

$$\frac{v^2}{2g} = \frac{(\text{ft/sec})^2}{\text{ft/sec}^2} = \text{ft} \qquad \text{or} \qquad \frac{v^2}{2g} = \frac{(\text{in./sec})^2}{\text{in./sec}^2} = \text{in.}$$

Since each term of Bernoulli's equation has units of length, the use of the expression *head* has picked up widespread use as follows:

Z_1 is called *elevation head*.

P_1/γ is called *pressure head*.

$v_1^2/2g$ is called *velocity head*.

We can correct Eq. (3-23) to take into account that frictional losses (H_L) take place between stations 1 and 2. H_L represents the energy per pound of fluid loss in going from station 1 to station 2. In addition, we want to take into account that a pump or hydraulic motor may exist between stations 1 and 2. H_p will represent the energy per pound of fluid added by a pump, and H_m will represent the energy per pound of fluid removed by a motor.

This leads us to the complete Bernoulli equation based on balancing the energy terms on a 1-lb chunk of fluid: The total energy at station 1 plus the energy added by a pump minus the energy removed by a motor minus the energy loss due to friction equals the total energy at station 2:

$$Z_1 + \frac{P_1}{\gamma} + \frac{v_1^2}{2g} + H_p - H_m - H_L = Z_2 + \frac{P_2}{\gamma} + \frac{v_2^2}{2g} \qquad (3\text{-}24)$$

The pump head can be calculated using:

Magic
formula!

$$H_p = \frac{3950(HP)}{QS_g} \qquad\qquad\qquad (3\text{-}25)$$

where HP = pump horsepower,
$\quad Q$ = pump flow (gpm),
$\quad S_g$ = specific gravity of fluid,
$\quad H_p$ = pump head (ft).

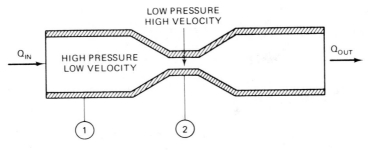

Figure 3-15.   Pressure reduction in a Venturi.

The use of a venturi in an automobile engine carburetor is a familiar application of Bernoulli's equation.

Figure 3-15 shows a venturi, which is a special pipe whose diameter is gradually reduced until a constant-diameter throat is reached. Then the pipe gradually increases in diameter until it reaches the original size. We know that inlet station 1 has a lower velocity than does the throat station 2 due to the continuity equation. Therefore, v_2 is greater than v_1.

Let's write Bernoulli's equation between stations 1 and 2 assuming ideal flow and equal elevations:

$$\frac{P_1}{\gamma} + \frac{v_1^2}{2g} = \frac{P_2}{\gamma} + \frac{v_2^2}{2g}$$

Solving for $P_1 - P_2$, we have

$$P_1 - P_2 = \frac{\gamma}{2g}(v_2^2 - v_1^2)$$

Since v_2 is greater than v_1, we know that P_1 must be greater than P_2. The reason is simple: In going from station 1 to 2 the fluid gained kinetic energy due to the continuity theorem. As a result, the fluid had to lose pressure energy in order not to create or destroy energy. This venturi effect is commonly called Bernoulli's principle.

Figure 3-16 shows how the venturi effect is used in an automobile carburetor. The volume of air flow is determined by the opening position of the butterfly valve. As the air flows through the venturi, it speeds up and loses some of its pressure. The pressure in the fuel bowl is equal to the pressure in the air horn above the venturi. This differential pressure between the fuel bowl and the venturi throat causes gasoline to flow into the air stream. The reduced pressure in the venturi helps the gasoline to vaporize.

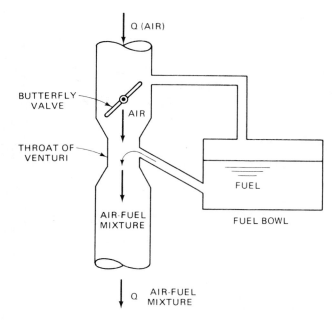

Figure 3-16. Use of Venturi to produce Bernoulli effect in automobile carburetor.

EXAMPLE 3-8

For the hydraulic system of Fig. 3-17, the following data are given:

1. The pump is adding 5 hp to the fluid.
2. Pump flow is 30 gpm.
3. The pipe has a 1-in. inside diameter.
4. The specific gravity of the oil is 0.9.

Find the pressure available at the inlet to the hydraulic motor (station 2). The pressure at station 1 in the hydraulic tank is atmospheric (0 psig). The head loss H_L due to friction between stations 1 and 2 is 30 ft of oil.

Solution   Writing Bernoulli's equation between stations 1 and 2, we have

$$Z_1 + \frac{P_1}{\gamma} + \frac{v_1^2}{2g} + H_p - H_m - H_L = Z_2 + \frac{P_2}{\gamma} + \frac{v_2^2}{2g}$$

Since there is no hydraulic motor between stations 1 and 2, $H_m = 0$. Also $v_1 = 0$ because the cross section of an oil tank is large. Thus, the velocity of the oil surface is negligible, and the value of v_1 approaches zero. Per Fig. 3-17, $Z_2 - Z_1 = 20$ ft. Also $H_L = 30$ ft, and $P_1 = 0$ per the given input data.

Substituting known values, we have

$$Z_1 + 0 + 0 + H_p - 0 - 30 = Z_2 + \frac{P_2}{\gamma} + \frac{v_2^2}{2g}$$

Solving for P_2/γ, we have

$$\frac{P_2}{\gamma} = (Z_1 - Z_2) + H_p - \frac{v_2^2}{2g} - 30$$

Since $Z_2 - Z_1 = 20$ ft, we have

$$\frac{P_2}{\gamma} = H_p - \frac{v_2^2}{2g} - 50$$

Using Eq. (3-25) yields

$$H_p = \frac{(3950)(5)}{(30)(0.9)} = 732 \text{ ft}$$

Then solve for

$$Q \text{ (ft}^3/\text{sec)} = \frac{Q \text{ (gpm)}}{448} = \frac{30}{448} = 0.0668$$

$$v_2 \text{ (ft/sec)} = \frac{Q \text{ (ft}^3/\text{sec)}}{A \text{ (ft}^2)}$$

$$A \text{ (ft}^2) = \frac{\pi}{4} (D \text{ ft})^2 = \frac{\pi}{4} \left(\frac{1}{12} \text{ ft}\right)^2 = 0.00546 \text{ ft}^2$$

$$v_2 = \frac{0.0668}{0.00546} \frac{\text{ft}^3/\text{sec}}{\text{ft}^2} = 12.2 \text{ ft/sec}$$

$$\frac{v_2^2}{2g} = \frac{(12.2 \text{ ft/sec})^2}{64.4 \text{ ft/sec}^2} = 2.4 \text{ ft}$$

Upon final substitution, we have

$$\frac{P_2}{\gamma} = 732 - 2.4 - 50 = 679.2 \text{ ft}$$

Solving for P_2 yields

$$P_2 \text{ (lb/ft}^2) = (679.2 \text{ ft})\gamma \text{ (lb/ft}^3)$$

where $\gamma = S_g \gamma_{water}$
$\gamma = (0.9)(62.4) = 56.2 \text{ lb/ft}^3$
$P_2 = (679.2)(56.2) = 38,200 \text{ lb/ft}^2$

Changing to units of psi yields

$$P_2 = \frac{38,200}{144} = 265 \text{ psi}$$

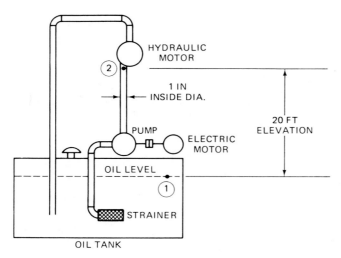

Figure 3-17. Hydraulic system for Example 3-8.

In Example 3-8, the velocity v_2 was calculated in units of ft/sec by dividing the flow rate Q in units of ft³/sec by the pipe cross-sectional area in units of ft². This required converting Q from gpm to ft³/sec and calculating the pipe area A in units of ft², knowing the pipe inside diameter D in units of in. Since flow rates are normally given in gpm and pipe diameters in units of in., this cumbersome series of calculations to find fluid velocities in pipes is frequently encountered. However, Eq. (3-26) can be used to provide answers quickly for fluid velocities:

$$v = \frac{0.408\,Q}{D^2}$$ (3-26)

where v = fluid velocity (ft/sec),
$\quad Q$ = flow rate (gpm),
$\quad D$ = pipe inside diameter (in.).

We can check the calculated value of v_2 from Example 3-8 by using Eq. (3-26):

$$v_2 = \frac{(0.408)(30)}{(1)^2} = 12.2 \text{ ft/sec}$$

As expected, the result is the same.

3.8 TORRICELLI'S THEOREM

Torricelli's theorem states that ideally the velocity of a free jet of fluid is equal to the square root of the product of two times the acceleration of gravity times the head producing the jet. As such, Torricelli's theorem is essentially a special case

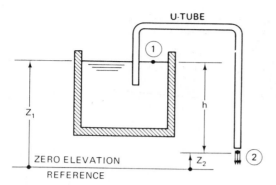

Figure 3-19. Siphon.

of a U-tube with one end submerged below the surface of the liquid in the container. The center portion of the U-tube rises above the level of the liquid surface, and the free end lies below it on the outside of the container.

For the fluid to flow out of the free end, two conditions must be met:

1. The elevation of the free end must be lower than the elevation of the liquid surface inside the container.

2. The fluid must initially be forced to flow up from the container into the center portion of the U-tube. This is normally done by temporarily providing a suction pressure at the free end of the siphon. For example, when siphoning gasoline from an automobile gas tank, a person can develop this suction pressure by momentarily "sucking" on the free end of the hose. This allows atmospheric pressure in the tank to push the gasoline up the U-tube hose, as required. For continuous flow operation, the free end of the U-tube hose must lie below the gasoline level in the tank.

We can analyze the flow through a siphon by applying Bernoulli's equation using points 1 and 2 in Figure 3-19:

$$Z_1 + \frac{P_1}{\gamma} + \frac{v_1^2}{2g} + H_p - H_m - H_L = Z_2 + \frac{P_2}{\gamma} + \frac{v_2^2}{2g}$$

The following conditions apply for the siphon:

1. $P_1 = P_2 =$ atmospheric pressure $= 0$ psig.
2. The area of the surface of the liquid in the container is large so that the velocity v_1 equals essentially zero.
3. There is no pump or motor ($H_p = H_m = 0$).
4. $Z_1 - Z_2 = h =$ differential head between liquid level and free end of U-tube.

Substituting known values, we have

$$Z_1 + 0 + 0 + 0 - H_L = Z_2 + 0 + \frac{v_2^2}{2g}$$

Solving for v_2 yields

SIPHON

$$v_2 = \sqrt{2g\,(Z_1 - Z_2 - H_L)} = \sqrt{2g\,(h - H_L)} \tag{3-29}$$

Equation (3-29) is identical to Torricelli's equation, and, as expected, the velocity inside the siphon tube is reduced as the head loss H_L due to viscosity increases. The flow rate of the siphon can be found by using the continuity equation.

EXAMPLE 3-10

For the siphon of Fig. 3-19, the following information is given

$$h = 30 \text{ ft}$$

$$H_L = 10 \text{ ft}$$

$$\text{U-tube inside diameter} = 1 \text{ in.}$$

Find the velocity and flow rate through the siphon.

Solution   Substitute known values into Eq. (3-29):

$$v_2 = \sqrt{(2)(32.2)(30 - 10)} = 35.8 \text{ ft/sec}$$

Use Eq. (3-25) to obtain the flow rate:

$$Q = \frac{(35.8)(1)^2}{0.408} = 87.8 \text{ gpm}$$

3.10 ENERGY, POWER, AND FLOW RATES IN THE METRIC-SI SYSTEM

It is important to be able to convert energy, power, and flow-rate parameters between the English and metric-SI systems. In this section we provide the necessary information to perform these conversions.

Energy

In the SI system, the joule (J) is the work done when a force of 1 N acts through a distance of 1 m. Since work equals force times distance, we have

$$1 \text{ J} = 1 \text{ N} \times 1 \text{ m} \tag{3-30}$$

The following equations allow conversion of energy or work values between the English and SI systems:

$$J(\text{N} \cdot \text{m}) = 0.113E \text{ (in.} \cdot \text{lb)} \tag{3-31}$$

$$J(\text{N} \cdot \text{m}) = 1.356E \text{ (ft} \cdot \text{lb)} \tag{3-32}$$

In other words 1 ft · lb of energy equals 1.356 J.

Power

Power is the rate of doing work. In the SI system, 1 watt (W) of power is the rate of 1 J of work per second:

$$\text{power} = \frac{\text{work}}{\text{time}} \tag{3-33}$$

$$1 \text{ W} = \frac{1 \text{ J}}{\text{s}} = 1 \text{ N} \times \frac{1 \text{ m}}{\text{s}} \tag{3-34}$$

where s = seconds.

In the metric-SI system all forms of power are expressed in watts. Thus, an electric motor has an electrical power input of watts and a mechanical power output of watts.

Conversions of power values between the English and SI systems can be made using the following equations:

$$W\left(\frac{\text{N} \cdot \text{m}}{\text{s}}\right) = 1.36P \text{ (ft} \cdot \text{lb/sec)} \tag{3-35}$$

$$W\left(\frac{\text{N} \cdot \text{m}}{\text{s}}\right) = 746P \text{ (hp)} \tag{3-36}$$

As a result we can conclude that 1 hp = 746 W.

Flow Rates

Volumes (V) can be converted between the English and SI systems as follows:

$$V \text{ (m}^3) = 0.0284V \text{ (ft}^3) \tag{3-37}$$

$$V \text{ (m}^3) = 0.00379V \text{ (gal)} \tag{3-38}$$

Thus, as an example, 1000 gal = 3.79 m³.
Thus, flow rates can be converted by using the following equations:

$$Q \text{ (m}^3\text{/s)} = 0.0284 Q \text{ (ft}^3\text{/sec)} \tag{3-39}$$

$$Q \text{ (m}^3\text{/s)} = 0.00379 Q \text{ (gal/sec)} \tag{3-40}$$

$$Q \text{ (m}^3\text{/s)} = 0.0000632 Q \text{ (gpm)} \tag{3-41}$$

So, for example, 1000 gpm = 0.0632 m³/s.

3.11 ILLUSTRATIVE EXAMPLE USING THE METRIC-SI SYSTEM

In this section we show how to solve a hydraulic system problem using the metric-SI system units. To provide a comparison with the English system of units, let's use the same hydraulic system analyzed in Example 3-8 (see Fig. 3-17). The metric-SI units solution is presented in the following example.

EXAMPLE 3-11

For the hydraulic system of Fig. 3-17, the following metric-SI data (which are equivalent to the English system of units data of Example 3-8) are given:

1. The pump is adding 5 hp (3730 W) to the fluid.
2. Pump flow is 0.001896 m³/s.
3. The pipe has a 0.0254-m inside diameter. Note that this size can also be represented in units of centimeters or millimeters as 2.54 cm or 25.4 mm, respectively.
4. The specific gravity of the oil is 0.9.
5. The elevation difference between stations 1 and 2 is 6.096 m.

Find the pressure available at the inlet to the hydraulic motor (station 2). The pressure at station 1 in the hydraulic tank is atmospheric (0 Pa or 0 N/m²). The head loss H_L due to friction between stations 1 and 2 is 9.144 m of oil.

Solution   Writing Bernoulli's equation between stations 1 and 2, we have

$$Z_1 + \frac{P_1}{\gamma} + \frac{v_1^2}{2g} + H_p - H_m - H_L = Z_2 + \frac{P_2}{\gamma} + \frac{v_2^2}{2g}$$

Since there is no hydraulic motor between stations 1 and 2, $H_m = 0$. Also, $v_1 = 0$ because the cross section of an oil tank is large. As per Fig. 3-17, $Z_2 - Z_1 = 20$ ft = 6.096 m. Also, $H_L = 9.144$ m and $P_1 = 0$ per the given input data.

Substituting known values, we have

$$Z_1 + 0 + 0 + H_p - 0 - 9.144 = Z_2 + \frac{P_2}{\gamma} + \frac{v_2^2}{2g}$$

Solving for P_2/γ, we have

$$\frac{P_2}{\gamma} = (Z_1 - Z_2) + H_p - \frac{v_2^2}{2g} - 9.144$$

Since $Z_2 - Z_1 = 6.096$ m, we have

$$\frac{P_2}{\gamma} = H_p - \frac{v_2^2}{2g} - 15.24$$

Equation (3-25) can be modified to allow solving for the pump head in metric units of meters as follows:

$$H_p(\text{ft}) = \frac{3950 \, (\text{HP})}{Q \, (\text{gpm}) \cdot S_g} \tag{3-25}$$

Per Eq. (3-41), we can solve for volumetric flow rates in units of gpm in terms of m^3/s:

$$Q \, (\text{gpm}) = \frac{Q \, (\text{m}^3/\text{s})}{0.0000632} = 15,820 Q \, (\text{m}^3/\text{s}) \tag{3-41}$$

Substituting Eq. (3-41) into Eq. (3-25) yields an intermediate relationship:

$$H_p \, (\text{ft}) = \frac{3950 \, (\text{HP})}{15,820 Q \, (\text{m}^3/\text{s}) \cdot S_g} = \frac{0.250 \, (\text{HP})}{Q \, (\text{m}^3/\text{s}) \cdot S_g} \tag{3-42}$$

Also, since 1 m = 3.28 ft, we have

$$H_p \, (\text{ft}) = 3.28 H_p \, (\text{m}) \tag{3-43}$$

Substituting Eq. (3-43) into Eq. (3-42) yields the desired result:

$$H_p \, (\text{m}) = \frac{H_p \, (\text{ft})}{3.28} = \frac{0.250 \, (\text{HP})}{3.28 Q \, (\text{m}^3/\text{s}) \cdot S_g} = \frac{0.0762 \, (\text{HP})}{Q \, (\text{m}^3/\text{s}) \cdot S_g} \tag{3-44}$$

Using values from Exercise 3-11, we have

$$H_p \, (\text{m}) = \frac{0.0762(5)}{0.001896(0.9)} = 223.3 \text{ m}$$

Next we solve for v_2 and $v_2^2/2g$:

$$v_2 \, (\text{m/s}) = \frac{Q \, (\text{m}^3/\text{s})}{A \, (\text{m}^2)} = \frac{0.001896}{(\pi/4)(0.0254 \text{ m})^2} = 3.74 \text{ m/s}$$

$$\frac{v_2^2}{2g} = \frac{(3.74 \text{ m/s})^2}{(2)(9.80 \text{ m/s}^2)} = 0.714 \text{ m}$$

Upon final substitution, we have

$$\frac{P_2}{\gamma} = 223.3 - 0.714 - 15.24 = 207.3 \text{ m}$$

Solving for P_2 yields

$$P_2 \text{ (N/m}^2) = (207.3 \text{ m})\gamma \text{ (N/m}^3)$$

where $\gamma = S_g \gamma_{\text{water}} = (0.9)(9797 \text{ N/m}^3) = 8817 \text{ N/m}^3$

$$P_2 \text{ (N/m}^2) = (207.3)(8817) = 1,828,000 \text{ Pa} = 1828 \text{ kPa}$$

where kPa = kilopascals = 1000 Pa.

Now since 1 Pa = 0.000145 psi, we have

$$P_2 = (1,828,000)(0.000145) = 265 \text{ psi}$$

This value checks with the answer to Example 3-8, which was solved using the English system of units.

EXERCISES

3-1. Explain the meaning of Bernoulli's equation and how it affects the flow of a liquid in a hydraulic circuit.

3-2. What is the continuity equation, and what are its implications relative to fluid flow?

3-3. At a velocity of 10 ft/sec, how many gpm of fluid will flow through a 1-in.-inside-diameter pipe?

3-4. How large an inside-diameter pipe is required to keep the velocity at 15 ft/sec if the flow rate is 20 gpm?

3-5. Fluid is flowing horizontally at 200 gpm from a 2-in.-diameter pipe to a 1-in.-diameter pipe, as shown in Fig. 3-20. If the pressure at point 1 is 10 psi, find the pressure at point 2. The specific gravity of the fluid is 0.9.

3-6. State Torricelli's theorem in your own words.

3-7. Explain how a siphon operates.

3-8. A hydraulic system is powered by a 5-hp motor and operates at 1000 psi. Assuming no losses, what is the flow rate through the system in gpm?

3-9. What is the potential energy of 1000 gal of water at an elevation of 100 ft?

3-10. What is the kinetic energy of 1 gal of water traveling at 20 ft/sec?

3-11. A siphon made of a 1-in.-inside-diameter pipe is used to maintain a constant level in a 20-ft-deep tank. If the siphon discharge is 30 ft below the top of the tank, what will be the flow rate if the fluid level is 5 ft below the top of the tank?

3-12. State the law of conservation of energy.

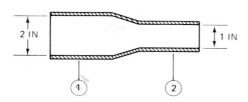

Figure 3-20.   System for Exercise 3-5.

3-13. Explain how a venturi is used to produce the Bernoulli effect in an automobile carburetor.

3-14. What is meant by the terms *elevation head, pressure head,* and *velocity head?*

3-15. State Newton's three laws of motion.

3-16. Differentiate between energy and power.

3-17. Define the term *torque.*

3-18. What is meant by the term *efficiency?*

3-19. An operator makes 20 complete cycles during a 15-sec interval using the hand pump of Fig. 3-2. Each complete cycle consists of two pump strokes (intake and power). The pump has a 2-in.-diameter piston and the load cylinder has a 4-in.-diameter piston. If the average hand force is 20 lb during each power stroke,

    (a) How much load can be lifted?

    (b) Through what distance will the load be moved during the 15-sec interval assuming no oil leakage? The pump piston has a 3-in. stroke.

    (c) What is the output *HP* assuming 90% efficiency?

3-20. For the pressure booster of Fig. 3-10, the following data are given:

$$\text{Inlet oil pressure } (P_1) = 125 \text{ psi.}$$

$$\text{Air piston area } (A_1) = 20 \text{ in.}^2$$

$$\text{Oil piston area } (A_2) = 1 \text{ in.}^2$$

$$\text{Load-carrying capacity } (F) = 75{,}000 \text{ lb.}$$

Find the required load piston area A_3.

3-21. Relative to power, there is an analogy between mechanical, electrical, and hydraulic systems. Describe this analogy.

3-22. A hydraulic cylinder is to compress a car body down to bale size in 8 sec. The operation requires an 8-ft stroke and a 10,000-lb force. If a 1000-psi pump has been selected, find

    (a) The required piston area.

    (b) The necessary pump flow rate.

    (c) The hydraulic horsepower delivered by the cylinder.

3-23. For the hydraulic system of Fig. 3-17, the following data are given:

    1. The pump is adding 4 hp to the fluid.

    2. Pump flow is 25 gpm.

    3. The pipe has an 0.75-in. inside diameter.

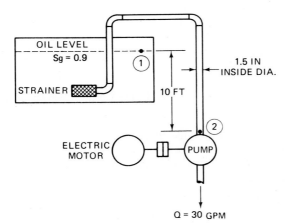

Figure 3-21. Hydraulic system for Exercise 3-24.

4. The specific gravity of the oil is 0.9.

5. The oil tank is vented to the atmosphere.

6. The head loss H_L between stations 1 and 2 is 40 ft of oil.

Find the pressure available at the inlet to the hydraulic motor (station 2).

3-24. The oil tank for the hydraulic system of Fig. 3-21 is air-pressurized at 10 psig. The inlet line to the pump is 10 ft below the oil level. The pump flow rate is 30 gpm. Find the pressure at station 2 if

(a) There is no head loss between stations 1 and 2.

(b) There is a 25-ft head loss between stations 1 and 2.

3-25. Solve Exercise 3-23, using the metric-SI system of units. The equivalent data are given in metric units as follows:

1. The pump is adding 4 hp (2984 W) to the fluid.

2. Pump flow is 0.00158 m³/s.

3. The pipe has an 0.01905-m inside diameter.

4. The specific gravity of the oil is 0.9.

5. The oil tank is vented to the atmosphere.

6. The head loss H_L between stations 1 and 2 is 12.19 m of oil.

7. The elevation difference between stations 1 and 2 is 6.096 m.

Find the pressure available at the inlet to the hydraulic motor (station 2).

3-26. Solve Exercise 3-24 using the metric-SI system of units. The equivalent data are given in metric units as follows:

1. Pump flow is 0.001896 m³/s.

2. The air pressure at station 1 in the hydraulic tank is 68,970-Pa or 68.97-kPa gage pressure.

3. The inlet line to the pump is 3.048 m below the oil level.

4. The pipe has an 0.0381-m inside diameter.

Find the pressure at station 2 if

(a) There is no head loss between stations 1 and 2.

(b) There is a 7.622-m head loss between stations 1 and 2.

3-27. The following relationship containing metric units as shown is analogous to equation (3-26).

$$v(\text{m/s}) = \frac{CQ(\text{m}^3/\text{s})}{[D(\text{m})]^2}$$

For the above equation, determine the numerical value of C. Using this value of C, calculate $v_2(\text{m/s})$ for the system of Example 3-11 and resolve any numerical difference.

3-28. Show that equation (3-44) can be rewritten as follows:

$$H_p(\text{m}) = \frac{0.1021\ \text{Power(kW)}}{Q(\text{m}^3/\text{s}) \cdot S_g}$$

where Power (kW) is the power in units of kilowatts (1 kW = 1000 W).

3-29. The power and load-carrying capacity of a hydraulic cylinder (extension direction) are 10 HP and 5000 lb, respectively. Find the piston velocity in units of ft/sec.

3-30. The power and load-carrying capacity of a hydraulic cylinder (extension direction) are 10 kW and 20,000 N, respectively. Find the piston velocity in units of m/s. Note the following applicable equation:

$$\text{Power(kW)} = F(\text{kN})v(\text{m/s})$$

3-31. At a velocity of 3 m/s, how many m^3/s of fluid will flow through a 0.10 m inside diameter pipe?

3-32. A hydraulic system is powered by a 5 kW motor and operates at 10 MPa. Assuming no losses, what is the flow rate through the system in units of m^3/s. Note the following applicable equation:

$$\text{Power(kW)} = \text{Pressure(kN/m}^2)\text{Flow(m}^3/\text{s})$$

$$= \text{Pressure(kPa)Flow(m}^3/\text{s})$$

3-33. A hydraulic cylinder is to compress a car body down to bale size in 8s. The operation requires a 3 m stroke and a 40,000 N force. If a 10 MPa pump has been selected, find

(a) The required piston area (m^2).

(b) The necessary pump flow rate (m^3/s).

(c) The hydraulic power (kW) delivered by the cylinder.

3-34. For the pressure booster of Fig. 3-10, the following data are given:

1. Inlet oil pressure (P_1) = 1 MPa.

2. Air piston area (A_1) = 0.02 m^2.

3. Oil piston area (A_2) = 0.001 m^2.

4. Load-carrying capacity (F) = 300,000 N.

Find the required load piston area A_3.

3-35. An automotive lift raises a 3500-lb car 7 ft above floor level. If the hydraulic cylinder contains an 8-in. diameter piston and 4-in. diameter rod, determine the

  (a) Work necessary to lift the car.

  (b) Required pressure.

  (c) Power if the lift raises the car in 10 sec.

  (d) Descending speed of the lift for a 10 gpm flow rate.

  (e) Flow rate for the auto to descend in 10 sec.

3-36. For the system in Exercise 3-35, change the data to metric units and solve for parts a, b, c, d, and e.

3-37. How long would it take a 20-gpm pump to extend a 4-in. diameter cylinder through a 20-in. stroke?

3-38. In Exercise 3-37, if the rod diameter is 2 in., how long would it take to retract the cylinder?

3-39. A hydraulic system has a 100-gallon reservoir mounted above the pump to produce a positive pressure (above atmospheric) at the pump inlet as shown in Fig. 3-22. The purpose of the positive pressure is to prevent the pump from cavitating when operating, especially start-up. If the pressure at the pump inlet is to be 5 psi prior to turning the pump on and the oil has a specific gravity of 0.90, what should the oil level be above the pump inlet?

3-40. For the hydraulic system of Figure 3-23, what would be the pressure at the pump inlet if the reservoir were located below the pump so that the oil level were 4 ft below the pump inlet. Ignore frictional losses and changes in kinetic energy. What effect would frictional losses and changes in kinetic energy have on the pressure at the pump inlet? Why? Would this increase or decrease the chances for having pump cavitation? Why?

3-41. A double-rod cylinder is one in which the rod extends out of the cylinder at both ends. Such a cylinder, with a 3-in. diameter piston and 2-in. diameter rod, cycles through a 10-in. stroke at 60 cycles per minute. What gpm size pump is required?

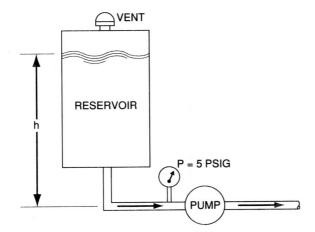

Figure 3-22. System for Exercise 3-39.

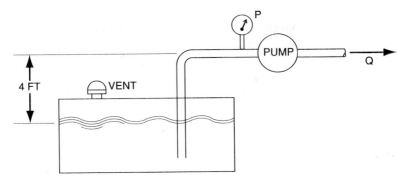

Figure 3-23.   System for Exercise 3-40.

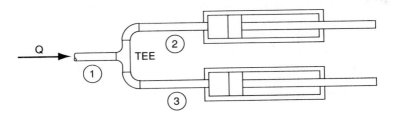

Figure 3-24.   System for Exercise 3-44.

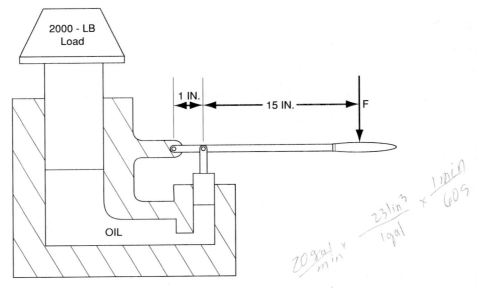

Figure 3-25.   System for Exercise 3-45.

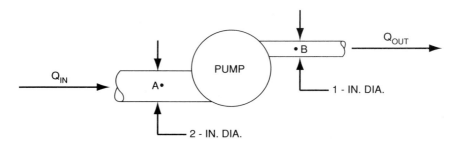

Figure 3-26. System for Exercise 3-48.

3-42. A hydraulic cylinder with a 3-in. diameter piston and 1.5-in. diameter rod extends while receiving fluid at 10 gpm through a $\frac{3}{4}$-inch Schedule 40 pipe. What is the velocity in this pipe when the cylinder is retracting?

3-43. A cylinder with a 8-cm diameter piston and 3-cm diameter rod receives fluid at 30 liters per min. If the cylinder has a stroke of 35 cm, what is the maximum cycle rate that can be accomplished?

3-44. Oil with specific gravity 0.9 enters a tee as shown in Fig. 3-24 with velocity $v_1 = 5$ m/s. The diameter at section 1 is 10 cm, the diameter at section 2 is 7 cm, and the diameter at section 3 is 6 cm. If equal flow rates are to occur at sections 2 and 3, find v_2 and v_3.

3-45. The hydraulic jack shown in Fig. 3-25 is filled with oil. The large and small pistons have diameters of 3 in. and 1 in. respectively. What force F on the handle is required to support the 2000-lb weight? If the force moves down 5 in., how far will the weight be lifted?

3-46. For the system of Exercise 3-45 as shown in Fig. 3-25, change the data to metric units. Then calculate the force F in units of N and determine how far the weight will be lifted in units of cm.

3-47. What is the significance of each term in Bernoulli's equation?

3-48. For the pump in Fig. 3-26, $Q_{out} = 30$ gpm. What is Q_{in}? What is the pressure difference between points A and B if (a) the pump is turned off and (b) input power to pump is 2 HP?

4

The Distribution System

4.1 INTRODUCTION

In a fluid power system, the fluid flows through a distribution system consisting of conductors and fittings, which carry the fluid from the reservoir through operating components and back to the reservoir. Since power is transmitted throughout the system by means of these conducting lines (conductors and fittings used to connect system components), it follows that they must be properly designed in order for the total system to function properly.

Today's fluid power systems use primarily four types of conductors:

1. Steel pipes
2. Steel tubing
3. Plastic tubing
4. Flexible hoses

The choice of which type of conductor to use depends primarily on the system's operating pressures and flow rates. In addition, the selection depends on environmental conditions such as the type of fluid, operating temperatures, vibration, and whether or not there is relative motion between connected components.

Conducting lines are available for handling working pressures up to 10,000 psi or greater. In general, steel tubing provides greater plumbing flexibility and neater appearance and requires fewer fittings than piping. However, piping is less expensive than steel tubing. Plastic tubing is finding increased industrial usage because it is not costly and circuits can be very easily hooked up due to its flexibility. Flexible hoses are used primarily to connect components that experience relative motion. They are made from a large number of elastomeric (rubber-like) compounds and are capable of handling pressures exceeding 10,000 psi.

Stainless steel conductors and fittings are used if extremely corrosive environments are expected. However, they are very expensive and should be used only if necessary. Copper conductors should not be used in hydraulic systems because the copper promotes the oxidation of petroleum oils. Zinc, magnesium, and cadmium conductors should not be used either because they are rapidly corroded by water-glycol fluids. Galvanized conductors should also be avoided because the galvanized surface has a tendency to flake off into the hydraulic fluid. When using steel pipe or steel tubing, hydraulic fittings should be made of steel except for inlet, return, and drain lines where malleable iron may be used.

Conductors and fittings must be designed with human safety in mind. They must be strong enough not only to withstand the steady-state system pressures but also the instantaneous pressure spikes resulting from hydraulic shock. Whenever control valves are closed suddenly, this stops the fluid, which possesses large amounts of kinetic energy. This produces shock waves whose pressure levels can be two to four times the steady-state system design values. Pressure spikes can also be caused by sudden stopping or starting of heavy loads. These high-pressure pulses are taken into account by the application of an appropriate factor of safety.

4.2 CONDUCTOR SIZING FOR FLOW-RATE REQUIREMENTS

A conductor must have a large enough cross-sectional area to handle the flow-rate requirements without producing excessive fluid velocity. Whenever we speak of fluid velocity in a conductor such as a pipe, we are referring to the average velocity. The concept of average velocity is important since we know that the velocity profile is not constant. Recall from Chapter 3 that the velocity is zero at the pipe wall and reaches a maximum value at the centerline of the pipe. The average velocity is defined as the volumetric flow rate divided by the pipe cross-sectional area:

$$v = v_{avg} = \frac{Q}{A} \tag{4-1}$$

In other words, the average velocity is that velocity which when multiplied by the pipe area equals the volumetric flow rate. It is also understood that the term *diameter* by itself always means inside diameter and that the pipe area is that area that corresponds to the pipe inside diameter. The maximum recommended velocity for pump suction lines is 4 ft/sec in order to prevent excessively low suction pressures and resulting pump cavitation. The maximum recommended velocity for pressure lines is 20 ft/sec in order to prevent turbulent flow and the corresponding excessive head losses and elevated temperatures. Note that these design values of 4 ft/sec and 20 ft/sec are average velocities.

EXAMPLE 4-1

A pipe handles a flow rate of 30 gpm. Find the minimum inside diameter that will provide an average fluid velocity not to exceed 20 ft/sec.

Solution   Rewrite Eq. (3-26), solving for D:

$$D = \sqrt{\frac{0.408 Q}{v}} = \sqrt{\frac{(0.408)(30)}{20}} = \sqrt{0.612} = 0.782 \text{ in.}$$

4.3 PRESSURE RATING OF CONDUCTORS

Conductors must be strong enough to prevent bursting under the operating fluid pressures. A factor of safety is applied to determine the maximum safe level of working pressure. Once the inside diameter has been established based on flow-rate requirements, the required wall thickness can be obtained using the following equations:

$$BP = \frac{2tS}{D_o} \tag{4-2}$$

$$WP = \frac{BP}{FS} \tag{4-3}$$

$$t = \frac{D_o - D_i}{2} \tag{4-4}$$

where BP = burst pressure (psi),
t = wall thickness (in.),
S = tensile strength of conductor material (psi),
D_o = outside diameter (in.),
WP = working pressure (psi),
FS = factor of safety (dimensionless),
D_i = inside diameter (in.).

The working pressure is the maximum safe operating pressure and is defined as the conductor bursting pressure divided by the appropriate factor of safety. This ensures the integrity of the conductor. Industry standards recommend the following factors of safety based on corresponding operating pressures:

$$FS = 8 \text{ for pressures from 0 to 1000 psi}$$

$$FS = 6 \text{ for pressures from 1000 to 2500 psi}$$

$$FS = 4 \text{ for pressures above 2500 psi}$$

For systems where severe pressure shocks are expected, a factor of safety of 10 is recommended.

EXAMPLE 4-2

A steel tubing has a 1.250-in. outside diameter and a 1.060-in. inside diameter. It is made of SAE 1010 dead soft cold-drawn steel having a tensile strength of 55,000 psi. What would be the safe working pressure for this tube assuming a factor of safety of 8?

Solution First, calculate the wall thickness of the tubing:

$$t = \frac{1.250 - 1.060}{2} = 0.095 \text{ in.}$$

Next, find the burst pressure for the tubing:

$$BP = \frac{(2)(0.095)(55,000)}{1.250} = 8370$$

Finally, calculate the working pressure at which the tube can safely operate:

$$WP = \frac{8370}{8} = 1050 \text{ psi}$$

4.4 STEEL PIPES

Pipes and pipe fittings are classified by nominal size and schedule number, as illustrated in Fig. 4-1. The schedules provided are 40, 80, and 160, which are the ones most commonly used for hydraulic systems. Notice that for each nominal size the outside diameter does not change. To increase wall thickness the next larger schedule number is used. Also observe that the nominal size is neither the outside nor the inside diameter. Instead the nominal pipe size indicates the thread size for the mating connections.

Pipes have tapered threads as opposed to tube and hose fittings, which have straight threads. As shown in Fig. 4-2, the joints are sealed by an interference fit between the male and female threads as the pipes are tightened. This causes one of the major problems in using pipe. When a joint is taken apart, the pipe must be

NOMINAL PIPE SIZE	PIPE OUTSIDE DIAMETER	PIPE INSIDE DIAMETER		
		SCHEDULE 40	SCHEDULE 80	SCHEDULE 160
1/8	0.405	0.269	0.215	—
1/4	0.540	0.364	0.302	—
3/8	0.675	0.493	0.423	—
1/2	0.840	0.622	0.546	0.466
3/4	1.050	0.824	0.742	0.614
1	1.315	1.049	0.957	0.815
1-1/4	1.660	1.380	1.278	1.160
1-1/2	1.900	1.610	1.500	1.338
2	2.375	2.067	1.939	1.689

Figure 4-1.   Common pipe sizes.

tightened farther to reseal. This frequently requires replacing some of the pipe with slightly longer sections, although this problem has been overcome somewhat by using Teflon tape to reseal the pipe joints. Hydraulic pipe threads are the *dry-seal* type. They differ from standard pipe threads because they engage the roots and crests before the flanks. In this way, spiral clearance is avoided.

Pipes can have only male threads, and they cannot be bent around obstacles. There are, of course, various required types of fittings to make end connections and change direction, as shown in Fig. 4-3. The large number of pipe fittings required in a hydraulic circuit presents many opportunities for leakage, especially as pressure increases. Threaded-type fittings are used in sizes up to $1\frac{1}{4}$ in. in diameter. Where larger pipes are required, flanges are welded to the pipe, as illustrated in Fig. 4-4. As shown, flat gaskets or O-rings are used to seal the flanged fittings.

4.5 STEEL TUBING

Seamless steel tubing is the most widely used type of conductor for hydraulic systems as it provides significant advantages over pipes. The tubing can be bent into almost any shape, thereby reducing the number of required fittings. Tubing is easier to handle and can be reused without any sealing problems. For low-volume systems, tubing can handle the pressure and flow requirements with less bulk and weight. However, tubing and its fittings are more expensive. A tubing size designation always refers to the outside diameter. Available sizes include $\frac{1}{16}$ in. increments from $\frac{1}{8}$-in. outside diameter up to 1-in. outside diameter. For sizes beyond 1

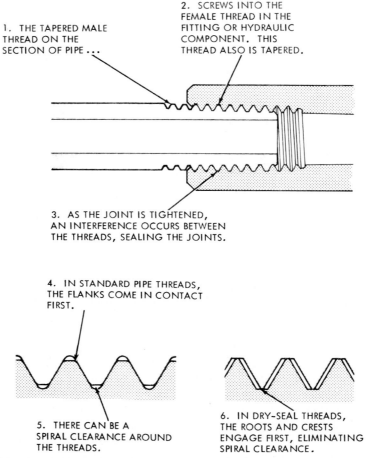

Figure 4-2. Hydraulic pipe threads are the dry-seal tapered type. (*Courtesy of Sperry Vickers, Sperry Rand Corp., Troy, Michigan.*)

in., the increments are $\frac{1}{4}$ in. Figure 4-5 shows some of the more common tube sizes used in fluid power systems.

SAE 1010 dead soft cold-drawn steel is the most widely used material for tubing. This material is easy to work with and has a tensile strength of 55,000 psi. If greater strength is required, the tube can be made of AISI 4130 steel, which has a tensile strength of 75,000 psi.

Tubing is not sealed by threads but by special kinds of fittings, as illustrated in Fig. 4-6. Some of these fittings are known as compression fittings. They seal by metal-to-metal contact and may be either the flared or flareless type. Other fittings may use O-rings for sealing purposes. The 37° flare fitting is the most widely used

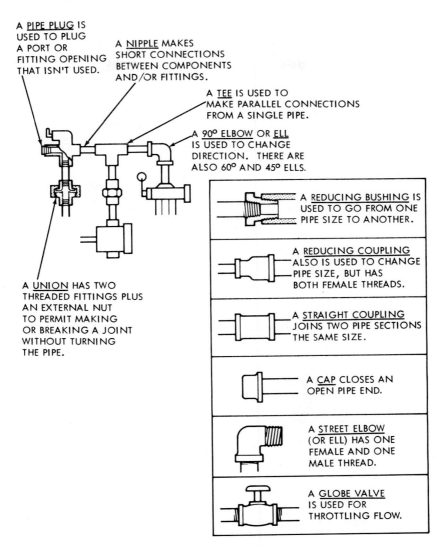

A <u>PIPE PLUG</u> IS USED TO PLUG A PORT OR FITTING OPENING THAT ISN'T USED.

A <u>NIPPLE</u> MAKES SHORT CONNECTIONS BETWEEN COMPONENTS AND/OR FITTINGS.

A <u>TEE</u> IS USED TO MAKE PARALLEL CONNECTIONS FROM A SINGLE PIPE.

A <u>90° ELBOW</u> OR <u>ELL</u> IS USED TO CHANGE DIRECTION. THERE ARE ALSO 60° AND 45° ELLS.

A <u>UNION</u> HAS TWO THREADED FITTINGS PLUS AN EXTERNAL NUT TO PERMIT MAKING OR BREAKING A JOINT WITHOUT TURNING THE PIPE.

A <u>REDUCING BUSHING</u> IS USED TO GO FROM ONE PIPE SIZE TO ANOTHER.

A <u>REDUCING COUPLING</u> ALSO IS USED TO CHANGE PIPE SIZE, BUT HAS BOTH FEMALE THREADS.

A <u>STRAIGHT COUPLING</u> JOINS TWO PIPE SECTIONS THE SAME SIZE.

A <u>CAP</u> CLOSES AN OPEN PIPE END.

A <u>STREET ELBOW</u> (OR ELL) HAS ONE FEMALE AND ONE MALE THREAD.

A <u>GLOBE VALVE</u> IS USED FOR THROTTLING FLOW.

Figure 4-3. Fittings make the connections between pipes and components. (*Courtesy of Sperry Vickers, Sperry Rand Corp., Troy, Michigan.*)

fitting for tubing that can be flared. The fittings shown in Fig. 4-6(a) and (b) seal by squeezing the flared end of the tube against a seal as the compression nut is tightened. A sleeve inside the nut supports the tube to dampen vibrations. The standard 45° flare fitting is used for very high pressures. It is also made in an inverted design with male threads on the compression nut. When the hydraulic component has straight thread ports, straight thread O-ring fittings can be used, as

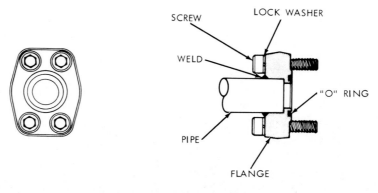

SOCKET WELD PIPE CONNECTIONS
STRAIGHT TYPE

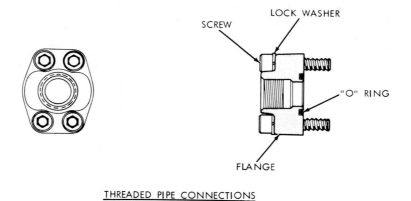

THREADED PIPE CONNECTIONS
STRAIGHT TYPE

Figure 4-4. Flanged connections for large pipes. (*Courtesy of Sperry Vickers, Sperry Rand Corp., Troy, Michigan.*)

shown in Fig. 4-6(c). This type is ideal for high pressures since the seal gets tighter as pressure increases.

For tubing that can't be flared, or if flaring is to be avoided, sleeve, ferrule, or O-ring compression fittings can be used [see Fig. 4-6(d), (e), (f)]. The O-ring fitting permits considerable variations in the length and squareness of the tube cut.

Figure 4-7 shows a Swagelok tube fitting which can contain any pressure up to the bursting strength of the tubing without leakage. This type of fitting can be

TUBE OD	WALL THICKNESS	TUBE ID	TUBE OD	WALL THICKNESS	TUBE ID	TUBE OD	WALL THICKNESS	TUBE ID
1/8	0.035	0.055	1/2	0.035	0.430	7/8	0.049	0.777
				0.049	0.402		0.065	0.745
3/16	0.035	0.118		0.065	0.370		0.109	0.657
				0.095	0.310			
1/4	0.035	0.180				1	0.049	0.902
	0.049	0.152	5/8	0.035	0.555		0.065	0.870
	0.065	0.120		0.049	0.527		0.120	0.760
				0.065	0.495			
5/16	0.035	0.243		0.095	0.435	1-1/4	0.065	1.120
	0.049	0.215					0.095	1.060
	0.065	0.183	3/4	0.049	0.652		0.120	1.010
				0.065	0.620			
3/8	0.035	0.305		0.109	0.532	1-1/2	0.065	1.370
	0.049	0.277					0.095	1.310
	0.065	0.245						

Figure 4-5.   Common tube sizes.

repeatedly taken apart and reassembled and remain perfectly sealed against leakage. Assembly and disassembly can be done easily and quickly using standard tools. In the illustration, notice that the tubing is supported ahead of the ferrules by the fitting body. Two ferrules grasp tightly around the tube with no damage to the tube wall. There is virtually no constriction of the inner wall, ensuring minimum flow restriction. Exhaustive tests have proven that the tubing will yield before a Swagelok tube fitting will leak. The secret of the Swagelok fitting is that all the action in the fitting moves along the tube axially instead of with a rotary motion. Since no torque is transmitted from the fitting to the tubing, there is no initial strain that might weaken the tubing. The double ferrule interaction overcomes variation in tube materials, wall thickness, and hardness.

In Fig. 4-8 we see the 45° flare fitting. The flared-type fitting was developed before the compression type and for some time was the only type that could successfully seal against high pressures.

Four additional types of tube fittings are depicted in Fig. 4-9: (a) union elbow, (b) union tee, (c) union, and (d) 45° male elbow. With fittings such as these, it is easy to install steel tubing as well as remove it for maintenance purposes.

EXAMPLE 4-3

Select the proper size steel tube for a flow rate of 30 gpm and an operating pressure of 1000 psi. The maximum recommended velocity is 20 ft/sec, and the tube material is SAE 1010 dead soft cold-drawn steel having a tensile strength of 55,000 psi.

Solution   The minimum inside diameter based on the fluid velocity limitation of 20 ft/sec is the same as that found in Example 4-1 (0.782 in.).

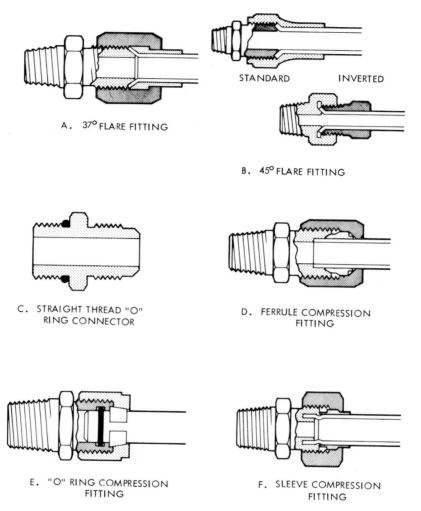

Figure 4-6. Threaded fittings and connectors used with tubing. (*Courtesy of Sperry Vickers, Sperry Rand Corp., Troy, Michigan.*)

From Fig. 4-5, the two smallest acceptable tube sizes based on flow-rate requirements are

1-in. OD, 0.049-in. wall thickness, 0.902-in. ID

1-in. OD, 0.065-in. wall thickness, 0.870-in. ID

Figure 4-7. Swagelok tube fitting. (*Courtesy of Swagelok Co., Solon, Ohio, copyright 1992 by Swagelok Co.*)

Figure 4-8. The 45° flare fitting. (*Courtesy of Gould, Inc., Valve and Fittings Division, Chicago, Illinois.*)

Let's check the 0.049-in. wall thickness tube first since it provides the smaller velocity:

$$BP = \frac{(2)(0.049)(55{,}000)}{1} = 5380 \text{ psi}$$

$$WP = \frac{5380}{8} = 673 \text{ psi}$$

This working pressure is not adequate, so let's next examine the 0.065-in. wall thickness tube:

$$BP = \frac{(2)(0.065)(55{,}000)}{1} = 7140 \text{ psi}$$

$$WP = \frac{7140}{8} = 890 \text{ psi}$$

Figure 4-9. Various steel tube fittings. (a) Union elbow, (b) union tee, (c) union, (d) 45° male elbow. (*Courtesy of Gould, Inc., Valve and Fittings Division, Chicago, Illinois.*)

This is still not adequate. Referring back to Fig. 4-5, the next larger size is 1¼-in. OD. Checking the 0.095-in. wall thickness size yields

$$BP = \frac{(2)(0.095)(55,000)}{1.25} = 8370 \text{ psi}$$

$$WP = \frac{8370}{8} = 1050 \text{ psi}$$

This is acceptable.

4.6 PLASTIC TUBING

Plastic tubing has gained rapid acceptance in the fluid power industry because it is relatively inexpensive. Also, it can be readily bent to fit around obstacles, it is easy to handle, and it can be stored on reels. Another advantage is that it can be color-coded to represent different parts of the circuit because it is available in many colors. Since plastic tubing is flexible, it is less susceptible to vibration damage than steel tubing.

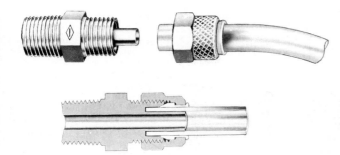

Figure 4-10. Poly-Flo Flareless Plastic Tube Fitting. (*Courtesy of Gould, Inc., Valve and Fittings Division, Chicago, Illinois.*)

Fittings for plastic tubing are almost identical to those designed for steel tubing. In fact many steel tube fittings can be used on plastic tubing as is the case for the Swagelok fitting of Fig. 4-7. In another design, a sleeve is placed inside the tubing to give it resistance to crushing at the area of compression, as illustrated in Fig. 4-10. In this particular design (called the Poly-Flo Flareless Tube Fitting), the sleeve is fabricated onto the fitting so it cannot be accidentally left off.

Plastic tubing is used universally in pneumatic systems because air pressures are low, normally less than 100 psi. Of course, plastic tubing is compatible with most hydraulic fluids and hence is used in low-pressure hydraulic applications.

Materials for plastic tubing include polyethylene, polyvinyl chloride, polypropylene, and nylon. Each material has special properties that are desirable for specific applications. Manufacturers' catalogs should be consulted to determine which material should be used for a particular application.

4.7 FLEXIBLE HOSES

The fourth major type of hydraulic conductor is the flexible hose, which is used when hydraulic components such as actuators are subjected to movement. Examples of this are found in portable power units, mobile equipment, and hydraulically powered machine tools. Hose is fabricated in layers of elastomer (synthetic rubber) and braided fabric or braided wire, which permits operation at higher pressures.

As illustrated in Fig. 4-11, the outer layer is normally synthetic rubber and serves to protect the braid layer. The hose can have as few as three layers (one being braid) or can have multiple layers to handle elevated pressures. When multiple wire layers are used, they may alternate with synthetic rubber layers, or the wire layers may be placed directly over one another.

Figure 4-12 gives some typical hose sizes and dimensions for single-wire braid and double-wire braid designs. Size specifications for a single-wire braid hose represent the outside diameter in sixteenths of an inch of standard tubing, and the hose will have about the same inside diameter as the tubing. For example,

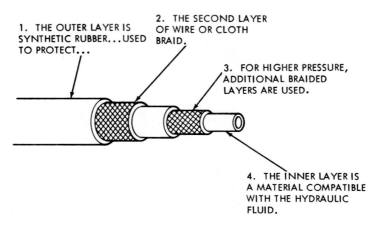

Figure 4-11. Flexible hose is constructed in layers. (*Courtesy of Sperry Vickers, Sperry Rand Corp., Troy, Michigan.*)

a size 8 single-wire braid hose will have an inside diameter very close to a $\frac{8}{16}$- or $\frac{1}{2}$- in. standard tubing. For double-braided hose, the size specification equals the actual inside diameter in sixteenths of an inch. For example, a size 8 double-wire braid hose will have a $\frac{1}{2}$-in. inside diameter. The minimum bend radii values provide the smallest values for various hose sizes to prevent undue strain or flow interference.

Figure 4-13 illustrates five different flexible hose designs whose constructions are described as follows:

a. *FC194:* Elastomer inner tube, single-wire braid reinforcement, and elastomer cover. Working pressures vary from 375 to 2750 psi depending on the size.

HOSE SIZE	O.D. TUBE SIZE (IN)	SINGLE-WIRE BRAID			DOUBLE-WIRE BRAID		
		HOSE I.D. (IN)	HOSE O.D. (IN)	MINIMUM BEND RADIUS (IN)	HOSE I.D. (IN)	HOSE O.D. (IN)	MINIMUM BEND RADIUS (IN)
4	1/4	3/16	33/64	1-15/16	1/4	11/16	4
6	3/8	5/16	43/64	2-3/4	3/8	27/32	5
8	1/2	13/32	49/64	4-5/8	1/2	31/32	7
12	3/4	5/8	1-5/64	6-9/16	3/4	1-1/4	9-1/2
16	1	7/8	1-15/64	7-3/8	1	1-9/16	11
20	1-1/4	1-1/8	1-1/2	9	1-1/4	2	16

Figure 4-12. Typical hose sizes.

(a)

(b)

(c)

(d)

(e)

Figure 4-13.   Various flexible hose designs. (a) FC194: single-wire braid, (b) FC195: double-wire braid, (c) FC300: single-wire braid, polyester inner braid, (d) 1525: single-textile braid, (e) 2791: four heavy spiral wires, partial textile braid. (*Courtesy of Aeroquip Corp., Jackson, Michigan.*)

b. *FC195:* Elastomer inner tube, double-wire braid reinforcement, and elastomer cover. Working pressures vary from 1125 to 5000 psi depending on the size.

c. *FC300:* Elastomer inner tube, polyester inner braid, single-wire braid reinforcement, and polyester braid cover. Working pressures vary from 350 to 3000 psi depending on the size.

d. *1525:* Elastomer inner tube, textile braid reinforcement, oil and mildew resistant, and textile braid cover. Working pressure is 250 psi for all sizes.

e. *2791:* Elastomer inner tube, partial textile braid, four heavy spiral wire reinforcements, and elastomer cover. Working pressure is 2500 psi for all sizes.

Hose assemblies of virtually any length and with various end fittings are available from manufacturers. See Fig. 4-14 for examples of hoses with the following permanently attached end fittings: (a) straight fitting, (b) 45° elbow fitting, and (c) 90° elbow fitting.

The elbow-type fittings allow access to hard-to-get-at connections. They also permit better flexing and improve the appearance of the system.

Figure 4-15 shows the three corresponding reusable-type end fittings. These types can be detached from a damaged hose and reused on a replacement hose. The renewable fittings idea had its beginning in 1941. With the advent of World War II, it was necessary to get aircraft with failed hydraulic lines back into operation as quickly as possible.

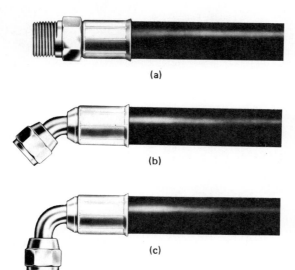

(a)

(b)

(c)

Figure 4-14. Flexible hoses with permanently attached end fittings. (a) Straight fitting, (b) 45° elbow fitting, (c) 90° elbow fitting. (*Courtesy of Aeroquip Corp., Jackson, Michigan.*)

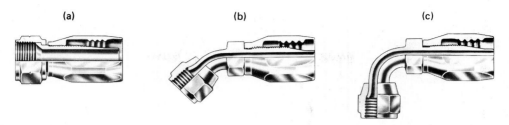

Figure 4-15. Flexible hose reusable-type end fittings. (a) Straight fitting, (b) elbow fitting, (c) 90° elbow fitting. (*Courtesy of Aeroquip Corp., Jackson, Michigan.*)

Care should be taken in changing fluid in hoses since the hose and fluid materials must be compatible. Flexible hose should be installed so there is no kinking during operation of the system. There should always be some slack to relieve any strain and allow for the absorption of pressure surges. It is poor practice to twist the hose and use long loops in the plumbing operation. It may be necessary to use clamps to prevent chafing or tangling of the hose with moving parts. If the hose is subject to rubbing, it should be encased in a protective sleeve. Figure 4-16 gives basic information on hose routing and installation procedures.

4.8 QUICK DISCONNECT COUPLINGS

One additional type of fitting is the quick disconnect coupling used for both plastic tubing and flexible hose. It is used mainly where a conductor must be disconnected frequently from a component. This type of fitting permits assembly and disassembly in a matter of a second or two.

1. Straight through: This type offers minimum restriction to flow but does not prevent fluid loss from the system when the coupling is disconnected.

2. One-way shutoff: This design locates the shutoff at the fluid source connection but leaves the actuator component unblocked. Leakage from the system is not excessive in short runs, but system contamination due to the entrance of dirt in the open end of the fitting can be a problem, especially with mobile equipment located at the work site.

3. Two-way shutoff: This design provides positive shutoff of both ends of pressurized lines when disconnected. See Fig. 4-17 for a cutaway of this type of quick disconnect coupling. Figure 4-18 shows an external view of the same coupling. Such a coupling puts an end to the loss of fluids. As soon as you release the locking sleeve, valves in both the socket and plug close, shutting off flow. When connecting, the plug contacts an O-ring in the socket, creating a positive seal.

Hose routing and installation

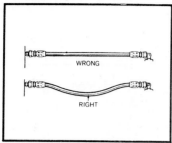

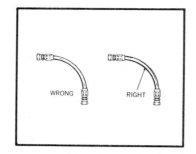

Under pressure, a hose may change in length. The range is from −4% to +2%. Always provide some slack in the hose to allow for this shrinkage or expansion. (However, excessive slack in hose lines is one of the most common causes of poor appearance.)

If a hose is installed with a twist in it, high operating pressures tend to force it straight. This can loosen the fitting nut or even burst the hose at the point of strain.

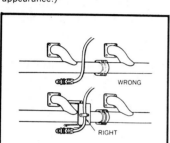

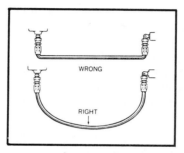

When hose lines pass near an exhaust manifold, or other heat source, they should be insulated by a heat resistant boot, firesleeve or a metal baffle. In any application, brackets and clamps keep hoses in place and reduce abrasion.

At bends, provide enough hose for a wide radius curve. Too tight a bend pinches the hose and restricts the flow. The line could even kink and close entirely. In many cases, use of the right fittings or adapters can eliminate bends or kinks.

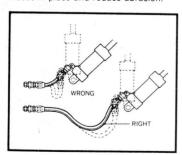

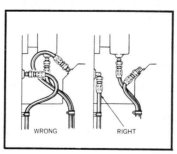

In applications where there is considerable vibration or flexing, allow additional hose length. The metal hose fittings, of course, are not flexible, and proper installation protects metal parts from undue stress, and avoids kinks in the hose.

When 90° adapters were used, this assembly became neater-looking and easier to inspect and maintain. It uses less hose, too!

Four basic adapter functions

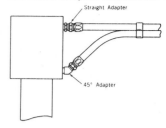

1. To join a hose to a component. Example, a valve might have a ½″ female pipe thread and a hose a ¾″ S.A.E. 37° swivel nut. The right Aeroquip Adapter fits both.

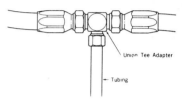

2. To connect two or more pieces of hose and tubing. Here, a T-shaped adapter connects two hoses with a length of tubing. Each end of the adapter may have a different thread.

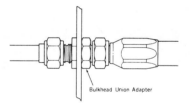

3. To provide both connection and anchor at a bulkhead. In this example, it provides an anchor in addition to connecting a hose to a tube.

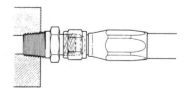

4. To eliminate the need for a bushing. Example, one end of the adapter is ¾″ pipe thread, connected to the assembly, and the other is ½″ S.A.E. 37° flare, which connects to an S.A.E. 37° swivel fitting. The adapter itself replaces the bushing.

Figure 4-16. Hose routing and installation information (*Courtesy of Aeroquip Corp., Jackson, Michigan.*)

Figure 4-17. Quick disconnect coupling (cross-sectional view). (*Courtesy of Hansen Manufacturing Co., Cleveland, Ohio.*)

Figure 4-18. Quick disconnect coupling (external view). (*Courtesy of Hansen Manufacturing Co., Cleveland, Ohio.*)

There is no chance of premature flow or waste due to a partial connection. The plug must be fully seated in the socket before the valves will open.

4.9 METRIC STEEL TUBING

In this section we examine common metric tube sizes and show how to select the proper size tube based on flow rate requirements and strength considerations.

Figure 4-19 shows the common tube sizes used in fluid power systems. Note that the smallest OD size is 4 mm (0.158 in.), whereas the largest OD size is 42 mm (1.663 in.). These values compare to 0.125 in. and 1.500 in., respectively, from Figure 4-5 for common English units tube sizes. It should be noted that since 1 m = 39.6 in., then 1 mm = 0.0396 in.

Tube OD mm	Wall Thickness mm	Tube ID mm	Tube OD mm	Wall Thickness mm	Tube ID mm	Tube OD mm	Wall Thickness mm	Tube ID mm
4	0.5	3	14	2.0	10	25	3.0	19
6	1.0	4	15	1.5	12	25	4.0	17
6	1.5	3	15	2.0	11	28	2.0	24
8	1.0	6	16	2.0	12	28	2.5	23
8	1.5	5	16	3.0	10	30	3.0	24
8	2.0	4	18	1.5	15	30	4.0	22
10	1.0	8	20	2.0	16	35	2.0	31
10	1.5	7	20	2.5	15	35	3.0	29
10	2.0	6	20	3.0	14	38	4.0	30
12	1.0	10	22	1.0	20	38	5.0	28
12	1.5	9	22	1.5	19	42	2.0	38
12	2.0	8	22	2.0	18	42	3.0	36

Figure 4-19. Common metric tube sizes.

Factors of safety based on corresponding operating pressures become:

FS = 8 for pressures from 0 to 1000 psi (0 to 7 MPa or 0 to 70 bars).

FS = 6 for pressures from 1000 to 2500 psi (7 to 17.5 MPa or 70 to 175 bars).

FS = 4 for pressures above 2500 psi (17.5 MPa or 175 bars).

The corresponding tensile strengths for SAE 1010 dead soft cold-drawn steel and AISI 4130 steel are:

SAE 1010   55,000 psi or 379 MPa
AISI 4130   75,000 psi or 517 MPa

EXAMPLE 4-4

Select the proper metric size steel tube for a flow rate of 0.00190 m³/s and an operating pressure of 70 bars. The maximum recommended velocity is 6.1 m/s and the tube material is SAE 1010 dead soft cold-drawn steel having a tensile strength of 379 MPa.

Solution   The minimum inside diameter based on the fluid velocity limitation of 6.1 m/s is found using Eq. (3-18).

$$Q(\text{m}^3/\text{s}) = A(\text{m}^2)v(\text{m/s})$$

Solving for A we have:

$$A = Q/v$$

Since $A = \frac{\pi}{4}(ID)^2$ we have the final resulting equation:

$$ID = \sqrt{\frac{4Q}{\pi v}} \qquad (4\text{-}5)$$

Substituting values we have:

$$ID = \sqrt{\frac{(4)(0.00190)}{\pi(6.1)}} = 0.0199 \text{ m} = 19.9 \text{ mm}$$

From Fig. 4-19, the smallest acceptable OD tube size is:

22-mm OD, 1.0-mm wall thickness, 20-mm ID

From equation 4-2 we obtain the burst pressure.

$$BP = \frac{2tS}{D_o} = \frac{(2)(0.001 \text{ m})(379 \text{ MN/m}^2)}{0.022 \text{ m}} = 34.5 \text{ MN/m}^2 = 34.5 \text{ MPa}$$

Then we calculate the working pressure using equation 4-3.

$$WP = \frac{BP}{FS} = \frac{34.5 \text{ MPa}}{8} = 4.31 \text{ MPa} = 43.1 \text{ bars}$$

This pressure is not adequate (less than operating pressure of 70 bars), so let's examine the next larger size OD tube having the necessary ID.

28-mm OD, 2.0-mm wall thickness, 24-mm ID

$$BP = \frac{(2)(0.002)(379)}{0.028} = 54.1 \text{ MPa}$$

$$WP = \frac{54.1}{8} = 6.76 \text{ MPa} = 67.6 \text{ bars}$$

This is still not adequate. Let's check the 28-mm OD tube having a 2.5-mm wall thickness.

$$BP = \frac{(2)(0.0025)(379)}{0.028} = 67.7 \text{ MPa}$$

$$WP = \frac{67.7}{8} = 8.46 \text{ MPa} = 84.6 \text{ bars}$$

This is acceptable.

EXERCISES

4-1. What is the primary purpose of the fluid distribution system?

4-2. What flow velocity is generally recommended for the discharge side of a pump?

4-3. What is the recommended flow velocity for the inlet side of the pump?

4-4. Why should conductors or fittings not be made of copper?

4-5. What metals cannot be used with water-glycol fluids?

4-6. What effect does hydraulic shock have on system pressure?

4-7. What variables determine the wall thickness and safety factor of a conductor for a particular operating pressure?

4-8. Why should conductors have greater strength than the system working pressure requires?

4-9. What size inlet line would you select for a 20-gpm pump?

4-10. What size discharge line would you select for a 20-gpm pump?

4-11. Name the major disadvantages of steel pipes.

4-12. Name the four primary types of conductors.

4-13. What is meant by the term *average fluid velocity?*

4-14. A steel tubing has a 1.250-in. outside diameter and a 1.060-in. inside diameter. It is made of AISI 4130 steel having a tensile strength of 75,000 psi. What would be the safe working pressure for this tube assuming a factor of safety of 8?

4-15. Why is malleable iron sometimes used for steel pipe fittings?

4-16. Why is steel tubing more widely used than steel pipe?

4-17. What principal advantage does plastic tubing have over steel tubing?

4-18. Select the proper-sized steel tube for a flow rate of 20 gpm and an operating pressure of 1000 psi. The maximum recommended velocity is 20 ft/sec and the factor of safety is 8.

    (a) Material is SAE 1010 with a tensile strength of 55,000 psi.

    (b) Material is AISI 4130 with a tensile strength of 75,000 psi.

4-19. Explain the purpose of a quick disconnect fitting.

4-20. What disadvantage do threaded fittings have?

4-21. What is the difference between a flared fitting and a compression fitting?

4-22. Under what conditions would flexible hoses be used in hydraulic systems?

4-23. Name three factors that should be considered when installing flexible hoses.

4-24. What is the basic construction of a flexible hose?

4-25. What metric size inlet line would you select for a 0.002 m^3/s pump?

4-26. What metric size discharge line would you select for a 0.002 m^3/s pump?

4-27. A steel tubing has a 30-mm outside diameter and a 24-mm inside diameter. It is made of AISI 4130 steel having a tensile strength of 517 MPa. What would be the safe working pressure in units of bars for this tube assuming a factor of safety of 8?

4-28. Select the proper-sized metric steel tube for a flow rate of 0.001 m^3/s and an operating pressure of 70 bars. The maximum recommended velocity is 6.1 m/s and the factor of safety is 8.

     (a) Material is SAE 1010.

     (b) Material is AISI 4130.

4-29. A steel tube of 1-in. ID has a burst pressure of 8000 psi. If the tensile strength of the tube material is 55,000 psi, find the minimum acceptable OD.

4-30. A steel tube of 25-mm ID has a burst pressure of 50 MPa. If the tensile strength is 379 MPa, find the minimum acceptable OD.

4-31. Relative to steel pipes, for a given nominal size, does the wall thickness increase or decrease as the schedule number is increased?

4-32. For liquid flow in a pipe, the velocity of the liquid varies inversely as the _____ of the pipe inside diameter.

4-33. For liquid flow in a pipe, doubling the pipe's inside diameter reduces the fluid velocity by a factor of ____.

4-34. For liquid flow in a pipe, derive the constants C_1 and C_2 in the following equations:

$$A(\text{in.}^2) = \frac{C_1 Q(\text{gpm})}{v(\text{ft/sec})} \qquad A(\text{m}^2) = \frac{C_2 Q(\text{m}^3/\text{s})}{v(\text{m/s})}$$

4-35. How is a pipe size classified?

4-36. What is meant by the *schedule number* of standard pipe?

4-37. What minimum commercial size tubing with a wall thickness of 0.095 in. would be required at the inlet and outlet of a 30-gpm pump if the inlet and outlet velocities are limited to 5 ft/sec and 20 ft/sec respectively? See table in Fig. 4-5.

4-38. Change the data in Exercise 4-37 to metric units and solve for the minimum commercial size tubing at the pump inlet and outlet.

5

Basics of Hydraulic Flow in Pipes

5.1 INTRODUCTION

Up to now we have not investigated the mechanism of energy losses due to friction associated with the flow of a fluid inside a pipe. It is intuitive that liquids, such as water or gasoline, flow much more readily than do heavier liquids such as oil. The resistance to flow is essentially a measure of the viscosity of the fluid. The greater the viscosity of a fluid, the less readily it flows and the more energy is required to move it. This energy is loss because it is dissipated into heat and thus represents wasted energy.

Energy losses also occur in pipeline restrictions called fittings. A fitting is a component (other than a straight pipe) that is used to carry or control the fluid. Examples are valves, tees, elbows, and orifices. The nature of the flow path through a fitting determines the amount of energy losses. Generally speaking, the more torturous the path, the greater the losses. In many fluid power applications, energy losses due to fittings exceed those due to viscous flow in pipes.

It is very important to keep all energy losses in a fluid power system to a minimum acceptable level. This requires the proper selection of the sizes of the pipes and fittings that make up the system. In general, the smaller the pipe diameter or fitting size, the greater the losses. However, using large-diameter pipes and fittings results in greater cost and poor space utilization. Thus, the selection of component sizes represents a compromise between energy losses and component cost and space requirements.

The resistance of pipes and fittings can be determined using empirical formulas that have been developed by experimentation. This permits the calculation of energy losses for any system component. Bernoulli's equation and the continuity equation can then be used to perform a complete analysis of a fluid power system. This includes calculating the pressure drops, flow rates, and horsepower losses for all components of the fluid power system.

5.2 LAMINAR AND TURBULENT FLOW

In our discussions of fluid flow in pipes in Chapter 3, we assumed a constant velocity at any one station (see Fig. 3-12). However, when a fluid flows through a pipe, the layer of fluid at the wall has zero velocity. This is due to viscosity, which causes fluid particles to cling to the wall. Layers of fluid at the progressively greater distances from the pipe surface have higher velocities, with the maximum velocity occurring at the pipe centerline, as illustrated in Fig. 5-1.

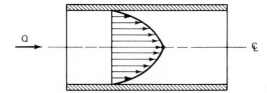

Figure 5-1.   Velocity profile in pipe.

Actually there are two basic types of flow in pipes, depending on the nature of the different factors that affect the flow. The first type is called laminar flow, which is characterized by the fluid flowing in smooth layers or laminae. In this type of flow, a particle of fluid in a given layer stays in that layer, as shown in Fig. 5-2. This type of fluid motion is called streamline flow because all the particles of fluid are moving in parallel paths. Therefore, laminar flow is smooth with essentially no collision of particles. For laminar flow, the friction is caused by the sliding of one layer or particle of fluid over another in a smooth continuous fashion.

If the velocity of flow reaches a high enough value, the flow ceases to be laminar and becomes turbulent. As shown in Fig. 5-3, in turbulent flow the movement of a particle becomes random and fluctuates up and down in a direction perpendicular as well as parallel to the mean flow direction. This mixing action generates turbulence due to the colliding fluid particles. This causes considerably

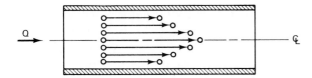

Figure 5-2.   Straight-line path of fluid particles in laminar flow.

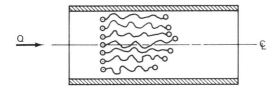

Figure 5-3. Random fluctuation of fluid particles in turbulent flow.

more resistance to flow and thus greater energy losses than that produced by laminar flow.

The difference between laminar and turbulent flow can be seen when using a water faucet. When the faucet is turned only partially open, with just a small amount of flow, the flow pattern observed is a smooth laminar one. However, when the faucet is opened wide, the flow mixes and becomes turbulent, as illustrated in Fig. 5-4.

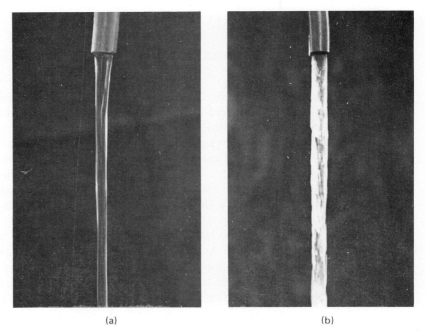

(a)   (b)

Figure 5-4. Flow patterns from water faucet. (a) Laminar flow, (b) turbulent flow. (Reprinted from *Introduction to Fluid Mechanics* by J. E. John and W. L. Haberman, Prentice Hall, Englewood Cliffs, N.J., 1988)

5.3 REYNOLDS NUMBER

It is important to know whether the flow pattern inside a pipe is laminar or turbulent. This brings us to the experiments performed by Osborn Reynolds in 1833 to determine the conditions governing the transition from laminar to turbulent flow. Using the test setup in Fig. 5-5, Reynolds allowed the fluid in the large tank to flow through a bell-mouthed entrance and along a smooth glass tube. He controlled the flow rate by means of a valve at the end of the tube. A capillary tube, connected to a reservoir of dye, allowed the flow of a fine jet of dye into the main flow stream.

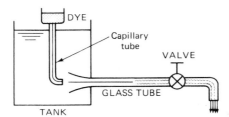

Figure 5-5. Reynolds' experiment.

If the flow in the tube was laminar, the dye jet flowed smoothly. However, when turbulent flow occurred in the tube, the dye jet would mix with the main fluid.

Reynolds came to a very significant conclusion as a result of his experiments: *The nature of the flow depends on the dimensionless parameter $vD\rho/\mu$,* where v = fluid velocity, D = pipe inside diameter, ρ = fluid mass density, and μ = absolute viscosity of the fluid.

This parameter has been named the *Reynolds number N_R* and (as Reynolds discovered from his tests) has the following significance:

1. If N_R is less than 2000, the flow is laminar.
2. If N_R is greater than 4000, the flow is turbulent.
3. Reynolds numbers between 2000 and 4000 cover a critical zone between laminar and turbulent flow.

It is not possible to predict the type of flow that will exist within the critical zone. However, since turbulent flow results in greater losses, fluid power systems should be designed to operate in the laminar flow region.

The Reynolds number can be calculated in several ways depending on the units chosen, e.g.,

$$N_R = \frac{v(\text{ft/sec}) \cdot D(\text{ft}) \cdot \rho(\text{slug/ft}^3)}{\mu(\text{lb} \cdot \text{sec/ft}^2)} \tag{5-1}$$

A more convenient set of units gives

$$N_R = \frac{7740v(\text{ft/sec}) \cdot D(\text{in.}) \cdot S_g}{\mu(\text{cP})} \tag{5-2}$$

A final relationship using kinematic viscosity is also desirable:

P.lee of crop formula

$$N_R = \frac{7740v(\text{ft/sec}) \cdot D(\text{in.})}{v(\text{cS})} \tag{5-3}$$

It should be noted that if turbulent flow is allowed to exist, the temperature of the fluid increases due to greater energy losses. Therefore, turbulent flow systems suffering from excessive temperatures can be helped by slightly increasing the pipe size to establish laminar flow.

EXAMPLE 5-1

The kinematic viscosity of a hydraulic oil is 100 cS. If the fluid is flowing in a 1-in.-diameter pipe at a velocity of 10 ft/sec, what is the Reynolds number?

Solution   Substitute directly into Eq. (5-3):

$$N_R = \frac{(7740)(10)(1)}{100} = 774$$

5.4 DARCY'S EQUATION

Friction is the main cause of energy losses in fluid power systems. The energy loss due to friction is transferred into heat, which is given off to the surrounding air. The result is a loss of potential energy in the system, and this shows up as a loss in pressure or head. In Chapter 3 we included this head loss in Bernoulli's equation as an H_L term. However, we did not discuss how the magnitude of this head loss term could be evaluated. The head loss (H_L) in a system actually consists of two components:

1. Losses in pipes
2. Losses in fittings

Head losses in pipes can be found by using Darcy's equation:

$$H_L = f\left(\frac{L}{D}\right)\left(\frac{v^2}{2g}\right) \tag{5-4}$$

where f = friction factor (dimensionless),
$\quad$ L = length of pipe (ft),
$\quad$ D = pipe inside diameter (ft),
$\quad$ v = average fluid velocity (ft/sec),
$\quad$ g = acceleration of gravity (ft/sec^2).

Darcy's equation can be used to calculate the head loss due to friction in pipes for both laminar and turbulent flow. The difference between the two lies in the evaluation of the friction factor f. The technique is discussed for laminar and turbulent flow in Secs. 5.5 and 5.6, respectively.

5.5 FRICTIONAL LOSSES IN LAMINAR FLOW

Darcy's equation can be used to find head losses in pipes experiencing laminar flow by noting that for laminar flow the friction factor equals the constant 64 divided by the Reynolds number:

$$f = \frac{64}{N_R} \tag{5-5}$$

Substituting Eq. (5-5) into Eq. (5-4) yields the Hagen-Poiseuille equation, which is valid for laminar flow only:

$$H_L = \frac{64}{N_R}\left(\frac{L}{D}\right)\left(\frac{v^2}{2g}\right) \tag{5-6}$$

The following example illustrates the use of the Hagen-Poiseuille equation.

EXAMPLE 5-2

For the system of Example 5-1, find the head loss due to friction in units of psi for a 100-ft length of pipe. The oil has a specific gravity of 0.90.

Solution   From Eq. (5-6) we solve for the head loss in units of feet of oil:

$$H_L = \frac{64}{774}\left(\frac{100}{\frac{1}{12}}\right)\left(\frac{10^2}{64.4}\right) = 154 \text{ ft}$$

Note that the units for H_L are really ft · lb/lb. Thus we can conclude that 154 ft · lb of energy is lost by each pound of oil as it flows through the 100-ft length of pipe.

Using Eq. (2-8), we convert head loss in units of feet of oil to pressure loss in units of psi:

$$P = (0.433)(154)(0.90) = 60 \text{ psi}$$

Thus there is a 60-psi pressure loss as the oil flows through the 100-ft length of pipe. This pressure loss is due to friction.

5.6 FRICTIONAL LOSSES IN TURBULENT FLOW

Darcy's equation will be used for calculating energy losses in turbulent fluid flow. However, the friction factor cannot be represented by a simple formula as was the case for laminar flow. This is due to the random and fluctuating movement of the fluid particles.

For turbulent flow, experiments have shown that the friction factor is a function of not only the Reynolds number but also the relative roughness of the pipe. The relative roughness is defined as the pipe inside surface roughness ε (Greek letter epsilon) divided by the pipe inside diameter D:

$$\text{relative roughness} = \frac{\varepsilon}{D} \tag{5-7}$$

Figure 5-6 illustrates the physical meaning of the pipe inside surface roughness ε, which is called the absolute roughness.

Pipe roughness values depend on the pipe material as well as the method of manufacture. Figure 5-7 gives typical values of absolute roughness for various types of pipes.

It should be noted that the values given in Fig. 5-7 are average values for new clean pipe. After the pipes have been in service for a time, the roughness values may change significantly due to the buildup of deposits on the pipe walls.

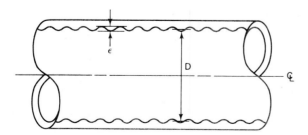

Figure 5-6.   Pipe absolute roughness (ε).

TYPE OF PIPE	ABSOLUTE ROUGHNESS ϵ(FT)
GLASS OR PLASTIC	SMOOTH
DRAWN TUBING	0.000005
COMMERCIAL STEEL OR WROUGHT IRON	0.00015
ASPHALTED CAST IRON	0.0004
GALVANIZED IRON	0.0005
CAST IRON	0.00085
RIVETED STEEL	0.006

Figure 5-7.   Typical values of absolute roughness.

To determine the value of the friction factor for use in Darcy's equation, we use the Moody diagram shown in Fig. 5-8. This diagram contains curves that were determined by data taken by L. F. Moody. The curves represent values of friction factor as a function of Reynolds number and relative roughness. Thus if we know the Reynolds number and relative roughness, we can quickly determine the friction factor.

The following important characteristics should be noted about the Moody diagram:

1. It is plotted on logarithmic paper because of the large range of values encountered for f and N_R.
2. At the left end of the chart (Reynolds numbers less than 2000) the straight-line curve gives the relationship for laminar flow: $f = 64/N_R$.
3. No curves are drawn in the critical zone (2000 $<$ N_R $<$ 4000), because it is not possible to predict whether the flow is laminar or turbulent in this region.
4. For Reynolds numbers greater than 4000, each curve plotted represents a particular value of ε/D. For intermediate values of ε/D, interpolation is required.
5. Once complete turbulence is reached (region to the right of the dashed line), increasing values of N_R have no effect on the value of f.

Example 5-3 illustrates the use of the Moody diagram for finding values of friction factor f for laminar and turbulent flow.

EXAMPLE 5-3

The kinematic viscosity of a hydraulic oil is 50 cS. If the oil flows in a 1-in.-diameter commercial steel pipe, find the friction factor if

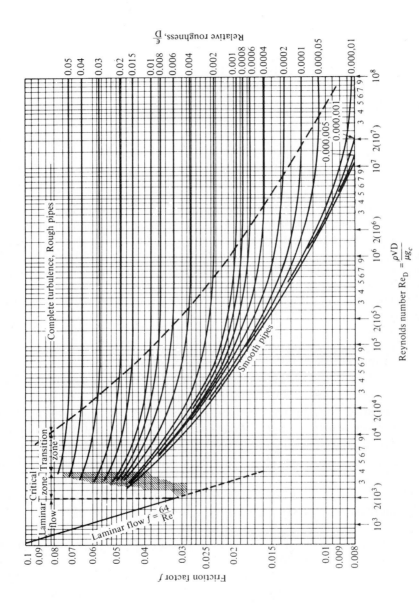

Figure 5-8. The Moody diagram. (Reprinted from *Introduction to Fluid Mechanics* by J. E. John and W. L. Haberman, Prentice Hall, Englewood Cliffs, N.J., 1988.)

a. The velocity is 10 ft/sec.

b. The velocity is 40 ft/sec.

Solution

a. Find N_R from Eq. (5-3):

$$N_R = \frac{(7740)(10)(1)}{50} = 1548 = 1.548 \times 10^3$$

Since the flow is laminar, we do not need to know the relative roughness value. To find f, locate 1.548×10^3 on the N_R axis of the Moody diagram (approximate value = 1.5×10^3). Then project vertically up until the straight line curve ($f = 64/N_R$) is reached. Then project horizontally to the f axis to obtain a value of 0.042.

b.
$$N_R = \frac{(7740)(40)(1)}{50} = 6192 = 6.192 \times 10^3$$

Since the flow is turbulent, we need the value of ε/D. First the relative roughness (a dimensionless parameter) is found using Fig. 5-7 to get the value of ε:

$$\frac{\varepsilon}{D} = \frac{0.00015 \text{ (ft)}}{\frac{1}{12} \text{ (ft)}} = 0.0018$$

Now locate the value 6.192×10^3 on the N_R axis (approximate value = 6.2×10^3). Then project vertically up until you are between the ε/D cuves of 0.0010 and 0.0020 (where the 0.0018 curve would approximately exist if it were drawn). Then project horizontally to the f axis to obtain a value of 0.036.

Since approximate values of N_R are used and interpolation of ε/D values is required, variations in the determined value of f is expected when using the Moody diagram. However, this normally produces variations of less than ±0.001, which is generally acceptable for calculating frictional losses in piping systems. This type of analysis is done for a complete system in Sec. 5.9.

5.7 LOSSES IN VALVES AND FITTINGS

In addition to losses due to wall friction in pipes, there also are energy losses in valves and fittings such as tees, elbows, and bends. For many fluid power applications, the majority of the energy losses occur in these valves and fittings in which there is a change in the cross section of the flow path and a change in the direction of flow. Thus, the nature of the flow through valves and fittings is very complex.

As a result, experimental techniques are used to determine losses. Tests have shown that head losses in valves and fittings are proportional to the square of the velocity of the fluid:

$$H_L = \frac{Kv^2}{2g} \tag{5-8}$$

The constant of proportionality (K) is called the K factor of the valve or fitting. Figure 5-9 gives typical K-factor values for several common types of valves and fittings.

VALVE OR FITTING		K FACTOR
GLOBE VALVE:	WIDE OPEN	10.0
	1/2 OPEN	12.5
GATE VALVE:	WIDE OPEN	0.19
	3/4 OPEN	0.90
	1/2 OPEN	4.5
	1/4 OPEN	24.0
RETURN BEND		2.2
STANDARD TEE		1.8
STANDARD ELBOW		0.9
45° ELBOW		0.42
90° ELBOW		0.75
BALL CHECK VALVE		4.0

Figure 5-9. K-factors of common valves and fittings.

Illustrations of several common valves and fittings are given as follows:

1. *Globe valve:* See Fig. 5-10. In this design, the fluid changes direction when flow occurs between the globe and seat. This construction increases resistance to fluid flow but also permits close regulation of fluid flow. Figure 5-10

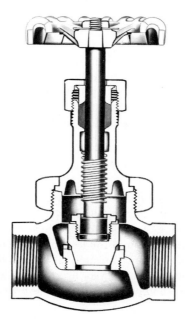

Figure 5-10. Globe value. (*Courtesy of Crane Co., New York, New York.*)

Figure 5-11. Gate valve (insertion-type stuffing box). (*Courtesy of Crane Co., New York, New York.*)

shows the globe valve in its fully closed position. The stem in a globe valve not only raises the globe (disk) but also helps guide it squarely to its seat.

2. *Gate valve:* See Figs. 5-11 and 5-12. Fluids flow through gate valves in a straight line, and thus there is little resistance to flow and the resulting pressure

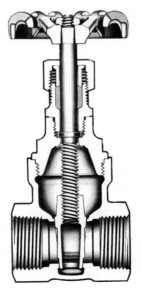

Figure 5-12. Gate valve (conventional stuffing box). (*Courtesy of Crane Co., New York, New York.*)

Figure 5-13. 45° elbow. (*Courtesy of Crane Co., New York, New York.*)

Figure 5-14. 90° elbow. (*Courtesy of Crane Co., New York, New York.*)

Figure 5-15. Tee. (*Courtesy of Crane Co., New York, New York.*)

Figure 5-16. Return bend. (*Courtesy of Crane Co., New York, New York.*)

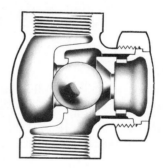

Figure 5-17. Ball check valve. (*Courtesy of Crane Co., New York, New York.*)

drops are small. A gatelike disk (actuated by a stem screw and handwheel) moves up and down at right angles to the path of flow and seats against two seat faces to shut off flow. Gate valves are best for services that require infrequent valve operation and where the disk is kept either fully opened or closed. They are not practical for throttling. With the usual type of gate valve, close regulation is impossible. Velocity of flow against a partly opened disk may cause vibration and chattering and result in damage to the seating surfaces.

    3. *45° elbow:* See Fig. 5-13.

    4. *90° elbow:* See Fig. 5-14.

    5. *Tee:* See Fig. 5-15.

    6. *Return bend:* See Fig. 5-16.

7. *Ball check valve:* See Fig. 5-17. The function of a check valve is to allow flow to pass through in only one direction. Thus, check valves are used to prevent backflow in hydraulic lines.

For some fluid power valves, *K* factors are not specified. Instead, an empirical curve of pressure drop versus flow rate is given by the valve manufacturer for the particular valve. Thus if the flow rate through the valve is known, the pressure drop can be determined by referring to the curve. This is normally done for directional control valves and also for flow control valves for various opening positions.

Figure 5-18 shows a cutaway of a directional control valve whose pressure drop versus flow-rate characteristics are provided by curves (see Fig. 5-19) rather than by *K*-factor values. As expected from Eq. (5-8), the curves show that the pressure drop increases approximately as the square of the flow rate.

Example 5-4

What is the head loss across a 1-in.-wide open globe valve when oil ($S_g = 0.9$) flows through it at a rate of 30 gpm?

Solution   Find the fluid velocity using Eq. (3-26):

$$v = \frac{(0.408)(30)}{(1)^2} = 12.2 \text{ ft/sec}$$

From Fig. 5-9, K for a wide open globe valve equals 10:

$$H_L = \frac{(10)(12.2)^2}{64.4} = 23.1 \text{ ft of oil}$$

The pressure drop (ΔP) across the valve can now be found:

$$\Delta P = 0.433 H_L S_g = (0.433)(23.1)(0.9) = 9.00 \text{ psi}$$

Figure 5-18.   Cutaway of a directional control valve. (*Courtesy of Continental Hydraulics, Division of Continental Machines, Inc., Savage, Minnesota.*)

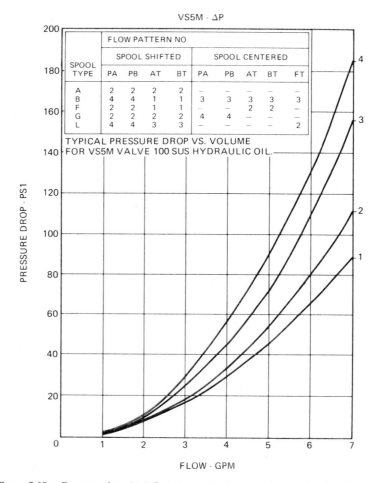

Figure 5-19.   Pressure drop (vs.) flow curves for directional control valve. (*Courtesy of Continental Hydraulics, Division of Continental Machines, Inc., Savage, Minnesota.*)

5.8 EQUIVALENT LENGTH TECHNIQUE

Darcy's equation shows that the head loss in a pipe, due to fluid friction, is proportional not only to the square of the fluid velocity but also to the length of the pipe. There is a similarity between Darcy's equation and Eq. (5-8), which states that the head loss in a valve or fitting is proportional to the square of the fluid velocity.

This suggests that it might be possible to find a length of pipe that for the same flow rate would produce the same head loss as a valve or fitting. This length of pipe, which is called the equivalent length of the valve or fitting, can be found by equating the head losses across the valve or fitting and the pipe:

$$H_{L(\text{valve or fitting})} = H_{L(\text{pipe})}$$

Substituting the corresponding expressions, we have

$$\frac{Kv^2}{2g} = f\frac{(L)}{D}\frac{v^2}{2g}$$

Since the velocities are equal, we can cancel the $v^2/2g$ terms from both sides of the equation. The result is

$$L_e = \frac{KD}{f} \qquad\qquad (5\text{-}9)$$

where L_e is the equivalent length of a valve or fitting whose K factor is K. Note that parameters K and f are both dimensionless. Therefore, L_e and D will have the same dimensions. For example, if D is measured in units of feet, then L_e will be calculated in units of feet.

   Equation (5-9) permits the convenience of examining each valve or fitting of a fluid power system as though it were a pipe of length L_e. This provides a convenient method of analyzing hydraulic circuits where frictional energy losses are to be taken into account. Section 5-9 deals with this type of problem. Example 5-5 shows how to find the equivalent length of a hydraulic component.

EXAMPLE 5-5

Hydraulic oil ($S_g = 0.9$, $v = 100$ cS) flows through a 1-in.-diameter commercial steel pipe at a rate of 30 gpm. What is the equivalent length of a 1-in.-wide open globe valve placed in the line?

Solution   We need to find the friction factor f, so let's first find the velocity v from Eq. (3-26):

$$v = \frac{0.408(30)}{(1)^2} = 12.2 \text{ ft/sec}$$

Using Eq. (5-3), we find the Reynolds number:

$$N_R = \frac{7740(12.2)(1)}{100} = 944$$

   Since the flow is laminar, we do not need to know the relative roughness to find the friction factor.

$$f = \frac{64}{N_R} = \frac{64}{944} = 0.0678$$

Finally, we determine the equivalent length using Eq. (5-9):

$$L_e = \frac{KD}{f} = \frac{(10)(\frac{1}{12}\text{ ft})}{0.0678} = 12.3 \text{ ft}$$

Thus a 1-in.-diameter pipe of length 12.3 ft would produce the same frictional energy loss as a 1-in.-wide open globe valve for a flow rate of 30 gpm.

5.9 HYDRAULIC CIRCUIT ANALYSIS

We are now ready to perform a complete analysis of a hydraulic circuit, taking into account energy losses due to friction. Let's analyze the hydraulic system of Fig. 5-20 by doing an example problem.

EXAMPLE 5-6

For the hydraulic system of Fig. 5-20, the following data are given:

1. The pump is adding 5 hp to the fluid.
2. Pump flow is 30 gpm.
3. The pipe has a 1-in. inside diameter.
4. The specific gravity of oil is 0.9.
5. The kinematic viscosity of oil is 100 cS.

Find the pressure available at the inlet to the hydraulic motor (station 2). The pressure at the oil top surface level in the hydraulic tank is atmospheric (0 psig). The head loss H_L due to friction between stations 1 and 2 is not given.

Solution   Writing Bernoulli's equation between stations 1 and 2, we have

$$Z_1 + \frac{P_1}{\gamma} + \frac{v_1^2}{2g} + H_p - H_m - H_L = Z_2 + \frac{P_2}{\gamma} + \frac{v_2^2}{2g}$$

Since there is no hydraulic motor between stations 1 and 2, $H_m = 0$. Also $v_1 = 0$ and $P_1/\gamma = 0$ (the oil tank is vented to the atmosphere). Also $Z_2 - Z_1 = 20$ ft per Fig. 5-20.

To make use of Bernoulli's equation, let's first solve for v_2 using Eq. (3-25):

$$v_2 = \frac{(0.408)(30)}{(1)^2} = 12.2 \text{ ft/sec}$$

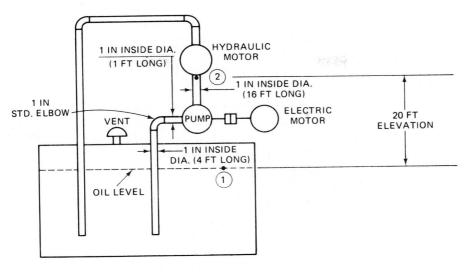

Figure 5-20.   Hydraulic system for Example 5-6.

Next, let's evaluate the velocity head at station 2:

$$\frac{v_2^2}{2g} = \frac{(12.2)^2}{64.4} = 2.4 \text{ ft}$$

The Reynolds number can now be found:

$$N_R = \frac{7740v(\text{ft/sec}) \cdot D (\text{in.})}{v(\text{cS})} = \frac{7740(12.2)(1)}{100} = 944$$

Since the flow is laminar, the friction factor can be found directly from the Reynolds number:

$$f = \frac{64}{N_R} = \frac{64}{944} = 0.0678$$

We can now determine the head loss due to friction between stations 1 and 2:

$$H_L = f\left(\frac{L}{D}\right) \frac{v^2}{2g}$$

where

$$L = 16 + 1 + 4 + \left(\frac{KD}{f}\right)_{\text{std elbow}}$$

$$L = 21 + \frac{(0.9)(\frac{1}{12} \text{ ft})}{0.0678} = 21 + 1.1 = 22.1 \text{ ft}$$

$$H_L = (0.0678)\frac{(22.1)}{\frac{1}{12}}(2.4) = 43 \text{ ft}$$

We can now substitute into Bernoulli's equation to solve for P_2/γ:

$$\frac{P_2}{\gamma} = (Z_1 - Z_2) + H_p + \frac{P_1}{\gamma} - H_L - \frac{v_2^2}{2g}$$

$$\frac{P_2}{\gamma} = -20 + H_p + 0 - 43 - 2.4$$

$$\frac{P_2}{\gamma} = H_p - 65.4$$

Using Eq. (3-24) allows us to solve for the pump head:

$$H_p = \frac{3950(HP)}{QS_g} = \frac{(3950)(5)}{(30)(0.9)} = 732 \text{ ft}$$

Thus we can solve for the pressure head at station 2:

$$\frac{P_2}{\gamma} = 732 - 65.4 = 666.6 \text{ ft}$$

Finally, we solve for the pressure at station 2:

$$P_2(\text{lb/ft}^2) = 666.6 \text{ (ft)} \cdot \gamma \text{ (lb/ft}^3)$$

where $\gamma = S_g \gamma_{\text{water}} = (0.9)(62.4) = 56.2$ lb/ft^3
Thus $P_2 = (666.6)(56.2) = 37,463$ lb/ft$^2 = 260$ psi

5.10 FLOW MEASUREMENT

Flow-rate measurements are frequently required to evaluate the performance of hydraulic components as well as to troubleshoot a hydraulic system. They can be used to check the volumetric efficiency of pumps and also to determine leakage paths within a hydraulic circuit.

Probably the most common type of flowmeter is the rotameter, which consists of a metering float in a calibrated vertical tube, as shown in Fig. 5-21. The operation of the rotameter is as follows (refer to Fig. 5-22):

The metering float is free to move vertically in the tapered glass tube. The fluid flows through the tube from bottom to top. When no fluid is flowing, the float rests at the bottom of the tapered tube, and its maximum diameter is usually so selected that it blocks the small end of the tube almost completely. When flow begins in the pipeline, the fluid enters the bottom of the meter and raises the float. This increases the flow area between the float and tube until an equilibrium position is reached. At this position, the weight of the float is balanced by the upward force of the fluid on the float. The greater the flow rate, the higher the float rises in the tube. The tube is graduated to allow a direct reading of the flow rate.

Figure 5-21.
Rotameter.
(*Courtesy of
Fischer & Porter
Co., Worminster,
Pennsylvania.*)

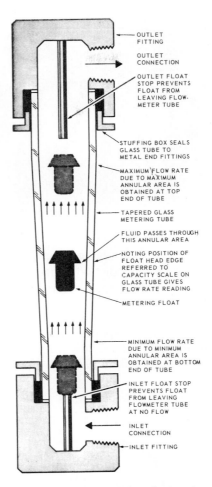

OUTLET FITTING

OUTLET CONNECTION

OUTLET FLOAT STOP PREVENTS FLOAT FROM LEAVING FLOW-METER TUBE

STUFFING BOX SEALS GLASS TUBE TO METAL END FITTINGS

MAXIMUM FLOW RATE DUE TO MAXIMUM ANNULAR AREA IS OBTAINED AT TOP END OF TUBE

TAPERED GLASS METERING TUBE

FLUID PASSES THROUGH THIS ANNULAR AREA

NOTING POSITION OF FLOAT HEAD EDGE REFERRED TO CAPACITY SCALE ON GLASS TUBE GIVES FLOW RATE READING

METERING FLOAT

MINIMUM FLOW RATE DUE TO MINIMUM ANNULAR AREA IS OBTAINED AT BOTTOM END OF TUBE

INLET FLOAT STOP PREVENTS FLOAT FROM LEAVING FLOWMETER TUBE AT NO FLOW

INLET CONNECTION

INLET FITTING

Figure 5-22.   Operation of rotameter.
(*Courtesy of Fischer & Porter Co.,
Worminster, Pennsylvania.*)

Sometimes it is desirable to determine whether or not fluid is flowing in a pipeline and to observe the flowing fluid visually. Such a device for accomplishing this is called a sight flow indicator. It does not measure the rate of flow but instead only indicates whether or not there is flow. The sight flow indicator shown in Fig. 5-23 has two windows located on opposite sides of the body fittings to give the best possible visibility.

Figure 5-23. Sight flow indicator. (*Courtesy of Fischer & Porter Co., Worminster, Pennsylvania.*)

Figure 5-24 shows another type of flowmeter, which incorporates what is called a disk piston. When the fluid passes through the measuring chamber, the disk piston develops a rotary motion, which is transmitted through gearing to a pointer on a dial.

A schematic drawing of a turbine-type flowmeter is given in Figure 5-25. This design incorporates a turbine rotor mounted in a housing connected in a pipeline whose fluid flow rate is to be measured. The fluid causes the turbine to rotate at a speed that is proportional to the flow rate. The rotation of the turbine

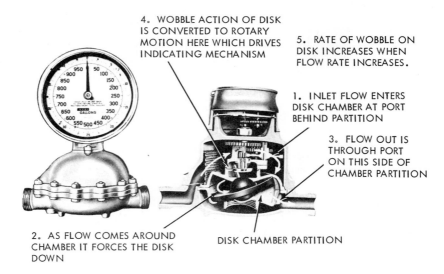

4. WOBBLE ACTION OF DISK IS CONVERTED TO ROTARY MOTION HERE WHICH DRIVES INDICATING MECHANISM

5. RATE OF WOBBLE ON DISK INCREASES WHEN FLOW RATE INCREASES.

1. INLET FLOW ENTERS DISK CHAMBER AT PORT BEHIND PARTITION

3. FLOW OUT IS THROUGH PORT ON THIS SIDE OF CHAMBER PARTITION

2. AS FLOW COMES AROUND CHAMBER IT FORCES THE DISK DOWN

DISK CHAMBER PARTITION

Figure 5-24. Flowmeter with disk piston. (*Courtesy of Sperry Vickers, Sperry Rand Corp., Troy, Michigan.*)

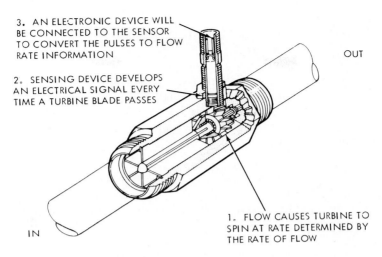

3. AN ELECTRONIC DEVICE WILL BE CONNECTED TO THE SENSOR TO CONVERT THE PULSES TO FLOW RATE INFORMATION

2. SENSING DEVICE DEVELOPS AN ELECTRICAL SIGNAL EVERY TIME A TURBINE BLADE PASSES

OUT

1. FLOW CAUSES TURBINE TO SPIN AT RATE DETERMINED BY THE RATE OF FLOW

IN

Figure 5-25. Turbine flowmeter. (*Courtesy of Sperry Vickers, Sperry Rand Corp. Troy, Michigan.*)

generates an electrical impulse every time a turbine blade passes a sensing device. An electronic device connected to the sensor converts the pulses to flow-rate information.

Figure 5-26 shows an orifice (a disk with a hole through which fluid flows) installed in a pipe. Such a device is an actual flowmeter when the pressure drop (ΔP) across the orifice is measured, because there is a unique relationship between ΔP and Q (flow rate) for a given orifice. Specifically, the greater the flow rate, the greater the pressure drop. It can be shown that the following equation relates the ΔP vs. Q relationship for an orifice installed in a pipe to measure liquid flow rate:

$$Q = 38.06CA \sqrt{\frac{\Delta P}{S_g}} \qquad (5\text{-}10)$$

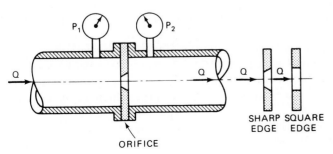

SHARP EDGE   SQUARE EDGE

ORIFICE

Figure 5-26. Orifice flowmeter.

where Q = flow rate (gpm),
   C = flow coefficient (C = 0.80 for sharp-edged orifice, C = 0.60 for square-edged orifice),
   A = area of orifice opening (in.^{2}),
   $\Delta P = P_2 - P_1$ = pressure drop across orifice (psi),
   S_g = specific gravity of flowing fluid.

The following example shows how an orifice flowmeter can be used to determine flow rate.

EXAMPLE 5-7

The pressure drop across the sharp-edged orifice of Fig. 5-26 is 100 psi. The orifice has a 1-in. diameter, and the fluid has a specific gravity of 0.9. Find the flow rate in units of gpm.

Solution   Substitute directly into Eq. (5-10):

$$Q = (38.06)(0.80)\left(\frac{\pi}{4} \times 1^2\right)\sqrt{\frac{100}{0.9}} = 252 \text{ gpm}$$

5.11 PRESSURE MEASUREMENT

Pressure-measuring devices are needed in hydraulic circuits for a number of reasons. In addition to testing and troubleshooting, they are used to adjust pressure settings of pressure control valves and to determine forces exerted by hydraulic cylinders and torques delivered by hydraulic motors.

One of the most widely used pressure-measuring devices is the Bourdon gage (see Fig. 5-27, which shows an assortment of Bourdon gages, each having different pressure ranges). The Bourdon gage contains a sealed tube formed in the shape of an arc (refer to Fig. 5-28). When pressure is applied at the port opening, the tube starts to straighten somewhat. This activates a linkage-gear system, which moves the pointer to indicate the pressure on the dial. The scale of most Bourdon gages reads zero when the gage is open to the atmosphere, because the gages are calibrated to read pressure above atmospheric pressure or gage pressure. Some Bourdon gages are capable of reading pressures below atmospheric or vacuum (suction) pressures, such as those existing in pump inlet lines. The range for vacuum gages is from 0 to 30 in. of mercury, which represents a perfect vacuum.

A second common type of pressure-measuring device is the Schrader gage. As illustrated in Fig. 5-29, pressure is applied to a spring-loaded sleeve and piston.

Figure 5-27. Bourdon gages with different pressure ranges. (*Courtesy of Span Instruments, Inc., Plano, Texas.*)

As the pressure moves the sleeve, it actuates the indicating pointer through mechanical linkages.

Figure 5-30 shows a digital electronic readout device that provides 5-digit displays for pressure, flow rate, and speed measurements accurate to ±0.15% of

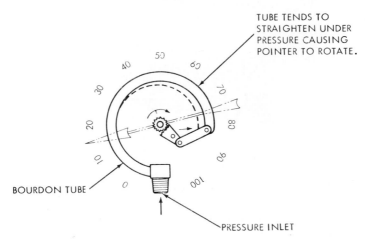

Figure 5-28. Operation of Bourdon gage. (*Courtesy of Sperry Vickers, Sperry Rand Corp., Troy, Michigan.*)

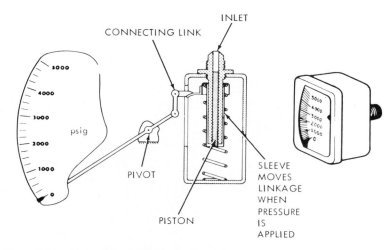

Figure 5-29. Operation of a Schrader gage. (*Courtesy of Sperry Vickers, Sperry Rand Corp., Troy, Michigan.*)

Figure 5-30. Digital Electronic Read-out. (*Courtesy of Flo-tech, Inc., Mundelein, Illinois.*)

full scale (maximum of 19999). The scale can be factory-calibrated to display values in units such as psi, Pascals, bars, gpm, liters/min, rpm, inches/min, meters/min, and so on via 10 independent input channels and memory. Data is updated $2\frac{1}{2}$ times per second and provides an over-range condition indication.

5.12 HYDRAULIC CIRCUIT ANALYSIS USING THE METRIC-SI SYSTEM

In this section we perform a complete analysis of a hydraulic circuit using the metric-SI system of units and taking into account energy losses due to friction. To provide a comparison with the English system of units, let's use the same hydraulic system analyzed in Example 5-6 (see Fig. 5-20). The metric-SI units solution is presented in the following example.

EXAMPLE 5-8

For the hydraulic system of Fig. 5-20, the following metric data (which are equivalent to the English system data of Example 5-6) are given:

1. The pump is adding 5 hp (3730 W) to the fluid.
2. Pump flow is 0.001896 m³/s.
3. The pipe has a 0.0254-m inside diameter.
4. The specific gravity of oil is 0.9.
5. The kinematic viscosity of oil is 100 cS.
6. The elevation difference between stations 1 and 2 is 6.096 m.
7. Pipe lengths are as follows: 1-ft length = 0.305 m, 4-ft length = 1.22 m, and 16-ft length = 4.88 m.

Find the pressure available at the inlet to the hydraulic motor (station 2). The pressure at the oil top surface level in the hydraulic tank is atmospheric (0 Pa). The head loss H_L due to friction between stations 1 and 2 is not given.

Solution   In the metric-SI system, absolute viscosity is given in units of newton-seconds per meter squared. Thus we have $\mu = $ N · s/m² = Pa · s. Our problem is to convert viscosity in cS to the appropriate units in the metric-SI system. This is accomplished as follows: The conversion between dyne · s/cm² or poise and N · s/m² is found first:

$$\mu \ (\text{N} \cdot \text{s/m}^2) = \mu \ (\text{dyne} \cdot \text{s/cm}^2) \times \frac{1 \ \text{N}}{10^5 \ \text{dynes}} \times \left(\frac{100 \ \text{cm}}{1 \ \text{m}}\right)^2$$

This yields a useful conversion equation dealing with absolute viscosities:

$$\mu \ (\text{N} \cdot \text{s/m}^2) = \frac{\mu \ (\text{dyne} \cdot \text{s/cm}^2)}{10} = \frac{\mu \ (\text{poise})}{10} \qquad (5\text{-}11)$$

Since a centipoise is one-hundredth of a poise, we can develop a converstion equation between $N \cdot s/m^2$ and centipose:

$$\mu \text{ (poise)} = \frac{\mu(cP)}{100} \tag{5-12}$$

Substituting Eq. (5-12) into Eq. (5-11) yields the desired result:

$$\mu \text{ (N } \cdot \text{ s/m}^2) = \frac{\mu(cP)}{100(10)} = \frac{\mu(cP)}{1000} \tag{5-13}$$

However, in this problem, we were given a kinematic viscosity value rather than an absolute viscosity value. Therefore, we need to convert viscosity in cS to the appropriate units in the metric-SI system. Kinematic viscosity in the metric-SI system is given in units of meters squared per second. Thus, we have $\nu = m^2/s$. The conversion between cm^2/s or stokes and m^2/s is found as follows:

$$\nu(m^2/s) = \nu(cm^2/s) \times \left(\frac{1 \text{ m}}{100 \text{ cm}}\right)^2 = \frac{\nu(cm^2/s)}{10,000} = \frac{\nu(\text{stokes})}{10,000} \tag{5-14}$$

Since a centistoke is one-hundredth of a stoke, we can now develop the desired conversion between m^2/s and centistokes:

$$\nu \text{ (stokes)} = \frac{\nu(cS)}{100} \tag{5-15}$$

Substituting Eq. (5-15) into Eq. (5-14) yields

$$\nu(m^2/s) = \frac{\nu \text{ (stokes)}}{10,000} = \frac{\nu(cS)}{10,000 \times 100} = \frac{\nu(cS)}{1,000,000} \tag{5-16}$$

Now back to the problem at hand: We write Bernoulli's equation between stations 1 and 2:

$$Z_1 + \frac{P_1}{\gamma} + \frac{v_1^2}{2g} + H_p - H_m - H_L = Z_2 + \frac{P_2}{\gamma} + \frac{v_2^2}{2g}$$

Since there is no hydraulic motor between stations 1 and 2, $H_m = 0$. Also $v_1 = 0$ and $P_1/\gamma = 0$ (the oil tank is vented to the atmosphere). Also $Z_2 - Z_1 = 6.096$ m per given input data. To make use of Bernoulli's equation, let's first solve for v_2:

$$v_2 \text{ (m/s)} = \frac{Q \text{ (m}^3/\text{s})}{A \text{ (m}^2)} = \frac{0.001896}{(\pi/4)(0.0254)^2} = 3.74 \text{ m/s}$$

Next, let's evaluate the velocity head at station 2:

$$\frac{v_2^2}{2g} = \frac{(3.74 \text{ m/s})^2}{2(9.80 \text{ m/s}^2)} = 0.714 \text{ m}$$

The Reynolds number can now be found:

$$N_R = \frac{v \text{ (m/s)} \cdot D \text{ (m)}}{\nu \text{ (m}^2/\text{s)}} = \frac{(3.74)(0.0254)}{100/1,000,000} = 944$$

As expected (since the Reynolds number is a dimensionless parameter), the value calculated here is the same as that found in Example 5-6. Since the flow is laminar, the friction factor can be found directly from the Reynolds number:

$$f = \frac{64}{N_R} = \frac{64}{944} = 0.0678$$

We can now determine the head loss due to friction between stations 1 and 2:

$$H_L = f\left(\frac{L}{D}\right) \frac{v^2}{2g}$$

where

$$L = 4.88 + 0.305 + 1.22 + \left(\frac{KD}{f}\right)_{\text{std elbow}}$$

$$L = 6.41 + \frac{(0.9)(0.0254)}{0.0678} = 6.41 + 0.34 = 6.75 \text{ m}$$

$$H_L = (0.0678)\frac{(6.75)}{0.0254}(0.714) = 12.9 \text{ m}$$

We can now substitute into Bernoulli's equation to solve for P_2/γ:

$$\frac{P_2}{\gamma} = (Z_1 - Z_2) + H_p + \frac{P_1}{\gamma} - H_L - \frac{v_2^2}{2g}$$

$$\frac{P_2}{\gamma} = -6.096 + H_p + 0 - 12.9 - 0.714 = H_p - 19.7$$

Using Eq. (3-44) allows us to solve for the pump head:

$$H_p = \frac{0.0762 \text{ (HP)}}{Q \text{ (m}^3/\text{s)} \cdot S_g} = \frac{0.0762(5)}{0.001896(0.9)} = 223.3 \text{ m}$$

Thus, we can now solve for the pressure head at station 2:

$$\frac{P_2}{\gamma} = 223.3 - 19.7 = 203.6 \text{ m}$$

Solving for P_2 yields

$$P_2 \text{ (N/m}^2) = (203.6 \text{ m}) \cdot \gamma \text{ (N/m}^3)$$

where $\qquad \gamma = S_g\gamma_{water} = (0.9)(9797 \text{ N/m}^3) = 8817 \text{ N/m}^3$

$P_2 \text{ (N/m}^2) = (203.6)(8817) = 1,795,000 \text{ Pa} = 1795 \text{ kPa}$

Since 1 Pa = 0.000145 psi,

$\qquad P_2 \text{ (psi)} = (1.795,000)(0.000145) = 260 \text{ psi}$

As expected, this value checks with the answer to Example 5-6, which was solved in the English system of units.

EXERCISES

5-1. Why is it important to select properly the size of pipes, valves, and fittings in hydraulic systems?

5-2. What is the physical difference between laminar and turbulent flow?

5-3. What are the important conclusions resulting from Reynolds' experiment?

5-4. The kinematic viscosity of a hydraulic oil is 75 cS. If it is flowing in a 1½-in.-diameter pipe at a velocity of 20 ft/sec, what is the Reynolds number? Is the flow laminar or turbulent?

5-5. For the system of Exercise 5.4, find the head loss due to friction in units of psi for a 100-ft length of pipe. The oil has a specific gravity of 0.90.

5-6. Define the term *relative roughness*.

5-7. The kinematic viscosity of a hydraulic oil is 100 cS. If it is flowing in a ¾-in.-diameter commercial steel pipe, find the friction factor if (a) the velocity is 15 ft/sec. (b) the velocity is 45 ft/sec.

5-8. What is meant by the expression "K factor of a valve or fitting"?

5-9. What is the head loss across a 1½-in.-wide open gate valve when oil ($S_g = 0.9$) flows through it at a rate of 100 gpm?

5-10. What is meant by the expression "equivalent length of a valve or fitting"?

5-11. Oil ($S_g = 0.9$, $\nu = 75$ cS) flows at a rate of 30 gpm through a ¾-in.-diameter commercial steel pipe. What is the equivalent length of a ¾-in.-wide open gate valve placed in the line?

5-12. For the hydraulic system of Fig. 5-20, the following data are given:

1. The pump is adding 4 hp to the fluid.

2. Pump flow is 25 gpm.

3. The pipe has an 0.75-in. inside diameter.

4. The specific gravity of oil is 0.9.

5. The kinematic viscosity of oil is 75 cS.

Find the pressure available at the inlet to the hydraulic motor (station 2).

5-13. The oil tank for the hydraulic system of Fig. 5-31 is air-pressurized at 10 psig. The inlet line to the pump is 10 ft below the oil level. The pump flow rate is 30 gpm. Find

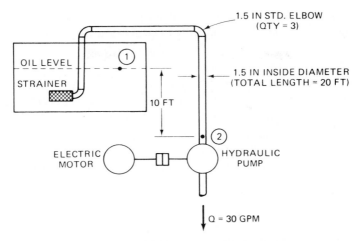

Figure 5-31. Hydraulic system for Exercise 5-13.

the pressure at station 2. S_g of oil $= 0.9$, and the kinematic viscosity of the oil $= 100$ cS. Assume that the pressure drop across the strainer is 1 psi.

5-14. Name two types of flow-measuring devices.

5-15. Name two types of pressure-measuring devices.

5-16. Why is it desirable to measure flow rates and pressures in a hydraulic system?

5-17. A 2-in.-diameter sharp-edged orifice is placed in a pipeline to measure flow rate. If the measured pressure drop is 50 psi and the fluid specific gravity is 0.90, find the flow rate in units of gpm.

5-18. For a given orifice and fluid, a graph can be generated showing the ΔP vs. Q relationship. For the orifice and fluid of Exercise 5-17, plot the curve and check the answer obtained mathematically. What advantage does the graph have over the equation? What is the disadvantage of the graph?

5-19. Solve Exercise 5-12 using the metric-SI system of units. The equivalent data are given in metric units as follows:

1. The pump is adding 4 hp (2984 W) to the fluid.

2. Pump flow is 0.00158 m³/s.

3. The pipe has an 0.01905-m inside diameter.

4. The specific gravity of oil is 0.9.

5. The kinematic viscosity of oil is 75 cS.

6. The elevation difference between stations 1 and 2 is 6.096 m.

7. Pipe lengths are as follows: 1-ft length $= 0.305$ m, 4-ft length $= 1.22$ m, and 16-ft length $= 4.88$ m.

Find the pressure available at the inlet to the hydraulic motor (station 2).

5-20. Solve Exercise 5-13 using the metric-SI system of units. The equivalent data are given in metric units as follows:

1. The oil tank is air-pressurized at 68,970-Pa gage pressure.
2. The inlet line to the pump is 3.048 m below the oil level.
3. The pump flow rate is 0.001896 m³/s.
4. The specific gravity of the oil is 0.9.
5. The kinematic viscosity of the oil is 100 cS.
6. Assume that the pressure drop across the strainer is 6897 Pa.
7. The pipe has an 0.0381-m inside diameter.
8. The total length of pipe = 6.097 m.

Find the pressure at station 2.

5-21. The kinematic viscosity of a hydraulic oil is 0.0001 m²/s. If it is flowing in a 30-mm diameter pipe at a velocity of 6 m/s, what is the Reynolds number? Is the flow laminar or turbulent?

5-22. For the system of Exercise 5-21, find the head loss due to friction in units of bars for a 100-m length of pipe. The oil has a specific gravity of 0.90.

5-23. The kinematic viscosity of a hydraulic oil is 0.0001 m²/s. If it is flowing in a 20-mm diameter commercial steel pipe, find the friction factor if

(a) The velocity is 2 m/s.

(b) The velocity is 10 m/s.

5-24. What is the head loss (in units of bars) across a 30-mm wide open gate valve when oil ($S_g = 0.9$) flows through at a rate of 0.004 m³/s?

5-25. Oil ($S_g = 0.9$, $\nu = 0.0001$ m²/s) flows at a rate of 0.002 m³/s through a 20-mm diameter commercial steel pipe. What is the equivalent length of a 20-mm-wide open gate valve placed in the line?

5-26. For the system of Fig. 5-32, solve for $P_2 - P_1$ in units of psi. The kinematic viscosity (ν) of the oil is 100 cS and the specific gravity (S_g) is 0.9.

5-27. For the system of Exercise 5-26, the following new data is applicable:

Pipe 1: length = 8 m, ID = 25 mm
Pipe 2: length = 8 m, ID = 25 mm

The globe valve is 25 mm in size and is wide open.

$S_g = 0.9$, $\nu = 0.0001$ m²/s, $Q = 0.0025$ m³/s

Find $P_2 - P_1$ in units of bars.

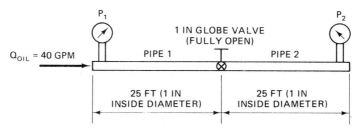

Figure 5-32. Hydraulic system for Exercise 5-26.

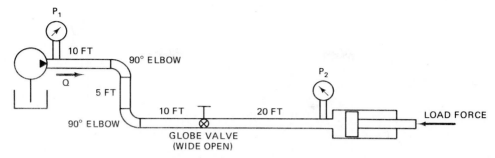

Figure 5-33. Hydraulic system for Exercise 5-28.

5-28. For the system of Fig. 5-33, if $P_1 = 100$ psi, solve for P_2 in units of psi. The pipe is 45 ft long, has a 1.5 in. *ID* throughout, and lies in a horizontal plane. $Q = 30$ gpm of oil, $S_g = 0.9$ and $\nu = 100$ cS.

5-29. For the system of Exercise 5-28, the following new data is applicable:

$P_1 = 7$ bars, $Q = 0.002$ m³/s

Pipe: L(total) = 15 m and ID = 38 mm

Oil: $S_g = 0.9$ and $\nu = 0.0001$ m²/s

Solve for P_2 in units of bars.

5-30. If the volumetric flowrate through a valve is doubled, by what factor does the pressure drop increase?

5-31. To minimize pressure losses, the K factor of a valve should be made as small as possible—true or false?

5-32. A hydraulic system is operating at a Reynolds number of 1000. If the temperature increases so that the oil viscosity decreases, the Reynolds number would (increase, decrease, or remain the same)? Specify one answer.

5-33. What advantage does a digital readout fluid parameter measuring device have over an analog device?

5-34. For laminar flow of a liquid in a pipe, frictional pressure losses are _____ to the liquid velocity.

5-35. For fully turbulent flow of a liquid in a pipe, frictional pressure losses vary as the _____ of the velocity.

5-36. What can be concluded about the pressures on the upstream and downstream sides of an orifice when oil is flowing through it?

5-37. Name two causes of turbulence in fluid flow.

5-38. A directional control valve with an effective area of 0.5 in.² provides a pressure drop of 40 psi at 60 gpm. If the fluid has a specific gravity of 0.90, what is the flow coefficient and K factor for the valve?

5-39. For the directional control valve of Exercise 5-38, if the data were converted to metric units, how would the flow coefficient and K-factor values calculated compare to those determined using English units? Explain your answer.

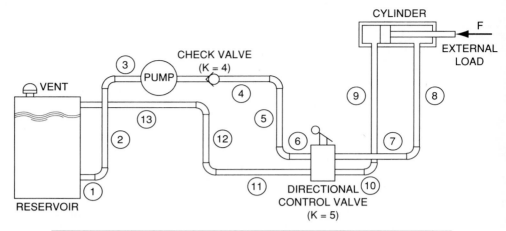

Pipe No.	Length(ft)	Dia(in.)	Pipe No.	Length(ft)	Dia(in.)
1	2	1.5	8	5	1.0
2	6	1.5	9	5	0.75
3	2	1.5	10	5	0.75
4	50	1.0	11	60	0.75
5	10	1.0	12	10	0.75
6	5	1.0	13	20	0.75
7	5	1.0			

Figure 5-34.   System for Exercise 5-40.

5-40. For the fluid power system shown in Fig. 5-34, determine the external load F that the hydraulic cylinder can sustain while moving in the extending direction. Take frictional pressure losses into account. The pump produces a pressure increase of 1000 psi from the inlet port to discharge port and a flow rate of 40 gpm. The following data is applicable:

Kinematic viscosity of oil = 0.001 ft²/sec

Weight density of oil = 50 lb/ft³

Cylinder piston diameter = 8 in.

Cylinder rod diameter = 4 in.

All elbows are 90° with a K factor = 0.75. Pipe lengths and inside diameters are given in Fig. 5-34.

5-41. For the system of Exercise 5-40 as shown in Fig. 5-34, determine the heat generation rate due to frictional pressure losses.

5-42. For the system of Exercise 5-40 as shown in Fig. 5-34, determine the extending and retracting speeds of the cylinder.

6

The Source of Hydraulic Power: Pumps

6.1 INTRODUCTION

A pump, which is the heart of a hydraulic system, converts mechanical energy into hydraulic energy. The mechanical energy is delivered to the pump via a prime mover such as an electric motor. Due to mechanical action, the pump creates a partial vacuum at its inlet. This permits atmospheric pressure to force the fluid through the inlet line and into the pump. The pump then pushes the fluid into the hydraulic system.

There are two broad classifications of pumps as identified by the fluid power industry.

1. Nonpositive displacement pumps: This type is generally used for low-pressure, high-volume flow applications. Because they are not capable of withstanding high pressures, they are of little use in the fluid power field. Normally their maximum pressure capacity is limited to 250–300 psi. This type of pump is primarily used for transporting fluids from one location to another.

2. Positive displacement pumps: This type is universally used for fluid power systems. As the name implies, a positive displacement pump ejects a fixed amount of fluid into the hydraulic system per revolution of pump shaft rotation. Such a pump is capable of overcoming the pressure resulting from the mechanical loads on the system as well as the resistance to flow due to friction. These are two features that are desired of fluid power pumps. These pumps have the following advantages over nonpositive displacement pumps:

low press.   high press

a. High-pressure capability (up to 10,000 psi or higher)

b. Small, compact size

c. High volumetric efficiency

d. Small changes in efficiency throughout the design pressure range

e. Great flexibility of performance (can operate over a wide range of pressure requirements and speed ranges)

There are three main types of positive displacement pumps: gear, vane, and piston. Many variations exist in the design of each of these main types of pumps. For example, vane and piston pumps can be of either fixed or variable displacement. A fixed displacement pump is one in which the amount of fluid ejected per revolution (displacement) cannot be varied. In a variable displacement pump, the displacement can be varied by changing the physical relationships of various pump elements. This change in pump displacement produces a change in pump flow output even though pump speed remains constant.

It should be understood that pumps do not pump pressure. Instead they produce fluid flow. The resistance to this flow, produced by the hydraulic system, is what determines the pressure. For example, if a positive displacement pump has its discharge line open to the atmosphere, there will be flow, but there will be no discharge pressure above atmospheric, because there is essentially no resistance to flow. However, if the discharge line is blocked, then we have theoretically infinite resistance to flow. Hence, there is no place for the fluid to go. The pressure will therefore rise until some component breaks unless pressure relief is provided. This is the reason a pressure relief valve is needed when a positive displacement pump is used. When the pressure reaches a set value, the relief valve will open to allow flow back to the oil tank. Thus, a pressure relief valve determines the maximum pressure level that the system will experience regardless of the magnitude of the load resistance. A discussion of hydraulic control components, such as pressure relief valves, is provided in Chapter 8.

Some pumps are made with variable displacement, pressure compensation capability. Such pumps are designed so that as system pressure builds up, they produce less flow. Finally at some predetermined maximum pressure level, the flow output goes to zero due to zero displacement. This prevents any additional pressure buildup. Pressure relief valves are not needed when pressure-compensated pumps are used.

The hydraulic power developed by pumps is converted back into mechanical energy by hydraulic actuators, which produce the useful work output. The subject of power output components is covered in Chapter 7.

It should be noted that pumps are used in hydraulic systems to provide flow of incompressible fluids (liquids) such as oil. In a pneumatic system where a compressible fluid such as air is used, the unit that produces fluid flow is called a compressor. Thus, in pneumatic systems, compressors perform functions similar to those performed by pumps in hydraulic systems. Compressors are discussed in Chapter 10.

6.2 PUMPING THEORY

All pumps operate on the principle whereby a partial vacuum is created at the pump inlet due to the internal operation of the pump. This allows atmospheric pressure to push the fluid out of the oil tank (reservoir) and into the pump intake. The pump then mechanically pushes the fluid out the discharge line.

This type of operation can be visualized by referring to the simple piston pump of Fig. 6-1. Notice that this pump contains two ball check valves, which are described as follows:

- Check valve 1 is connected to the pump inlet line and allows fluid to enter the pump only at this location.
- Check valve 2 is connected to the pump discharge line and allows fluid to leave the pump only at this location.

As the piston is pulled to the left, a partial vacuum is generated in pump cavity 3, because the close tolerance between the piston and cylinder (or the use of piston ring seals) prevents air inside cavity 4 from traveling into cavity 3. This flow of air, if allowed to occur, would destroy the vacuum. This vacuum holds check valve 2 against its seat (lower position) and allows atmospheric pressure to push fluid from the reservoir into the pump via check valve 1. This inlet flow occurs because the force of the fluid pushes the ball of check valve 1 off its seat.

When the piston is pushed to the right, the fluid movement closes inlet valve 1 and opens outlet valve 2. The quantity of fluid, displaced by the piston, is

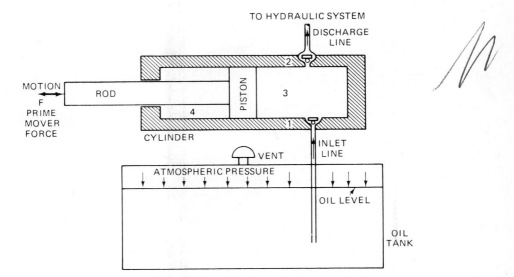

Figure 6-1. Pumping action of a simple piston pump.

forcibly ejected out the discharge line leading to the hydraulic system. The volume of oil displaced by the piston during the discharge stroke is called the displacement volume of the pump.

From the operation of the simple piston pump, it can be seen why a pump does not pump pressure. Pumps produce flow. The pressure developed is due to the resistance of the load, which is being driven by the system hydraulic actuators.

6.3 PUMP CLASSIFICATION

There are two broad classifications of pumps as identified by the fluid power industry. They are described as follows:

1. Hydrodynamic or nonpositive displacement pumps: Examples of this type are the centrifugal (impeller) and axial (propeller) pumps shown in Fig. 6-2. Although these pumps provide smooth continuous flow, their flow output is reduced as circuit resistance is increased. In fact, it is possible to completely block off the outlet to stop all flow, even while the pump is running at design speed. These pumps are typically used for low-pressure, high-volume flow applications.

Since there is a great deal of clearance between the rotating and stationary elements, these pumps are not self-priming. This is because there is too much clearance space to seal against atmospheric pressure, and thus the displacement between the inlet and outlet is not a positive one. Thus the pump flow rate depends not only on the rotational speed (rpm) at which it is driven but also on the resistance of the external system.

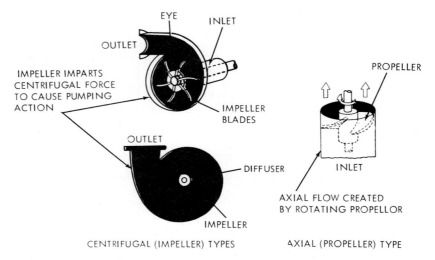

CENTRIFUGAL (IMPELLER) TYPES   AXIAL (PROPELLER) TYPE

Figure 6-2. Nonpositive displacement pumps. (*Courtesy of Sperry Vickers, Sperry Rand Corp., Troy, Michigan.*)

As the resistance of the external system starts to increase, some of the fluid slips back into the clearance spaces, causing a reduction in the discharge flow rate. This slippage is due to the fact that the fluid follows the path of least resistance. When the resistance of the external system becomes infinitely large (for example, a closed valve blocks the outlet line), the pump will produce no flow and thus its volumetric efficiency becomes zero.

For example, this dramatic drop in volumetric efficiency with increase in load resistance occurs when using a centrifugal pump. The operation of a centrifugal pump is simple. The fluid enters at the center of the impeller and is picked up by the rotating impeller. As the fluid rotates with the impeller, the centrifugal force causes the fluid to move radially outward. This causes the fluid to flow through the outlet discharge port of the housing. One of the interesting characteristics of a centrifugal pump is its behavior when there is no demand for fluid. In such a case, no harm occurs to the pump, and thus there is no need for safety devices to prevent pump damage. The tips of the impeller blades merely slosh through the fluid, and the rotational speed maintains a fluid pressure corresponding to the centrifugal force established. The fact that there is no positive internal seal against leakage is the reason that the centrifugal pump is not forced to produce flow against no demand. When demand for the fluid occurs (for example, the opening of a valve), the pressure delivers the fluid to the source of the demand. This is why centrifugal pumps are so desirable for pumping stations used for delivering water to homes and factories. The demand for water may go to near zero during the evening and reach a peak sometimes during the daytime. The centrifugal pump can readily handle these large changes in fluid demand.

Although hydrodynamic pumps provide smooth continuous flow (when a demand exists), their output flow rate is reduced as resistance to flow is increased. This is shown in Fig. 6-3 where pump pressure is plotted versus pump flow. The maximum pressure is called the shutoff head because all external circuit valves are closed and there is no flow. As the external resistance decreases, the flow increases at the expense of reduced pressure. Because the output flow changes

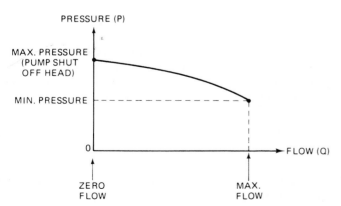

Figure 6-3.   Typical centrifugal pump pressure versus flow curve.

significantly with external circuit resistance, nonpositive displacement pumps are rarely used in hydraulic systems.

2. Hydrostatic or positive displacement pumps: This type of pump ejects a fixed quantity of fluid per revolution of the pump shaft. As a result, pump output flow, neglecting the small internal leakage, is constant and not dependent on system pressure. This makes them particularly well suited for fluid power systems. However, positive displacement pumps must be protected against overpressure if the resistance to flow becomes very large or infinite. This can happen if a valve is completely closed and there is no physical place for the fluid to go. The reason for this is that a positive displacement pump continues to eject fluid (even though it has no place to go), causing an extremely rapid buildup in pressure as the fluid is compressed. A pressure relief valve is used to protect the pump against overpressure by diverting pump flow back to the hydraulic tank where the fluid is stored for system use.

Positive displacement pumps can be classified by the type of motion of internal elements. The motion may be either rotary or reciprocating. Although these pumps come in a wide variety of different designs, there are essentially three basic types:

1. *Gear pumps* (fixed displacement only by geometrical necessity):
   a. External gear pumps
   b. Internal gear pumps
   c. Lobe pumps
   d. Screw pumps
2. *Vane pumps:*
   a. Unbalanced vane pumps (fixed or variable displacement)
   b. Balanced vane pumps (fixed displacement only)
3. *Piston pumps* (fixed or variable displacement):
   a. Axial design
   b. Radial design

In addition, vane pumps can be of the balanced or unbalanced design. The unbalanced design can have pressure compensation capability, which automatically protects the pump against overpressure. In Secs. 6.4, 6.5, and 6.6, we discuss the details of the construction and operation of gear, vane, and piston pumps, respectively.

6.4 GEAR PUMPS

Figure 6-4 illustrates the operation of an external gear pump, which develops flow by carrying fluid between the teeth of two meshing gears. One of the gears is connected to a drive shaft connected to the prime mover. The second gear is

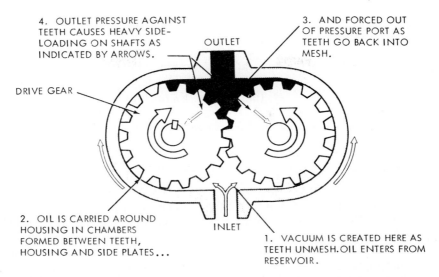

Figure 6-4.   External gear pump operation. (*Courtesy of Sperry Vickers, Sperry Rand Corp., Troy, Michigan.*)

driven as it meshes with the driver gear. Oil chambers are formed between the gear teeth, the pump housing, and the side wear plates. The suction side is where teeth come out of mesh, and it is here where the volume expands, bringing about a reduction in pressure to below atmospheric pressure. Fluid is pushed into this void by atmospheric pressure because the oil supply tank is vented to the atmosphere. The discharge side is where teeth go into mesh, and it is here where the volume decreases between mating teeth. Since the pump has a positive internal seal against leakage, the oil is positively ejected into the outlet port.

The following analysis permits us to evaluate the theoretical flow rate of a gear pump using specified nomenclature:

$$D_o = \text{outside diameter of gear teeth (in.)}$$

$$D_i = \text{inside diameter of gear teeth (in.)}$$

$$L = \text{width of gear teeth (in.)}$$

$$V_D = \text{displacement volume of pump (in.}^3/\text{rev)}$$

$$N = \text{rpm of pump}$$

$$Q_T = \text{theoretical pump flow rate}$$

From gear geometry, the volumetric displacement is found:

$$V_D = \frac{\pi}{4}(D_o^2 - D_i^2)L$$

The theoretical flow rate is determined next:

$$Q_T \text{ (in.}^3/\text{min)} = V_D \text{ (in.}^3/\text{rev)} \cdot N \text{ (rpm)}$$

Since 1 gal = 231 in.^{3}, we have

$$Q_T \text{ (gpm)} = \frac{V_D N}{231} \tag{6-1}$$

Equation (6-1) shows that the pump flow varies directly with speed [see Fig. 6-5(a)]. Hence, the theoretical flow is constant at a given speed, as shown by the solid line in Fig. 6-5(b).

There must be a small clearance (less than 0.001 in.) between the teeth tip and pump housing. As a result, some of the oil at the discharge port can leak directly back toward the suction. This means that the actual flow rate Q_A is less than the theoretical flow rate Q_T, which is based on volumetric displacement and pump speed. This internal leakage, which is called pump slippage, is identified by the term volumetric efficiency η, which is usually greater than 90% for positive displacement pumps, operating at design pressure:

$$\eta = \frac{Q_A}{Q_T} \times 100 \tag{6-2}$$

The higher the discharge pressure, the lower the volumetric efficiency because internal leakage increases with pressure. This is shown by the dashed line in Fig. 6-5(b). Pump manufacturers usually specify volumetric efficiency at the pump

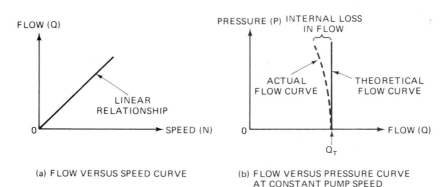

Figure 6-5.   Positive displacement pump Q versus N and P versus Q curves. (a) Flow versus speed curve, (b) flow versus pressure curve at constant pump speed.

rated pressure. The rated pressure of a positive displacement pump is that pressure below which no mechanical damage due to overpressure will occur to the pump and the result will be a long reliable service life. Too high a pressure not only produces excessive leakage but also can damage a pump by distorting the casing and overloading the shaft bearings. This brings to mind once again the need for overpressure protection. Also keep in mind that high pressures occur when a large load or resistance to flow is encountered.

EXAMPLE 6-1

A gear pump has a 3-in outside diameter a 2-in. inside diameter, and a 1-in. width. If the actual pump flow at 1800 rpm and rated pressure is 28 gpm, what is the volumetric efficiency?

Solution   Find the displacement volume:

$$V_D = \frac{\pi}{4}[(3)^2 - (2)^2](1) = 3.93 \text{ in.}^3$$

Next, use Eq. (6-1) to find the theoretical flow rate:

$$Q_T = \frac{V_D N}{231} = \frac{(3.93)(1800)}{231} = 30.6 \text{ gpm}$$

The volumetric efficiency is then found:

$$\eta = \frac{28}{30.6} \times 100 = 91.3\%$$

Figure 6-6 is a photograph showing detailed features of an external gear pump. Also shown is the hydraulic symbol used to represent fixed displacement pumps in hydraulic circuits. This external gear pump uses spur gears (teeth are parallel to the axis of the gear), which are noisy at relatively high speeds. To reduce noise and provide smoother operation, helical gears (teeth inclined at a small angle to the axis of the gear) are sometimes used. However, these helical gear pumps are limited to low-pressure applications (below 200 psi) because they develop excessive end thrust due to the action of the helical gears. Herringbone gear pumps eliminate this thrust action and thus can be used to develop much higher pressures (above 750 psi). Herringbone gears consist basically of two rows of helical teeth cut into one gear. One of the rows of each gear is right-handed and the other is left-handed to cancel out the axial thrust force. Herringbone gear pumps operate as smoothly as helical gear pumps and provide greater flow rates with much less pulsating action.

Figure 6-7 illustrates the configuration and operation of the internal gear pump. This design consists of an internal gear, a regular spur gear, a crescent-

Figure 6-6. Photograph showing detailed features of an external gear pump. (*Courtesy of Webster Electric Company, Inc., subsidiary of STA-RITE Industries, Inc., Racine, Wisconsin.*)

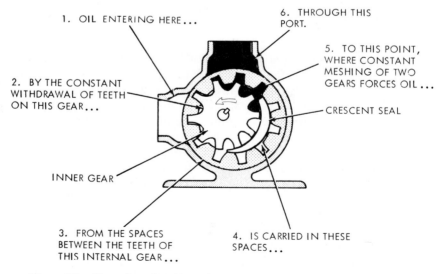

1. OIL ENTERING HERE...

6. THROUGH THIS PORT.

5. TO THIS POINT, WHERE CONSTANT MESHING OF TWO GEARS FORCES OIL...

2. BY THE CONSTANT WITHDRAWAL OF TEETH ON THIS GEAR...

CRESCENT SEAL

INNER GEAR

3. FROM THE SPACES BETWEEN THE TEETH OF THIS INTERNAL GEAR...

4. IS CARRIED IN THESE SPACES...

Figure 6-7. Operation of an internal gear pump. (*Courtesy of Sperry Vickers, Sperry Rand Corp., Troy, Michigan.*)

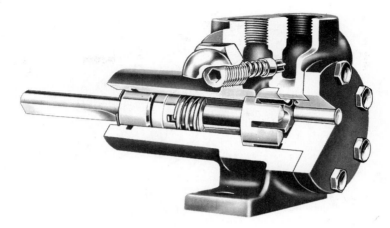

Figure 6-8. Cutaway view of an internal gear pump with built-in safety relief valve. (*Courtesy of Viking Pump Division of Houdaille Industries, Inc., Cedar Falls, Iowa.*)

shaped seal, and an external housing. As power is applied to either gear, the motion of the gears draws fluid from the reservoir and forces it around both sides of the crescent seal, which acts as a seal between the suction and discharge ports. When the teeth mesh on the side opposite to the crescent seal, the fluid is forced to enter the discharge port of the pump.

Figure 6-8 provides a cutaway view of an internal gear pump that contains its own built-in safety relief valve.

Also in the general family of gear pumps is the lobe pump, which is illustrated in Fig. 6-9. This pump operates in a fashion similar to the external gear

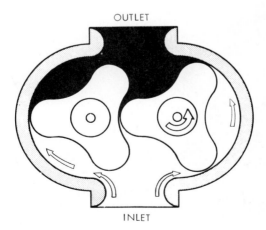

Figure 6-9. Operation of the lobe pump. (*Courtesy of Sperry Vickers, Sperry Rand Corp., Troy, Michigan.*)

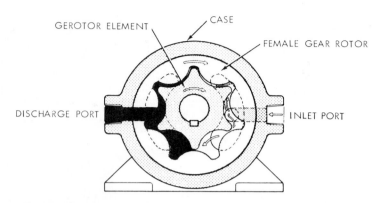

Figure 6-10. Operation of the Gerotor pump. (*Courtesy of Sperry Vickers, Sperry Rand Corp, Troy, Michigan.*)

pump. But unlike the external gear pump, both lobes are driven externally so that they do not actually contact each other. Thus, they are quieter than other types of gear pumps. Due to the smaller number of mating elements, the lobe pump output will have a somewhat greater amount of pulsation, although its volumetric displacement is generally greater than that for other types of gear pumps.

The Gerotor pump, shown in Fig. 6-10, operates very much like the internal gear pump. The inner gear rotor (Gerotor element) is power-driven and draws the outer gear rotor around as they mesh together. This forms inlet and discharge pumping chambers between the rotor lobes. The tips of the inner and outer rotors make contact to seal the pumping chambers from each other. The inner gear has

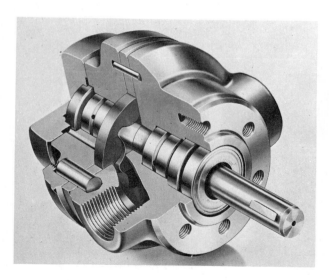

Figure 6-11. Gerotor pump. (*Courtesy of Brown & Sharpe Mfg. Co., Manchester, Michigan.*)

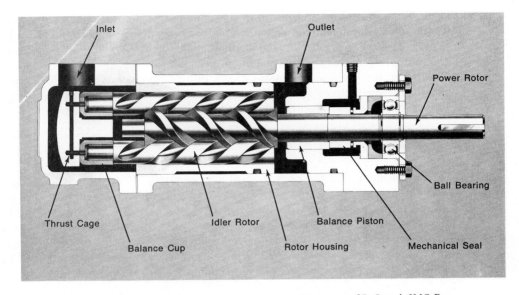

Figure 6-12. Nomenclature of a screw pump. (*Courtesy of DeLaval, IMO Pump Division, Trenton, New Jersey.*)

one tooth less than the outer gear, and the volumetric displacement is determined by the space formed by the extra tooth in the outer rotor.

Figure 6-11 is a photograph of an actual Gerotor pump. As can be seen, this is a simple type of pump since there are only two moving parts.

The screw pump (see Fig. 6-12) is an axial flow positive displacement unit. Three precision ground screws, meshing within a close-fitting housing, deliver nonpulsating flow quietly and efficiently. The two symmetrically opposed idler rotors act as rotating seals, confining the fluid in a succession of closures or

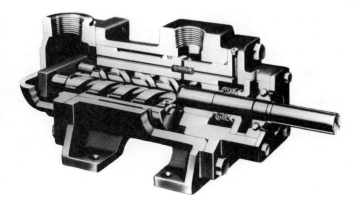

Figure 6-13. Screw pump. (*Courtesy of DeLaval, IMO Pump Division, Trenton, New Jersey.*)

stages. The idler rotors are in rolling contact with the central power rotor and are free to float in their respective housing bores on a hydrodynamic oil film. There are no radial bending loads. Axial hydraulic forces on the rotor set are balanced, eliminating any need for thrust bearings.

In Fig. 6-13, we see a cutaway view of an actual screw pump. It is rated at 500 psi and can deliver up to 123 gpm. High-pressure designs are available for 3500-psi operation with output flow rates up to 88 gpm.

6.5 VANE PUMPS

Figure 6-14 illustrates the operation of a vane pump. The rotor, which contains radial slots, is splined to the drive shaft and rotates inside a cam ring. Each slot contains a vane designed to mate with the surface of the cam ring as the rotor turns. Centrifugal force keeps the vanes out against the surface of the cam ring. During one-half revolution of rotor rotation, the volume increases between the rotor and cam ring. The resulting volume expansion causes a reduction of pressure. This is the suction process, which causes fluid to flow through the inlet port and fill the void. As the rotor rotates through the second half revolution, the surface of the cam ring pushes the vanes back into their slots, and the trapped volume is reduced. This positively ejects the trapped fluid through the discharge port.

Careful observation of Fig. 6-14 will reveal that there is an eccentricity between the centerline of the rotor and the centerline of the cam ring. If the

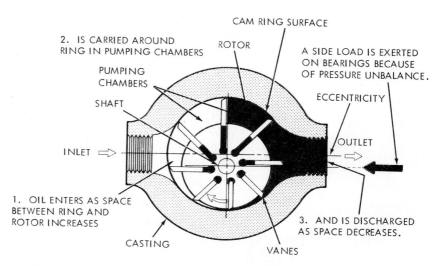

Figure 6-14. Vane pump operation. (*Courtesy of Sperry Vickers, Sperry Rand Corp., Troy, Michigan.*)

eccentricity is zero, there will be no flow. The following analysis and nomenclature is applicable to the vane pump:

$$D_C = \text{diameter of cam ring (in.)}$$

$$D_R = \text{diameter of rotor (in.)}$$

$$L = \text{width of rotor (in.)}$$

$$N = \text{rotor rpm}$$

$$V_D = \text{pump volumetric displacement (in.}^3\text{)}$$

$$e = \text{eccentricity (in.)}$$

$$e_{max} = \text{maximum possible eccentricity (in.)}$$

$$V_{D_{max}} = \text{maximum possible volumetric displacement (in.}^3\text{)}$$

From geometry, we can find the maximum possible eccentricity:

$$e_{max} = \frac{D_C - D_R}{2}$$

This maximum valve of eccentricity produces a maximum volumetric displacement:

$$V_{D_{max}} = \frac{\pi}{4}(D_C^2 - D_R^2)L$$

Rearranging, we have

$$V_{D_{max}} = \frac{\pi}{4}(D_C + D_R)(D_C - D_R)L$$

Substituting the expression for e_{max} yields

$$V_{D_{max}} = \frac{\pi}{4}(D_C + D_R)(2e_{max})L$$

The actual volumetric displacement occurs when $e_{max} = e$:

$$V_D = \frac{\pi}{2}(D_C + D_R)eL \qquad\qquad (6\text{-}3)$$

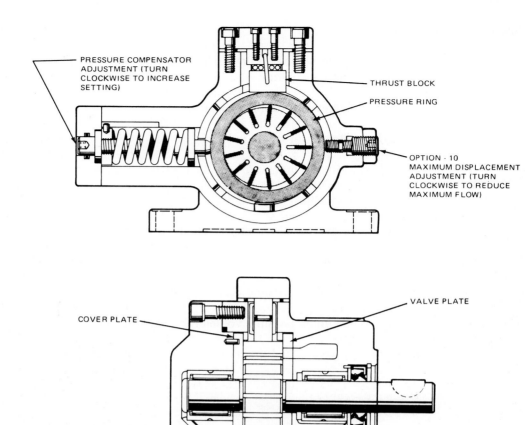

Figure 6-15. Variable displacement, press-compensated vane pump. (*Courtesy of Brown & Sharpe Mfg. Co., Manchester, Michigan.*)

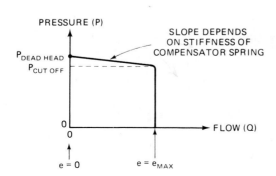

Figure 6-16. Pressure versus flow for pressure-compensated vane pump.

Some vane pumps have provisions for mechanically varying the eccentricity. Such a design is called a variable displacement pump and is illustrated in Fig. 6-15. A handwheel or a pressure compensator can be used to move the cam ring to change the eccentricity. The direction of flow through the pump can be reversed by movement of the cam ring on either side of center.

The design we see in Fig. 6-15 is a pressure-compensated one in which system pressure acts directly on the cam ring via a hydraulic piston on the right side (not shown). This forces the cam ring against the compensator spring-loaded piston on the left side of the cam ring. If the discharge pressure is large enough, it overcomes the compensator spring force and shifts the cam ring to the left. This reduces the eccentricity, which is maximum when discharge pressure is zero. As the discharge pressure continues to increase, zero eccentricity is finally achieved, and the pump flow becomes zero. Such a pump basically has its own protection against excessive pressure buildup, as shown in Fig. 6-16. When the pressure reaches a value called P_{cutoff}, the compensator spring force equals the hydraulic piston force. As the pressure continues to increase, the compensator spring is compressed until zero eccentricity is achieved. The maximum pressure achieved is called $P_{deadhead}$ at which point the pump is protected because it attempts to produce no more flow. As a result there is no horsepower wasted and fluid heating is reduced.

Figure 6-17 shows the internal configuration of an actual pressure-compensated vane pump. This design contains a cam ring that rotates slightly during use, thereby distributing wear over the entire inner circumference of the ring.

Figure 6-17. Cutaway photograph of pressure-compensated vane pump. (*Courtesy of Continental Hydraulics, Division of Continental Machines, Inc., Savage, Minnesota.*)

Notice in Figs. 6-14 and 6-15 that a side load is exerted on the bearings of the vane pump because of pressure unbalance. This same undesirable side load exists for the gear pump of Fig. 6-4. Such pumps are hydraulically unbalanced.

A balanced vane pump is one that has two intake and two outlet ports diametrically opposite each other. Thus, pressure ports are opposite each other, and a complete hydraulic balance is achieved. One disadvantage of a balanced vane pump is that it cannot be designed as a variable displacement unit. Instead of having a circular cam ring, a balanced design vane pump has an elliptical housing, which forms two separate pumping chambers on opposite sides of the rotor. This eliminates the bearing side loads and thus permits higher operating pressures. Figure 6-18 shows the balanced vane pump principle of operation.

Figure 6-19 is a cutaway view of a balanced vane pump containing 12 vanes and a spring-loaded end plate. The inlet port is in the body, and the outlet is in the cover, which may be assembled in any of four positions for convenience in piping.

EXAMPLE 6-2

A vane pump is to have a volumetric displacement of 5 in.[3] It has a rotor diameter of 2 in., a cam ring diameter of 3 in., and a vane width of 2 in. What must be the eccentricity?

Solution   Use Eq. (6-3):

$$e = \frac{2V_D}{\pi(D_C + D_R)L} = \frac{(2)(5)}{\pi(2 + 3)(2)} = 0.318 \text{ in.}$$

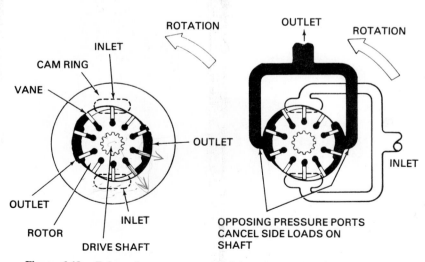

Figure 6-18. Balanced vane pump principles. (*Courtesy of Sperry Vickers, Sperry Rand Corp., Troy, Michigan.*)

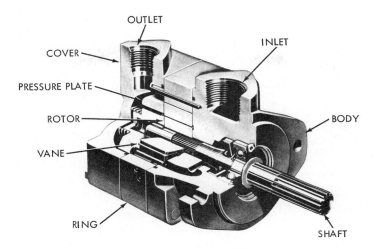

OUTLET

INLET

COVER

PRESSURE PLATE

ROTOR

BODY

VANE

RING

SHAFT

Figure 6-19.   Cutaway view of balanced vane pump. (*Courtesy of Sperry Vickers, Sperry Rand Corp., Troy, Michigan.*)

6.6 PISTON PUMPS

A piston pump works on the principle that a reciprocating piston can draw in fluid when it retracts in a cylinder bore and discharge it when it extends. The basic question is how to mechanize a series of reciprocating pistons. There are two basic types of piston pumps. One is the axial design, having pistons that are parallel to the axis of the cylinder block. Axial piston pumps can be either of the bent axis configuration or of the swash plate design. The second type of piston pump is the radial design, which has pistons arranged radially in a cylinder block.

Figure 6-20 shows an axial piston pump (bent-axis type) that contains a cylinder block rotating with the drive shaft. However, the centerline of the cylinder block is set at an offset angle relative to the centerline of the drive shaft. The cylinder block contains a number of pistons arranged along a circle. The piston rods are connected to the drive shaft flange by ball and socket joints. The pistons are forced in and out of their bores as the distance between the drive shaft flange and cylinder block changes. A universal link connects the block to the drive shaft to provide alignment and positive drive.

The volumetric displacement of the pump varies with the offset angle θ as shown in Fig. 6-21. No flow is produced when the cylinder block centerline is parallel to the drive shaft centerline. θ can vary from 0° to a maximum of about 30°. Fixed displacement units are usually provided with 23° or 30° offset angles.

Variable displacement units are available with a yoke and some external control to change the offset angle. One such design, which uses a stroking cylin-

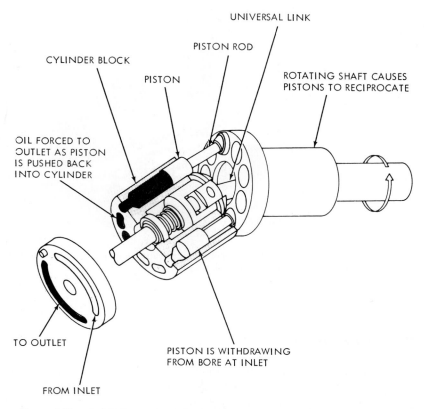

UNIVERSAL LINK

PISTON ROD

CYLINDER BLOCK

PISTON

ROTATING SHAFT CAUSES
PISTONS TO RECIPROCATE

OIL FORCED TO
OUTLET AS PISTON
IS PUSHED BACK
INTO CYLINDER

TO OUTLET

PISTON IS WITHDRAWING
FROM BORE AT INLET

FROM INLET

Figure 6-20.   Axial piston pump (bent-axis type). (*Courtesy of Sperry Vickers,
Sperry Rand Corp., Troy, Michigan.*)

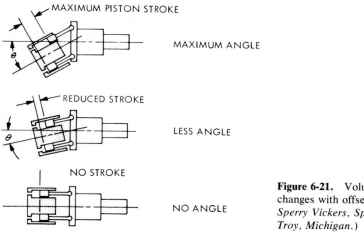

MAXIMUM PISTON STROKE

MAXIMUM ANGLE

REDUCED STROKE

LESS ANGLE

NO STROKE

NO ANGLE

Figure 6-21.   Volumetric displacement
changes with offset angle. (*Courtesy of
Sperry Vickers, Sperry Rand Corp.,
Troy, Michigan.*)

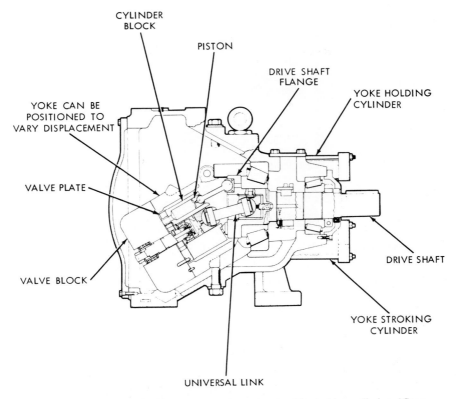

Figure 6-22.   Variable displacement piston pump with stroking cylinder. (*Courtesy of Sperry Vickers, Sperry Rand Corp., Troy, Michigan.*)

der, is shown in Fig. 6-22. Some designs have controls that move the yoke over the center position to reverse the direction of flow through the pump.

Figure 6-23 is a cutaway of a variable displacement piston pump in which an external handwheel can be turned to establish the desired offset angle. Also shown is the hydraulic symbol used to represent variable displacement pumps in hydraulic circuits.

The following nomenclature and analysis are applicable to an axial piston pump:

$$\theta = \text{offset angle (deg)}$$

$$S = \text{piston stroke (in.)}$$

$$D = \text{piston circle diameter (in.)}$$

$$Y = \text{number of pistons}$$

$$A = \text{piston area (in.}^2)$$

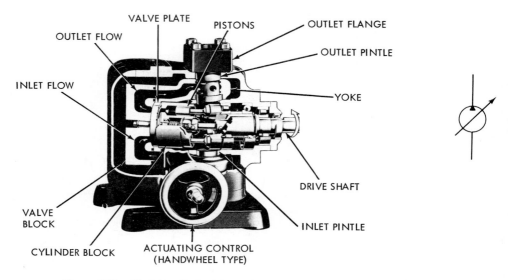

Figure 6-23. Variable displacement piston pump with handwheel. (*Courtesy of Sperry Vickers, Sperry Rand Corp., Troy, Michigan.*)

From trigonometry we have

$$\tan(\theta) = \frac{S}{D}$$

or

$$S = D \tan(\theta)$$

The total displacement volume equals the number of pistons multiplied by the displacement volume per piston:

$$V_D = YAS$$

Substituting, we have

$$V_D = YAD \tan(\theta) \tag{6-4}$$

From Eq. (6-1) we obtain, upon substitution,

$$Q = \frac{DANY \tan(\theta)}{231} \tag{6-5}$$

Example 6-3

Find the offset angle for an axial piston pump that delivers 16 gpm at 3000 rpm. The pump has nine $\frac{1}{2}$-in.-diameter pistons arranged on a 5-in. piston circle diameter.

Solution Use Eq. (6-5):

$$\tan(\theta) = \frac{231Q}{DANY} = \frac{(231)(16)}{(5)[(\pi/4)(\frac{1}{2})^2](3000)(9)} = 0.14$$

$$\theta = 8°$$

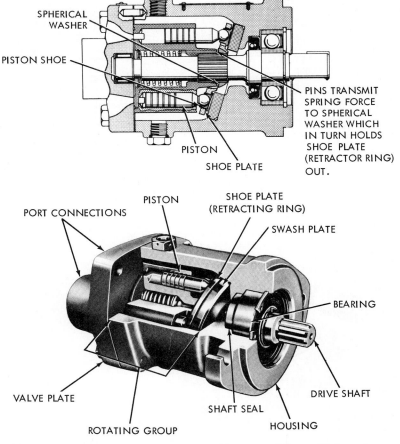

Figure 6-24. In-line design piston pump. (*Courtesy of Sperry Vickers, Sperry Rand Corp., Troy, Michigan.*)

Figure 6-24 provides a photograph and sketch illustrating the swash plate design in-line piston pump. In this type, the cylinder block and drive shaft are located on the same centerline. The pistons are connected to a shoe plate, which bears against an angled swash plate. As the cylinder rotates (see Fig. 6-25), the pistons reciprocate because the piston shoes follow the angled surface of the swash plate.

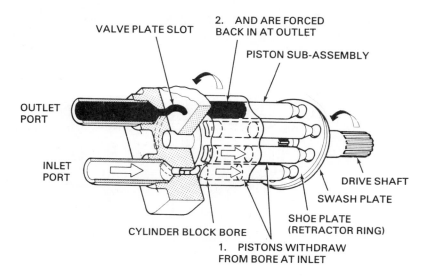

Figure 6-25. Swash plate causes pistons to reciprocate. (*Courtesy of Sperry Vickers, Sperry Rand Corp., Troy, Michigan.*)

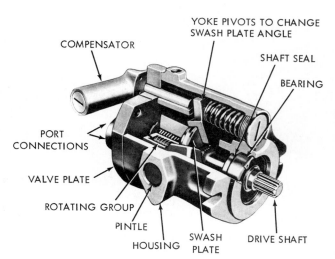

Figure 6-26. Variable displacement version of in-line piston pump. (*Courtesy of Sperry Vickers, Sperry Rand Corp., Troy, Michigan.*)

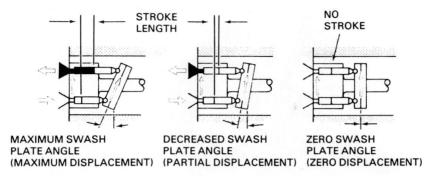

Figure 6-27. Variation in pump displacement. (*Courtesy of Sperry Vickers, Sperry Rand Corp., Troy, Michigan.*).

The outlet and inlet ports are located in the valve plate so that the pistons pass the inlet as they are being pulled out and pass the outlet as they are being forced back in. This type of pump can also be designed to have variable displacement capability. In such a design, the swash plate is mounted in a movable yoke, as depicted in Fig. 6-26. The swash plate angle can be changed by pivoting the yoke on pintles (see Fig. 6-27). Positioning of the yoke can be accomplished by manual operation, servo control, or a compensator control, as shown in Fig. 6-26. The maximum swash plate angle is limited to $17\frac{1}{2}°$ by construction.

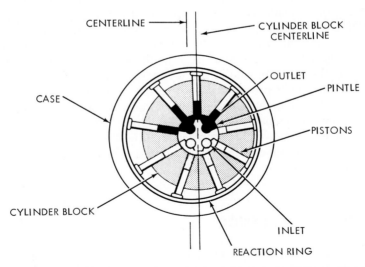

Figure 6-28. Operation of a radial piston pump. (*Courtesy of Sperry Vickers, Sperry Rand Corp., Troy, Michigan.*)

Figure 6-29. Cutaway view of a radial piston pump. (*Courtesy of Deere & Co., Moline, Illinois.*)

The operation and construction of a radial piston pump is illustrated in Fig. 6-28. This design consists of a pintle to direct fluid in and out of the cylinders, a cylinder barrel with pistons, and a rotor containing a reaction ring. The pistons remain in constant contact with the reaction ring due to centrifugal force and back pressure on the pistons. For pumping action, the reaction ring is moved eccentrically with respect to the pintle or shaft axis. As the cylinder barrel rotates, the pistons on one side travel outward. This draws in fluid as each cylinder passes the suction ports of the pintle. When a piston passes the point of maximum eccentricity, it is forced inward by the reaction ring. This forces the fluid to enter the discharge port of the pintle. In some models, the displacement can be varied by moving the reaction ring to change the piston stroke.

Figure 6-29 provides a photograph of a cutaway view of an actual radial piston pump that has variable displacement, pressure-compensated discharge. This pump is available in three sizes (2.40-, 3.00-, and 4.00-in.³ volumetric displacements) and weighs approximately 60 lb. Variable displacement is accomplished by hydraulic rather than mechanical means and is responsive to discharge line pressure.

6.7 PUMP PERFORMANCE

The performance delivered by a pump is primarily a function of the precision of its manufacture. Components must be made to close tolerances, which must be maintained while the pump is operating under design conditions. The maintenance of close tolerances is accomplished by designs that have mechanical integrity and balanced pressures.

Theoretically the ideal pump would be one having zero clearance between all mating parts. Although this is not feasible, working clearances should be as small as possible while maintaining proper oil films for lubrication between rubbing parts.

Pump manufacturers run tests to determine performance data for their various types of pumps. The overall efficiency of a pump can be computed by comparing the power available at the output of the pump to the power supplied at the input. Overall efficiency can be broken into two distinct components called volumetric and mechanical efficiencies.

1. *Volumetric efficiency* (η_v) indicates the amount of leakage that takes place within the pump. This involves considerations such as manufacturing tolerances and flexing of the pump casing under design pressure operating conditions:

$$\eta_v = \frac{\text{actual flow rate produced by pump}}{\text{theoretical flow rate pump should produce}} \times 100 = \frac{Q_A}{Q_T} \times 100 \qquad (6\text{-}6)$$

Volumetric efficiencies typically run from 80% to 90% for gear pumps, 82% to 92% for vane pumps, and 90% to 98% for piston pumps.

2. *Mechanical efficiency* (η_m) indicates the amount of energy losses that occur due to reasons other than leakage. This includes friction in bearings and between other mating parts. It also includes energy losses due to fluid turbulence. Mechanical efficiencies typically run from 90% to 95%:

$$\eta_m = \frac{\text{theoretical horsepower required to operate pump}}{\text{actual horsepower delivered to pump}} \times 100$$

or

$$\eta_m = \frac{\text{pump output horsepower assuming no leakage}}{\text{input horsepower delivered to pump}} \times 100$$

$$\eta_m = \frac{PQ_T/1714}{TN/63,000} \times 100 \qquad (6\text{-}7)$$

The parameters of Eq. (6-7) are defined as follows in conjunction with Fig. 6-30:

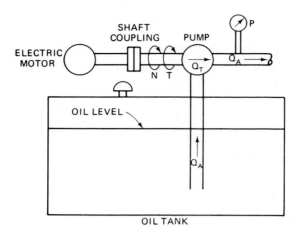

Figure 6-30. Terms involving pump mechanical efficiency.

P = measured pump discharge pressure (psi)

Q_T = calculated theoretical pump flow rate (gpm)

T = measured input torque in prime mover shaft of pump (in.·lb)

N = measured pump speed (rpm)

Mechanical efficiency can also be computed in terms of torques:

$$\eta_m = \frac{\text{theoretical torque required to operate pump}}{\text{actual torque supplied to pump}} \times 100 = \frac{T_T}{T_A} \times 100 \qquad (6\text{-}8)$$

Equations for evaluating T_T and T_A are given as follows:

$$T_T \text{ (in.·lb)} = \frac{V_D \text{ (in.}^3) \cdot P \text{ (psi)}}{2\pi} \qquad (6\text{-}9)$$

$$T_A = \frac{\text{horsepower supplied to pump} \times 63{,}000}{N \text{ (rpm)}} \qquad (6\text{-}10)$$

3. The overall efficiency (η_o) considers all energy losses and is defined mathematically as follows:

$$\text{overall efficiency} = \frac{\text{volumetric efficiency} \times \text{mechanical efficiency}}{100}$$

Substituting from Eq. (6-6) and (6-7), we have

$$\eta_o = \frac{\eta_v \eta_m}{100} = \frac{Q_A}{Q_T} \frac{100}{100} \frac{PQ_T/1714}{TN/63,000}$$

Canceling like terms yields

$$\eta_o = \frac{PQ_A/1714}{TN/63,000} \times 100 = \frac{\text{pump output horsepower}}{\text{pump input horsepower}} \times 100 \qquad (6\text{-}11)$$

EXAMPLE 6-4

A pump has a displacement volume of 5 in.^{3} It delivers 20 gpm at 1000 rpm and 1000 psi. If the prime mover input torque is 900 in.-lb,

    a. What is the overall efficiency of the pump?
    b. What is the theoretical torque required to operate the pump?

Solution

    a. Use Eq. (6-1) to find the theoretical flow rate:

$$Q_T = \frac{V_D N}{231} = \frac{(5)(1000)}{231} = 21.6 \text{ gpm}$$

Next solve for the volumetric efficiency:

$$\eta_v = \frac{Q_A}{Q_T} \times 100 = \frac{20}{21.6} \times 100 = 92.6\%$$

Then solve for the mechanical efficiency:

$$\eta_m = \frac{PQ_T/1714}{TN/63,000} \times 100 = \frac{[(1000)(21.6)]/1714}{[(900)(1000)]/63,000} \times 100 = 88.1\%$$

Finally we solve for the overall efficiency:

$$\eta_o = \frac{\eta_v \eta_m}{100} = \frac{92.6 \times 88.1}{100} = 81.6\%$$

    b. $\qquad T_T = T_A \eta_m = 900 \times 0.881 = 793 \text{ in.-lb}$

    Thus, due to mechanical losses within the pump, 900 in.-lb of torque are required to drive the pump instead of 793 in.-lb.

Pump manufacturers specify pump performance characteristics in the form of graphs. Test data are obtained initially in tabular form and then put in graphical form for better visual interpretation. Figure 6-31 represents typical performance

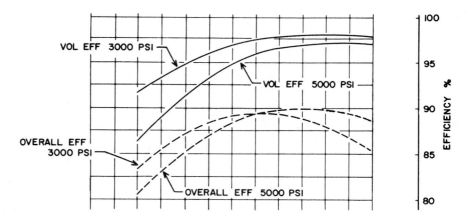

THESE CURVES INCLUDES LOSSES FROM INTEGRAL
SERVO/CHARGE PUMP & TRANSMISSION VALVE PACKAGE

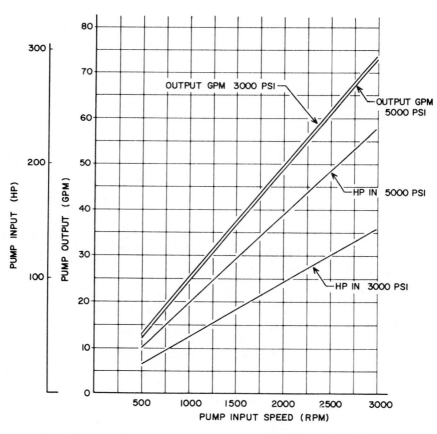

Figure 6-31. Performance curves for 6-in.³ variable displacement piston pump. (*Courtesy of Abex Corp., Dennison Division, Columbus, Ohio.*)

curves obtained for a 6-in.[3] variable displacement pump operating at full displacement. The upper graph gives curves of overall and volumetric efficiencies as a function of pump speed (rpm) for pressure levels of 3000 and 5000 psi. The lower graph gives curves of pump input horsepower (hp) and pump output flow (gpm) as a function of pump speed for the same two pressure levels.

Noise is another significant parameter used to determine the performance of a pump. Noise is measured in units of decibels [db(A)]. Any increase in the noise level normally indicates increased wear and imminent pump failure. Pumps are good generators but poor radiators of noise. As such, pumps are one of the main contributors to noise in a fluid power system. However, the noise we hear is not just the sound coming directly from the pump. It includes the vibration and fluid pulsations produced by the pump as well. Pumps are compact, and because of their relatively small size, they are poor radiators of noise, especially at lower frequencies. Reservoirs, electric motors, and piping, being large, are better radiators. Therefore, pump-induced vibrations or pulsations can cause them to radiate audible noise greater than that coming from the pump. In general, fixed displacement pumps are less noisy than variable displacement units because they have a more rigid construction.

Performance curves for the radial piston pump of Fig. 6-29 are presented in Fig. 6-32. Recall that this pump comes in three different sizes:

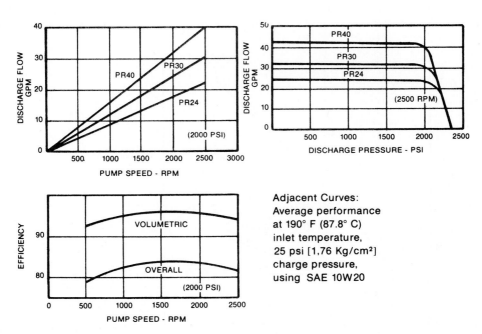

Figure 6-32.   Performance curves of radial piston pumps. (*Courtesy of Deere & Co., Moline, Illinois.*)

PR24: 2.40-in.³ displacement

PR30: 3.00-in.³ displacement

PR40: 4.00-in.³ displacement

Thus, there are three curves on two of the graphs. Observe the linear relationship between discharge flow (gpm) and pump speed (rpm). Also note that the discharge flow of these pumps is nearly constant over a broad pressure range. Discharge flow is infinitely variable between the point of inflection on the constant-discharge portion of the curve and zero flow. The volumetric and overall efficiency curves are based on a 2000-psi pump pressure.

As illustrated in Fig. 6-33, pump speed has a strong effect on noise, whereas pressure and pump size have about equal but smaller effects. Since these three factors determine horsepower, they provide a trade-off for noise. To achieve the lowest noise levels, use the lowest practical speed (1000 or 2000 rpm where electric motors are used, a reducer gear for engine prime movers) and select the most advantageous combination of size and pressure to provide the needed horsepower.

Another cause of noise is entrained air bubbles in the hydraulic fluid. A quiet pump is designed to operate using "solid" fluid. Entrained air bubbles, even if they represent less than 1% by volume, change the compressibility of the fluid so much that it can cause a quiet pump to operate with excessive noise.

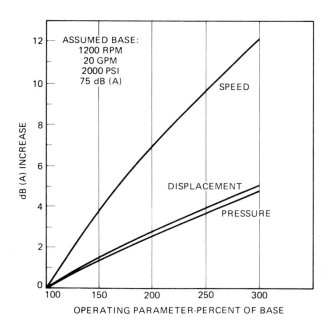

Figure 6-33. Data showing effect of changing size, pressure, and speed on noise. (*Courtesy of Sperry Vickers, Sperry Rand Corp., Troy, Michigan.*)

Still yet another noise problem called cavitation can occur due to entrained air bubbles. This occurs when suction lift is excessive and the inlet pressure falls below the vapor pressure of the fluid (usually about 5-psi suction). As a result, air bubbles, which form in the low-pressure inlet region of the pump, are collapsed when they reach the high-pressure discharge region. This produces high velocity and explosive forces, which severely erode the metallic components and shorten pump life.

The following rules will control or eliminate cavitation of a pump by keeping the suction pressure above the saturation pressure of the fluid:

1. Keep suction line velocities below 5 ft/sec.
2. Keep pump inlet lines as short as possible.
3. Minimize the number of fittings in the inlet line.
4. Mount the pump as close as possible to the reservoir.
5. Use low-pressure drop inlet filters or strainers. Use indicating-type filters and strainers so that they can be replaced at proper intervals as they become dirty.
6. Use the proper oil as recommended by the pump manufacturer. Figure 6-34 shows the preferred range of viscosities and temperatures for optimum pump operation. The importance of temperature control lies in the fact that increased temperatures tend to accelerate the liberation of air bubbles. Therefore, oil temperatures should be kept in the range of 120°F to 150°F to provide an optimum viscosity range and maximum resistance to liberation of air bubbles to reduce the possibility of cavitation.

Figure 6-35 contains a chart showing a comparison of various performance factors for hydraulic pumps. In general, gear pumps are the least expensive but also provide the lowest level of performance. In addition, gear pump efficiency is rapidly reduced by wear, which contributes to high maintenance costs. The volumetric efficiency is greatly affected by the following leakage losses, which can rapidly accelerate due to wear:

1. Leakage around the outer periphery of the gears
2. Leakage across the faces of the gears
3. Leakage at the points where the gear teeth make contact

Gear pumps are simple in design and compact in size. Therefore, they are the most common type of pump used in fluid power systems. The greatest number of applications of gear pumps are in the mobile equipment and machine tool fields.

Vane pump efficiencies and costs fall between those of gear and piston pumps. Vane pumps have good efficiencies and last for a reasonably long period of time. However, continued satisfactory performance necessitates clean oil with good lubricity. Excessive shaft speeds can cause operating problems. Leakage

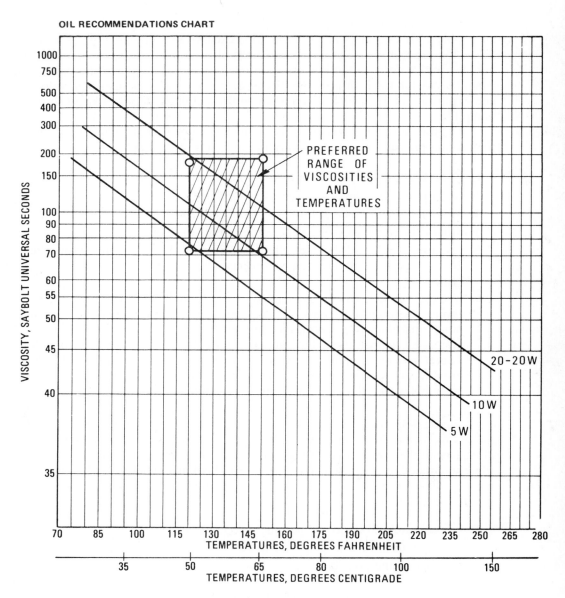

Figure 6-34. Preferred range of oil viscosities and temperatures. (*Courtesy of Sperry Vickers, Sperry Rand Corp., Troy, Michigan.*)

PUMP TYPE	PRESSURE RATING (PSI)	SPEED RATING (RPM)	OVERALL EFFICIENCY (PER CENT)	HP PER LB RATIO	FLOW CAPACITY (GPM)	COST (DOLLARS PER HP)
EXTERNAL GEAR	2000– 3000	1200– 2500	80-90	2	1–150	4-8
INTERNAL GEAR	500– 2000	1200- 2500	70-85	2	1–200	4-8
VANE	1000– 2000	1200– 1800	80-95	2	1–80	6-30
AXIAL PISTON	2000- 12,000	1200– 3000	90-98	4	1–200	6-50
RADIAL PISTON	3000- 12,000	1200– 1800	85-95	3	1–200	5-35

Figure 6-35.   Comparison of various performance factors for pumps.

losses in vane pumps occur across the faces of the rotor and between the bronze wear plates and the pressure ring.

Piston pumps are the most expensive and provide the highest level of overall performance. They can be driven at high speeds (up to 5000 rpm) to provide a high horsepower-to-weight ratio. They produce essentially a nonpulsating flow and can operate at the highest pressure levels. Due to very close-fitting pistons, they have the highest efficiencies. Since no side loads occur to the pistons, the pump life expectancy is at least several years. However, because of their complex design, piston pumps cannot normally be repaired in the field.

6.8 PUMP SELECTION

Pumps are selected by taking into account a number of considerations for a complete hydraulic system involving a particular application. Among these considerations are flow-rate requirements (gpm), operating speed (rpm), pressure rating (psi), performance, reliability, maintenance, cost, and noise. The selection of a pump typically entails the following sequence of operations:

1. Select the actuator (hydraulic cylinder or motor) that is appropriate based on the loads encountered.
2. Determine the flow-rate requirements. This involves the calculation of the flow rate necessary to drive the actuator to move the load through a specified distance within a given time limit.
3. Determine the pump speed and select the prime mover. This, together with the flow-rate calculation, determines the pump size (volumetric displacement).

4. Select the pump type based on the application (gear, vane, or piston pump and fixed or variable displacement).

5. Select the system pressure. This ties in with the actuator size and the magnitude of the resistive force produced by the external load on the system. Also involved here is the total amount of power to be delivered by the pump.

6. Select the reservoir and associated plumbing, including piping, valving, hydraulic cylinders, and motors and other miscellaneous components.

7. Calculate the overall cost of the system.

8. Consider factors such as noise levels, horsepower loss, need for a heat exchanger due to generated heat, pump wear, and scheduled maintenance service to provide a desired life of the total system.

Normally the sequence of operations is repeated several times with different sizes and types of components. After the procedure is repeated for several alternative systems, the best overall system is selected for the given application. This process is called optimization. It means determining the ultimate selection of a combination of system components to produce the most efficient overall system at minimum cost commensurate with the requirements of a particular application.

6.9 PRESSURE INTENSIFIERS

Although a pump is the primary power source for a hydraulic system, auxiliary units are frequently employed for special purposes. One such auxiliary unit is the pressure intensifier or booster.

A pressure intensifier is used to increase the pressure in a hydraulic system to a value above the pump discharge pressure. It accepts a high-volume flow at relatively low pump pressure and converts a portion of this flow to high pressure.

Figure 6-36 shows a cutaway view of a Racine pressure booster. The internal construction consists of an automatically reciprocating large piston that has two small rod ends (also see Fig. 6-37). This piston has its large area (total area of piston) exposed to pressure from a low-pressure pump. The force of the low-pressure oil moves the piston and causes the small area of the piston rod to force the oil out at intensified high pressure. This device is symmetrical about a vertical centerline. Thus, as the large piston reciprocates, the left- and right-hand halves of the unit duplicate each other during each stroke of the large piston.

The increase in pressure is in direct proportion to the ratio of the large piston area and the rod area. The volume output is inversely proportional to this same ratio.

$$\frac{\text{high discharge pressure}}{\text{low inlet pressure}} = \frac{\text{area of piston}}{\text{area of rod}} = \frac{\text{high inlet flow rate}}{\text{low discharge flow rate}} \quad (6\text{-}12)$$

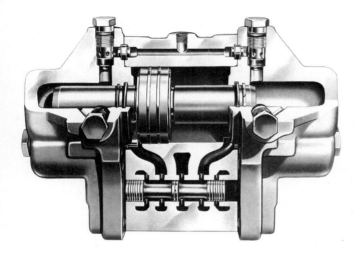

Figure 6-36. Photograph showing cutaway view of pressure intensifier. (*Courtesy of Rexnord Inc., Hydraulic Components Division, Racine, Wisconsin.*)

Racine pressure boosters are available with area ratios of 3:1, 5:1, and 7:1, developing pressures to 5000 psi and flows to 7 gpm. There are many applications for pressure intensifiers such as the elimination of a high-pressure/low-flow pump used in conjunction with a low-pressure/high-flow pump. In an application such as a punch press, it is necessary to extend a hydraulic cylinder rapidly using little pressure to get the ram near the sheet metal strip as quickly as possible. Then the cylinder must exert a large force using only a small flow rate. The large force is needed to punch the workpiece from the sheet metal strip. Since the strip is thin, only a small flow rate is required to perform the punching operation in a small period of time. The use of the pressure intensifier results in a significant cost

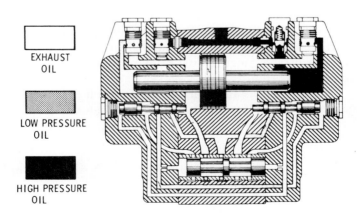

EXHAUST
OIL

LOW PRESSURE
OIL

HIGH PRESSURE
OIL

Figure 6-37. Sketch showing oil flow paths of pressure intensifier. (*Courtesy of Rexnord Inc., Hydraulic Components Division, Racine, Wisconsin.*)

savings in this application, because it replaces the expensive high-pressure pump that would normally be required.

EXAMPLE 6-5

Oil at 20 gpm and 500 psi enters the low-pressure inlet of a 5:1 Racine booster. Find the discharge flow and pressure.

Solution   Substitute directly into Eq. (6-12):

$$\frac{\text{high discharge pressure}}{500 \text{ psi}} = \frac{5}{1} = \frac{20 \text{ gpm}}{\text{low discharge flow rate}}$$

Solving for the unknown quantities, we have

$$\text{high discharge pressure} = 5(500 \text{ psi}) = 2500 \text{ psi}$$

$$\text{low discharge flow rate} = \frac{20}{5} = 4 \text{ gpm}$$

6.10 PUMP PERFORMANCE RATINGS IN METRIC UNITS

Performance data for hydraulic pumps are measured and specified in metric units as well as English units. Figure 6-38 shows actual performance data curves for a Vickers Model VVB20 variable displacement, pressure-compensated vane pump operating at 1200 rpm. The curves give values of flow rate (gpm), efficiency and power (HP and kW) versus output pressure (psi and bars). This particular pump (see Fig. 6-38) can operate at speeds between 1000 and 1800 rpm, is rated at 2540 psi (175 bars), and has a nominal displacement volume of 1.22 in.^{3}/rev (20 cm^3/r). Figure 6-39 shows a sectional view of this pump along with an outline drawing containing dimensions in inches and millimeters (mm). Although the curves give flow rates in gpm, metric flow rates of liters per second (1/s) are frequently specified.

One liter equals 1000 cm^3 or 0.001 m^3 (a liter is defined as the volume of a cube having sides of length equal to 10 cm). In terms of English units, the following conversion factor is applicable since 1 in. = 2.54 cm:

$$1 \text{ liter} = 1000 \text{ cm}^3 \times \left(\frac{1 \text{ in.}}{2.54 \text{ cm}}\right)^3 = 61.0 \text{ in.}^3$$

Since 1 quart equals $\frac{231 \text{ in.}^3}{4}$ or 57.7 in.^{3}, we can conclude that 1 liter equals $\frac{61.0}{57.7}$ or 1.06 quarts. Thus, a liter is about 6 percent larger than a quart.

Performance Data

Typical for single pumps or sections at 1200 rpm; full oil displacement at 150 SUS, 104°F (40°C). Note compensator characteristics in flow curve shown with each control diagram.

Model VVB020   1200 RPM

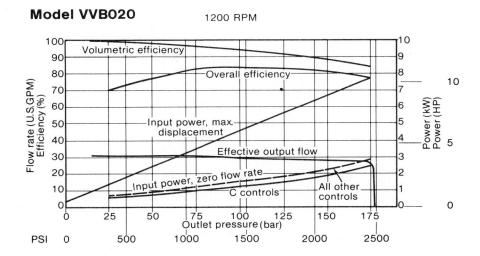

Figure 6-38.   English/metric performance curves for variable displacement, pressure-compensated vane pump at 1200 rpm. (*Courtesy of Vickers, Inc., Troy, Michigan.*)

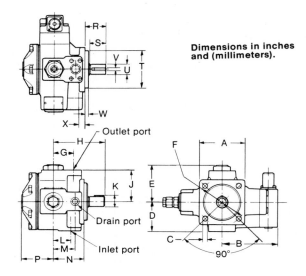

Dimensions in inches and (millimeters).

Model	A	B	C	D	E	F ⌀	G	H	J	K ⌀	L	M	N	P
VVB020*P	4.72 (120)	5.83 (147)	.56 (14)	3.15 (80)	3.94 (98)	5.005/4.995 (127)	2.13 (54)	5.04 (128)	3.35 (85)	1.111/1.101 (28.22/28.00)	1.81 (46)	2.32 (59)	3.23 (82)	3.23 (82)

Sectional view

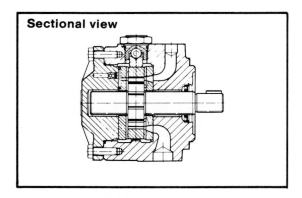

Figure 6-39. Sectional view of vane pump along with outline drawing containing dimensions in inches and millimeters. (*Courtesy of Vickers, Inc., Troy, Michigan.*)

EXAMPLE 6-6

A pump has a displacement volume of 100 cm³. It delivers 0.0015 m³/s at 1000 rpm and 70 bars. If the prime mover input torque is 120 N·m,

  a. What is the overall efficiency of the pump?
  b. What is the theoretical torque required to operate the pump?

Solution

  a. Use Eq. (6-1) in metric units to find the theoretical flow rate:

$$Q_T = V_D N = (0.000100 \text{ m}^3/\text{r}) \left(\frac{1000}{60} \text{ r/s} \right) = 0.00167 \text{ m}^3/\text{s}$$

Next solve for the volumetric efficiency:

$$\eta_v = \frac{Q_A}{Q_T} \times 100 = \frac{0.0015}{0.00167} \times 100 = 89.8\%$$

Then solve for the mechanical efficiency:

$$\eta_m = \frac{PQ_T}{TN} \times 100 = \frac{(70 \times 10^5 \text{ N/m}^2)(0.00167 \text{ m}^3/\text{s})}{(120 \text{ N·m}) \left(1000 \times \frac{2\pi}{60} \text{ radians/s} \right)} \times 100$$

$$\eta_m = \frac{11,690 \text{ N·m/s}}{12,570 \text{ N·m/s}} \times 100 = 93.0\%$$

Note that the product TN gives power in units of N·m/s (watts) where torque (T) has units of N·m and shaft speed has units of radians/s. Finally, we solve for the overall efficiency:

$$\eta_o = \frac{\eta_v \eta_m}{100} = \frac{89.8 \times 93.0}{100} = 83.5\%$$

  b. $$T_T = T_A \eta_m = (120)(0.93) = 112 \text{ N·m}$$

Thus, due to mechanical losses within the pump, 120 N·m of torque are required to drive the pump instead of 112 N·m.

EXERCISES

6-1. Name the three popular construction types of positive displacement pumps.

6-2. What is a positive displacement pump, and in what ways does it differ from a centrifugal pump?

6-3. How is the pumping action in positive displacement pumps accomplished?

6-4. How much hydraulic horsepower would a pump produce when operating at 2000 psi and delivering 10 gpm? What hp electric motor would be selected to drive this pump if its overall efficiency is 85%?

6-5. How is the volumetric efficiency of a positive displacement pump determined?

6-6. How is the mechanical efficiency of a positive displacement pump determined?

6-7. How is the overall efficiency of a positive displacement pump determined?

6-8. Explain how atmospheric pressure pushes hydraulic oil up into the inlet port of a pump.

6-9. What is the difference between a fixed displacement pump and a variable displacement pump?

6-10. Name three designs of external gear pumps.

6-11. Name two designs of internal gear pumps.

6-12. Why is the operation of a screw pump quiet?

6-13. Name the important considerations when selecting a pump for a particular application.

6-14. What is a pressure-compensated vane pump, and how does it work?

6-15. What is pump cavitation, and what is its cause?

6-16. What is the theoretical flow rate from a fixed displacement axial piston pump with a nine-bore cylinder operating at 2000 rpm? Each bore has an 0.5-in. diameter and the stroke is 0.75 in.

6-17. A positive displacement pump has an overall efficiency of 88% and a volumetric efficiency of 92%. What is the mechanical efficiency?

6-18. How is pressure developed in a hydraulic system?

6-19. Why should the suction head of a pump not exceed 5 psi?

6-20. Why must positive displacement pumps be protected by relief valves?

6-21. Why are centrifugal pumps so little used in fluid power systems?

6-22. What are the reasons for the popularity of external gear pumps?

6-23. What is meant by a balanced design hydraulic pump?

6-24. What is a pressure intensifier? List one application.

6-25. Name the two basic types of piston pumps.

6-26. What parameters affect the noise level of a positive displacement pump?

6-27. What is meant by the term "pressure rating of a positive displacement pump"?

6-28. Name four rules that will control or eliminate cavitation of a pump.

6-29. Comment on the relative comparison in performance among gear, vane, and piston pumps.

6-30. A gear pump has a $3\frac{1}{4}$-in. outside diameter, a $2\frac{1}{4}$-in. inside diameter, and a 1-in. width. If the actual pump flow at 1800 rpm and rated pressure is 29 gpm, what is the volumetric efficiency?

6-31. A vane pump is to have a volumetric displacement of 7 in.³ It has a rotor diameter of $2\frac{1}{2}$ in., a cam ring diameter of $3\frac{1}{2}$ in., and a vane width of 2 in. What must be the eccentricity?

6-32. Find the offset angle for an axial piston pump that delivers 30 gpm at 3000 rpm. The pump has nine ⅝-in.-diameter pistons arranged on a 5-in. piston circle diameter.

6-33. A pump has a displacement volume of 6 in.^{3} It delivers 24 gpm at 1000 rpm and 1000 psi. If the prime mover input torque is 1100 in.-lb,

  (a) What is the overall efficiency of the pump?

  (b) What is the theoretical torque required to operate the pump?

6-34. Oil at 21 gpm and 1000 psi enters the low-pressure inlet of a 3:1 Racine booster. Find the discharge flow and pressure.

6-35. How much hydraulic power would a pump produce when operating at 140 bars and delivering 0.001 m^3/s? What power-rated electric motor would be selected to drive this pump if its overall efficiency is 85%?

6-36. What is the theoretical flow rate from a fixed displacement axial piston pump with a nine-bore cylinder operating at 2000 rpm? Each bore has a 15 mm diameter and the stroke is 20 mm.

6-37. A gear pump has a 82.6 mm outside diameter, a 57.2 mm inside diameter, and a 25.4 mm width. If the actual pump flow at 1800 rpm and rated pressure is 0.00183 m^3/s, what is the volumetric efficiency?

6-38. A vane pump is to have a volumetric displacement of 115 cm^3. It has a rotor diameter of 63.5 mm, a cam ring diameter of 88.9 mm, and a vane width of 50.8 mm. What must be the eccentricity?

6-39. Find the offset angle for an axial piston pump which delivers 0.0019 m^3/s at 3000 rpm. The pump has nine 15.9 mm diameter pistons arranged on a 127 mm piston circle diameter.

6-40. A pump has a displacement volume of 98.4 cm^3. It delivers 0.0152 m^3/s at 1000 rpm and 70 bars. If the prime mover input torque is 124.3 N·m,

  (a) What is the overall efficiency of the pump?

  (b) What is the theoretical torque required to operate the pump?

6-41. Oil at 0.001 m^3/s and 70 bars enters the low-pressure inlet of a 3:1 Racine booster. Find the discharge flow and pressure.

6-42. A pump has a displacement volume of 6 in.^{3} and delivers 29 gpm at 1200 rpm and 500 psi. If the overall efficiency of the pump is 88%, find the actual torque required to operate the pump.

6-43. What pressure is typically available to push liquid into the inlet port of a pump? Why?

6-44. How is pressure created in a hydraulic system?

6-45. What are the basic characteristics of positive displacement pumps?

6-46. What are two ways of expressing pump size?

6-47. What types of pumps are available in variable displacement designs?

6-48. Explain the principle of a balanced vane design pump?

6-49. How can displacement be varied in an axial piston pump?

6-50. What limits the pressure capability of a gear pump?

6-51. What are two ways of altering displacement volumes in gear pumps?

6-52. A pump operates at 3000 psi and delivers 5 gpm. It requires 10 HP to drive the pump. Determine the overall efficiency of the pump. If the pump is driven at 1000 rpm, what is the input torque to the pump?

6-53. A pump has an overall efficiency of 88% and a volumetric efficiency of 92% while consuming 8 HP. Determine the mechanical efficiency and the frictional horsepower.

6-54. Determine the overall efficiency of a pump driven by a 10-HP prime mover if the pump delivers fluid at 40 liters per minute at a pressure of 10 MPa.

6-55. A 1.5 in.^{3}/rev displacement pump delivers 10 gpm at a pressure of 2000 psi. At 100% overall efficiency, what prime mover HP and speed are required?

6-56. A pump having a 96% volumetric efficiency delivers 29 liters per minute at 1000 rpm. What is the volumetric displacement of the pump?

6-57. Explain how the size of the pumping chamber of a variable displacement vane pump is changed.

6-58. How is the capability of a variable displacement pump affected by the addition of pressure compensation?

6-59. What is the metric equivalent of the equation $Q = \dfrac{V_D N}{231}$?

6-60. For the fluid power system of Fig. 6-40, the following data is given:

Cylinder piston diameter = 8 in.

Cylinder rod diameter = 4 in.

Extending speed of cylinder = 3 in./sec.

External load on cylinder = 40,000 lb.

Pump volumetric efficiency = 92%.

Pump mechanical efficiency = 90%.

Pump speed = 1800 rpm.

Pump inlet pressure = −4.0 psi.

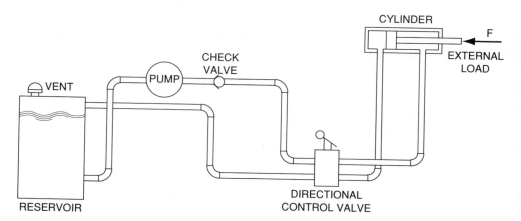

Figure 6-40. System for Exercise 6-60.

The total pressure drop in the line from the pump discharge port to the blank end of the cylinder is 75 psi. The total pressure drop in the return line from the rod end of the cylinder is 50 psi. Determine the:

(a) Volumetric displacement of the pump.

(b) Input HP required to drive the pump.

(c) Input torque required to drive the pump.

(d) Percentage of pump input power delivered to the load.

7

Fluid Power Actuators

7.1 INTRODUCTION

Pumps perform the function of adding energy to a hydraulic system for transmission to some remote point. Fluid power actuators do just the opposite. They extract energy from a fluid and convert it to a mechanical output to perform useful work.

Fluid power can be transmitted through either linear or rotary motion by using linear actuators called hydraulic cylinders or rotary actuators called hydraulic motors. Hydraulic cylinders extend and retract to perform a complete cycle of operation. They sometimes include cushions in their end plates to prevent shock loading, which can damage the moving piston or the stationary cylinder end plates. Rotary actuators can be of the limited rotation or the continuous rotation type. Limited rotation motors are frequently called oscillation fluid motors because they produce a reciprocating motion. Continuous rotary hydraulic motors (or simply hydraulic motors), in reality, are pumps that have been redesigned to withstand the different forces that are involved in motor applications. As a result, hydraulic motors are of the gear, vane, or piston configuration. A gear motor, like a gear pump, must be a fixed displacement unit. Also, as in the case of pumps, piston motors can be either fixed or variable displacement units.

Hydrostatic transmissions are hydraulic systems specifically designed to have a pump drive a hydraulic motor. Thus, a hydrostatic transmission simply transforms mechanical power into fluid power and then reconverts the fluid power back into shaft power. The advantages of hydrostatic transmissions include power transmission to remote areas, infinitely variable speed control, self-overload protection, reverse rotation capability, dynamic braking, and a high horsepower-to-weight ratio. They are used in applications where lifting, lowering, opening, closing, and indexing are required. Specific applications include materials handling equipment, farm tractors, railway locomotives, buses, automobiles, and machine tools.

7.2 *LINEAR HYDRAULIC ACTUATORS (HYDRAULIC CYLINDERS)*

The simplest type of linear actuator is the single-acting cylinder, which is shown schematically in Fig. 7-1(a). It consists of a piston inside a cylindrical housing called a barrel. Attached to one end of the piston is a rod, which extends outside one end of the cylinder (rod end). At the other end (blank end) is a port for the entrance and exit of oil. A single-acting cylinder can exert a force in only the extending direction as fluid from the pump enters the blank end of the cylinder. Single-acting cylinders do not retract hydraulically. Retraction is accomplished by using gravity or by the inclusion of a compression spring in the rod end. Figure 7-1(b) shows the symbolic representation of a single-acting cylinder.

Figure 7-2 shows the design of a double-acting hydraulic cylinder. Such a cylinder can be extended and retracted hydraulically. Thus, an output force can be applied in two directions (extension and retraction). This particular cylinder has a working pressure rating of 2000 psi for its smallest bore size of $1\frac{1}{8}$ in. and 800 psi for its largest bore size of 8 in.

The nomenclature of a double-acting cylinder is provided in Fig. 7-2. In this design, the barrel is made of seamless steel tubing, honed to a fine finish on the inside. The piston, which is made of ductile iron, contains U-cup packings to seal against leakage between the piston and barrel. The ports are located in the end caps, which are secured to the barrel by tie rods. The tapered cushion plungers provide smooth deceleration at both ends of the stroke. Therefore, the piston does not bang into the end caps with excessive impact, which could damage the hydraulic cylinder after a given number of cycles. The symbol for a double-acting

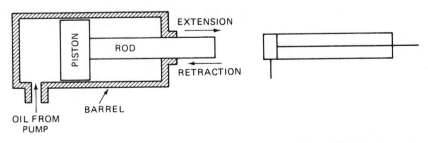

(a) SCHEMATIC REPRESENTATION   (b) SYMBOLIC REPRESENTATION

Figure 7-1.   Single-acting hydraulic cylinder.

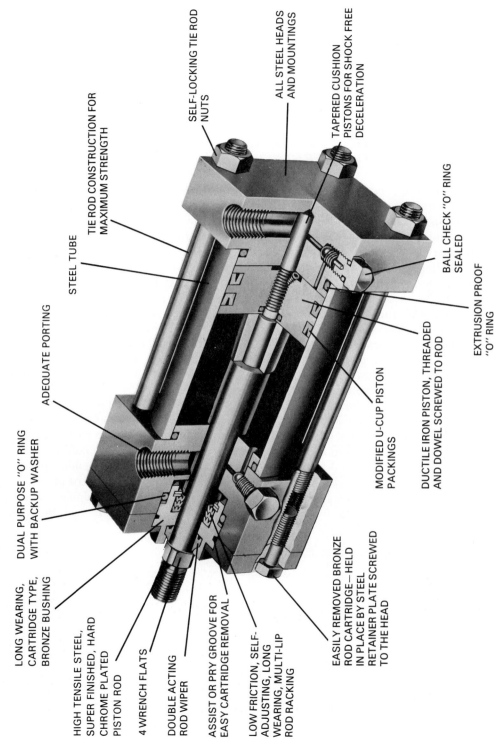

SELF-LOCKING TIE ROD NUTS

ALL STEEL HEADS AND MOUNTINGS

TAPERED CUSHION PISTONS FOR SHOCK FREE DECELERATION

TIE ROD CONSTRUCTION FOR MAXIMUM STRENGTH

STEEL TUBE

BALL CHECK "O" RING SEALED

EXTRUSION PROOF "O" RING

ADEQUATE PORTING

DUCTILE IRON PISTON, THREADED AND DOWEL SCREWED TO ROD

MODIFIED U-CUP PISTON PACKINGS

DUAL PURPOSE "O" RING WITH BACKUP WASHER

LONG WEARING, CARTRIDGE TYPE, BRONZE BUSHING

HIGH TENSILE STEEL, SUPER FINISHED, HARD CHROME PLATED PISTON ROD

4 WRENCH FLATS

DOUBLE ACTING ROD WIPER

ASSIST OR PRY GROOVE FOR EASY CARTRIDGE REMOVAL

LOW FRICTION, SELF-ADJUSTING, LONG WEARING, MULTI-LIP ROD RACKING

EASILY REMOVED BRONZE ROD CARTRIDGE—HELD IN PLACE BY STEEL RETAINER PLATE SCREWED TO THE HEAD

Figure 7-2. Double-acting cylinder design. (*Courtesy of Sheffer Corp., Cincinnati, Ohio.*)

222

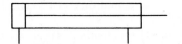

Figure 7-3. Symbolic representation of double-acting cylinder. (*Courtesy of Sheffer Corp., Cincinnati, Ohio.*)

cylinder is shown in Fig. 7-3. Notice that the symbol implies how the cylinder operates without showing any details. In drawing hydraulic circuits (as done in Chapter 9), symbolic representation of all components will be used. This facilitates circuit analysis and troubleshooting. Also, it would be too time-consuming to draw each component schematically. The symbols, which are merely combinations of simple geometric figures such as circles, rectangles, and lines, make no attempt to show the internal configuration of a component. However, symbols must clearly show the function of each component.

Various types of cylinder mountings are in existence, as illustrated in Fig. 7-4. This permits versatility in the anchoring of cylinders. The rod ends are usually

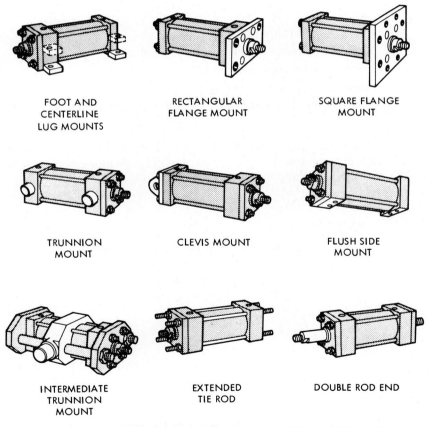

FOOT AND CENTERLINE LUG MOUNTS

RECTANGULAR FLANGE MOUNT

SQUARE FLANGE MOUNT

TRUNNION MOUNT

CLEVIS MOUNT

FLUSH SIDE MOUNT

INTERMEDIATE TRUNNION MOUNT

EXTENDED TIE ROD

DOUBLE ROD END

Figure 7-4. Various cylinder mountings. (*Courtesy of Sperry Vickers, Sperry Rand Corp., Troy, Michigan.*)

threaded so that they can be attached directly to the load, a clevis, a yoke, or some other mating device.

Through the use of various mechanical linkages, the applications of hydraulic cylinders are limited only by the ingenuity of the fluid power designer. As illustrated in Fig. 7-5, these linkages can transform a linear motion into either an oscillating or rotary motion. In addition, linkages can also be employed to increase or decrease the effective leverage and stroke of a cylinder.

Much effort has been made by manufacturers of hydraulic cylinders to reduce or eliminate the side loading of cylinders created as a result of misalignment. It is almost impossible to achieve perfect alignment even though the alignment of a hydraulic cylinder has a direct bearing on its life.

A universal alignment mounting accessory designed to reduce misalignment problems is illustrated in Fig. 7-6. By using one of these accessory components and a mating clevis at each end of the cylinder (see Fig. 7-6), the following benefits are obtained:

1. Freer range of mounting positions
2. Reduced cylinder binding and side loading
3. Allowance for universal swivel
4. Reduced bearing and tube wear
5. Elimination of piston blow-by caused by misalignment

The force output and piston velocity of double-acting cylinders is not the same for extension and retraction strokes. This phenomenon is due to the effect of the rod and is defined by Eqs. (7-1)–(7-4).

Extension stroke:

$$\text{force (lb)} = \text{pressure (psi)} \times \text{piston area (in.}^2) \qquad (7\text{-}1)$$

$$\text{velocity (ft/sec)} = \frac{\text{input flow (ft}^3/\text{sec)}}{\text{piston area (ft}^2)} \qquad (7\text{-}2)$$

Retraction stroke:

$$\text{force (lb)} = \text{pressure (psi)} \times [\text{piston area (in.}^2) - \text{rod area (in.}^2)]$$

$$(7\text{-}3)$$

$$\text{velocity (ft/sec)} = \frac{\text{input flow (ft}^3/\text{sec)}}{\text{piston area (ft}^2) - \text{rod area (ft}^2)} \qquad (7\text{-}4)$$

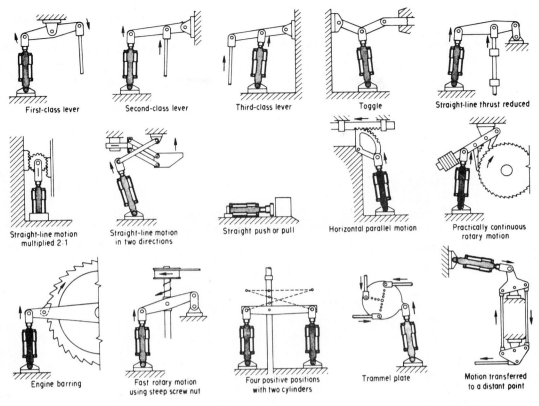

First-class lever

Second-class lever

Third-class lever

Toggle

Straight-line thrust reduced

Straight-line motion multiplied 2:1

Straight-line motion in two directions

Straight push or pull

Horizontal parallel motion

Practically continuous rotary motion

Engine barring

Fast rotary motion using steep screw nut

Four positive positions with two cylinders

Trammel plate

Motion transferred to a distant point

Figure 7-5. Typical mechanical linkages that can be combined with hydraulic cylinders. (*Courtesy of Rexnord Inc., Hydraulic Components Division, Racine, Wisconsin.*)

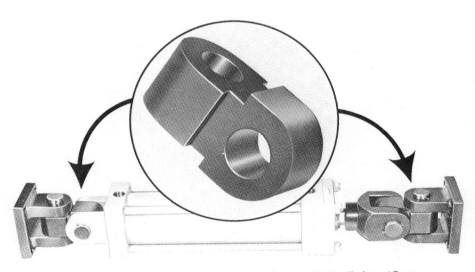

Figure 7-6. Universal alignment mounting accessory for fluid cylinders. (*Courtesy of Sheffer Corp., Cincinnati, Ohio.*)

The hydraulic horsepower developed by a hydraulic cylinder can be found using

$$\text{horsepower} = \frac{\text{piston velocity (ft/sec)} \times \text{hydraulic force (lb)}}{550}$$

$$= \frac{\text{input flow (gpm)} \times \text{pressure (psi)}}{1714} \qquad (7\text{-}5)$$

EXAMPLE 7-1

A pump supplies oil at 20 gpm to a 2-in.-diameter double-acting hydraulic cylinder. If the load is 1000 lb (extending and retracting) and the rod diameter is 1 in., find

    a. The hydraulic pressure during the extending stroke
    b. The piston velocity during the extending stroke
    c. The cylinder horsepower during the extending stroke
    d. The hydraulic pressure during the retraction stroke
    e. The piston velocity during the retraction stroke
    f. The cylinder horsepower during the retraction stroke

Solution

    a. $\text{pressure} = \dfrac{\text{force (lb)}}{\text{piston area (in.}^2)} = \dfrac{1000}{(\pi/4)(2)^2} = \dfrac{1000}{3.14} = 318$ psi

    b. $\text{velocity} = \dfrac{\text{input flow (ft}^3/\text{sec)}}{\text{piston area (ft}^2)} = \dfrac{20/448}{3.14/144} = \dfrac{0.0446}{0.0218} = 2.05$ ft/sec

    c. $\quad HP = \dfrac{\text{piston velocity (ft/sec)} \times \text{hydraulic force (lb)}}{550}$

$$= \frac{(2.05 \times (1000)}{550} = 3.72 \text{ hp}$$

or

$$HP = \frac{\text{input flow (gpm)} \times \text{pressure (psi)}}{1714} = \frac{(20)(318)}{1714} = 3.72 \text{ hp}$$

    d. $\text{pressure} = \dfrac{\text{force (lb)}}{\text{piston area (in.}^2) - \text{rod area (in.}^2)}$

$$= \frac{1000}{3.14 - (\pi/4)(1)^2} = \frac{1000}{2.355} = 425 \text{ psi}$$

Therefore, as expected, more pressure is required to retract than extend the same load due to the effect of the rod.

e. $$\text{velocity} = \frac{\text{input flow (ft}^3/\text{sec})}{\text{piston area (ft}^2) - \text{rod area (ft}^2)}$$

$$= \frac{0.0446}{2.355/144} = 2.73 \text{ ft/sec}$$

Therefore, as expected (for the same pump flow), the piston retraction velocity is greater than that for extension due to the effect of the rod.

f. $$HP = \frac{\text{piston velocity (ft/sec)} \times \text{hydraulic force (lb)}}{550}$$

$$= \frac{(2.73)(1000)}{550} = 4.96 \text{ hp}$$

or

$$HP = \frac{\text{input flow (gpm)} \times \text{pressure (psi)}}{1714} = \frac{(20)(425)}{1714} = 4.96 \text{ hp}$$

Thus, more horsepower is supplied by the cylinder during the retraction stroke because the piston velocity is greater during retraction and the load force remained the same during both strokes. This, of course, was accomplished by the greater pressure level during the retraction stroke. Recall that the pump output flow rate is constant, with a value of 20 gpm.

Double-acting cylinders sometimes contain cylinder cushions at the ends of the cylinder to slow the piston down near the ends of the stroke. This prevents excessive impact when the piston is stopped by the end caps, as illustrated in Fig. 7-7. As shown, deceleration starts when the tapered plunger enters the opening in the cap. This restricts the exhaust flow from the barrel to the port. During the last small portion of the stroke, the oil must exhaust through an adjustable opening. The cushion design also incorporates a check valve to allow free flow to the piston during direction reversal.

The maximum pressure developed by cushions at the ends of a cylinder must be considered since excessive pressure buildup would rupture the cylinder. Example 7-2 illustrates how to calculate this pressure, which decelerates the piston at the ends of its extension and retraction strokes.

EXAMPLE 7-2

A pump delivers oil at a rate of 18.2 gpm into the blank end of the 3-in.-diameter hydraulic cylinder shown in Fig. 7-8. The piston contains a 1-in. diameter cushion

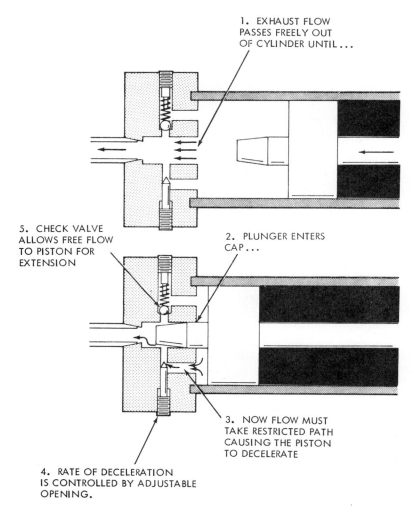

1. EXHAUST FLOW
PASSES FREELY OUT
OF CYLINDER UNTIL...

5. CHECK VALVE
ALLOWS FREE FLOW
TO PISTON FOR
EXTENSION

2. PLUNGER ENTERS
CAP...

3. NOW FLOW MUST
TAKE RESTRICTED PATH
CAUSING THE PISTON
TO DECELERATE

4. RATE OF DECELERATION
IS CONTROLLED BY ADJUSTABLE
OPENING.

Figure 7-7. Operation of cylinder cushions. (*Courtesy of Sperry Vickers, Sperry Rand Corp., Troy, Michigan.*)

plunger which is 0.75 in. long, and therefore the piston decelerates over a distance of 0.75 in. at the end of its extension stroke. The cylinder drives a 1500-lb load, which slides on a flat horizontal surface having a coefficient of friction (μ) equal to 0.12. The pump pressure relief valve setting equals 750 psi. Therefore, the maximum pressure (P_1) at the blank end of the cylinder equals 750 psi while the cushion is decelerating the piston. Find the maximum pressure (P_2) developed by the cushion.

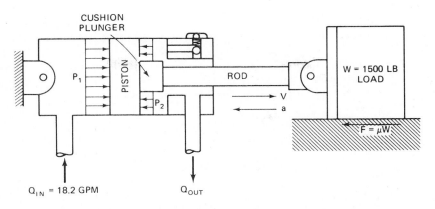

Figure 7-8. Cylinder cushion problem for Example 7-2.

Solution

Step 1: Calculate the steady-state piston velocity (V) prior to deceleration:

$$V = \frac{Q_{pump}}{A_{piston}} = \frac{(18.2/448) \text{ ft}^3/\text{sec}}{[(\pi/4)(3)^2/144] \text{ ft}^2} = \frac{0.0406}{0.049} = 0.83 \text{ ft/sec}$$

Step 2: Calculate the deceleration (a) of the piston during the 0.75-in. displacement (S) using the constant acceleration (or deceleration) equation:

$$V^2 = 2aS$$

Solving for acceleration, we have

$$a = \frac{V^2}{2S} \qquad\qquad (7\text{-}6)$$

Substituting known values, we obtain the value of deceleration:

$$a = \frac{(0.83 \text{ ft/sec})^2}{2(0.75/12 \text{ ft})} = 6.64 \text{ ft/sec}^2 \quad 5.511 \quad \text{ft/sec}^2$$

Step 3: Use Newton's law of motion: The sum of all external forces (ΣF) acting on a mass (m) equals the product of the mass (m) and its acceleration or deceleration (a):

$$F = ma$$

When substituting into Newton's equation, we shall consider forces that tend to slow down the piston as being positive forces. Also the mass m equals the mass of all the moving members (piston, rod, and load). Since the weight of the piston and rod is small compared to the weight of the load, the weight of the piston and rod will be ignored. Also note that mass (m) equals weight (W) divided by the acceleration of gravity (g).

The frictional retarding force (f) between the load (W) and its horizontal support surface equals μ times W.

Substituting into Newton's equation yields

$$P_2(A_{\text{piston}} - A_{\text{cushion plunger}}) + \mu W - P_1(A_{\text{piston}}) = \frac{W}{g} a$$

Solving for P_2 yields a usable equation:

$$P_2 = \frac{(W/g)a + P_1(A_{\text{piston}}) - \mu W}{A_{\text{piston}} - A_{\text{cushion plunger}}} \tag{7-7}$$

Substituting known values produces the desired result:

$$P_2 = \frac{[(1500)(6.64)/32.2] + 750(\pi/4)(3)^2 - (0.12)(1500)}{(\pi/4)(3)^2 - (\pi/4)(1)^2}$$

$$P_2 = \frac{309.3 + 5303 - 180}{7.07 - 0.785} = \frac{5432.3}{6.285} = 864 \text{ psi} \quad \boxed{855 \text{ psi}}$$

In Fig. 7-9 we see a unique, compact, self-contained hydraulic package called the Powr-Pak by its manufacturer. It supplies a force where a minimum size and maximum power are required. To provide for flexibility of operation, a selection of speeds and power is available. It consists of a heavy-duty hydraulic cylinder, a reversible electric motor, a reversible Gerotor pump, a reservoir, and automatic valving. This package is a complete hydraulic power system that is simple and easy to put into operation. You mount it and connect three wires through a reversing switch to an electrical power source. Because manifold porting is utilized, the customer does not have to do any plumbing. The mounting operation is simplified because 11 standard mountings are available. The hydraulic cylinder comes in bore sizes from $2\frac{1}{2}$ to 14 in. Operating pressures up to 2000 psi are obtainable. Figure 7-10 shows the Powr-Pak mounted to a press-type machine. Applications that have also been found for this system (due to its versatility, low cost, and reliability) include forming, bending, clamping, raising and lowering, and tilting operations.

Figure 7-11 illustrates a double-rod cylinder in which the rod extends out of the cylinder at both ends. For such a cylinder, the words "extend" and "retract" have no meaning. Since the force and speed are the same for either end, this type of cylinder is typically used when the same task is to be performed at either end. Since each end contains the same size rod, the velocity of the piston is the same for both strokes.

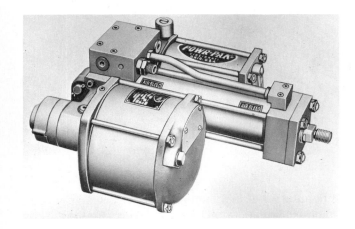

Figure 7-9.   Powr-Pak hydraulic power package. (*Courtesy of Sheffer Corp., Cincinnati, Ohio.*)

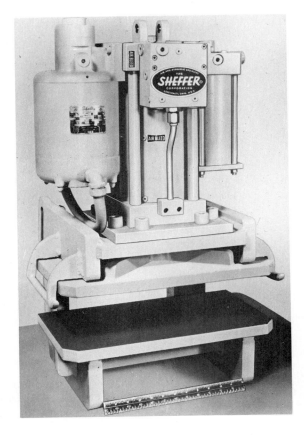

Figure 7-10.   Powr-Pak driving a press-type machine. (*Courtesy of Sheffer Corp., Cincinnati, Ohio.*)

Figure 7-11.   Double-rod cylinder. (*Courtesy of Allenair Corp., Mineola, New York.*)

Figure 7-12 illustrates the internal design features of a telescopic cylinder. This type actually contains multiple cylinders that slide inside each other. They are used where long work strokes are required but the full retraction length must be minimized. One application for a telescopic cylinder is the high-lift fork truck, illustrated in Fig. 7-13.

Figure 7-12.   Telescopic cylinder. (*Courtesy of Commercial Shearing, Inc., Youngstown, Ohio.*)

Figure 7-13. High-lift fork truck. (*Courtesy of Eaton Corp., Industrial Truck Division, Philadelphia, Pennsylvania.*)

7.3 LIMITED ROTATION HYDRAULIC ACTUATOR

A limited rotation hydraulic actuator (or rotary actuator) provides rotary output motion over a finite angle. This device produces high instantaneous torque in either direction and requires only small space and simple mountings. Rotary actuators consist of a chamber or chambers containing the working fluid and a movable surface against which the fluid acts. The movable surface is connected to an output shaft to produce the output motion. A direct-acting vane-type actuator is shown schematically in Fig. 7-14. Fluid under pressure is directed to one side of the moving vane, causing it to rotate. This type provides about 280° of rotation. Vane unit capacity ranges from 3 to 1 million in. · lb.

Rotary actuators are available with working pressures up to 5000 psi. They are typically mounted by foot, flange, and end mounts. Cushioning devices are available in most designs. Figure 7-15 shows an actual rotary actuator similar to the design depicted schematically in Fig. 7-14. Since it contains two vanes, the

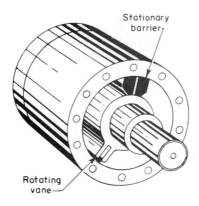

Figure 7-14. Limited rotation hydraulic actuator. (*Courtesy of Rexnord Inc., Hydraulic Components Division, Racine, Wisconsin.*)

Figure 7-15. Rotary actuator. (*Courtesy of Ex-cell-O Corp., Troy, Michigan.*)

maximum angle of rotation is reduced to about 100°. However, the torque-carrying capacity is twice that obtained by a single-vane design. This particular unit can operate with either air or oil at pressures up to 1000 psi.

The following nomenclature and analysis are applicable to a limited rotation hydraulic actuator containing a single rotating vane:

$$R_R = \text{outer radius of rotor (in.)}$$

$$R_V = \text{outer radius of vane (in.)}$$

$$L = \text{width of vane (in.)}$$

$$P = \text{hydraulic pressure (psi)}$$

F = hydraulic force acting on vane (lb)

A = surface area of vane in contact with oil (in.^{2})

T = torque developed (in. · lb)

The force on the vane equals the pressure times the vane surface area:

$$F = PA = P(R_V - R_R)L$$

The torque equals the vane force times the mean radius of the vane:

$$T = P(R_V - R_R)L\,\frac{(R_V + R_R)}{2}$$

Upon rearranging we have

$$T = \frac{PL}{2}(R_V^2 - R_R^2) \qquad (7\text{-}8)$$

A second equation for torque can be developed by noting the following relationship for volumetric displacement V_D:

$$V_D = \pi(R_V^2 - R_R^2)L \qquad (7\text{-}9)$$

Combining Eqs. (7-8) and (7-9) yields

$$T = \frac{PV_D}{6.28} \qquad (7\text{-}10)$$

Observe from Eq. (7-10) that torque output can be increased by increasing the pressure or volumetric displacement or both.

EXAMPLE 7-3

A single-vane rotary actuator has the following physical data:

Outer radius of rotor = 0.5 in.

Outer radius of vane = 1.5 in.

Width of vane = 1 in.

If the torque load is 1000 in. · lb, what pressure must be developed to overcome the load?

Solution

Use Eq. (7-9) to solve for the volumetric displacement:

$$V_D = \pi(1.5^2 - 0.5^2)(1) = 6.28 \text{ in.}^3$$

Then use Eq. (7-10) to solve for the pressure:

$$P = \frac{6.28T}{V_D} = \frac{6.28(1000)}{6.28} = 1000 \text{ psi}$$

Figure 7-16 illustrates a rotary actuator that uses a rack and pinion drive. Pressure applied to the tube chamber on one side moves the rack to the opposite chamber, causing the shaft to rotate. The shaft is double-ended, giving either clockwise or counter-clockwise rotation, depending on which end is driving. This type of rack and pinion rotary actuator is available to 1 million in. · lb of torque and can provide rotations up to 360°.

Still a third type of rotary actuator is illustrated in Fig. 7-17. Looking into the driver end, counterclockwise rotation is obtained with pressure applied on the driver end port. Clockwise rotation results when pressure is applied to the idler end port. Using simple valving, the piston and exclusive helix design allows the actuator to be stopped at any point in the rotation cycle, where it will hold indefinitely. The work load is thus firmly held and cannot back off under reverse tension, shock, or vibration—even if a complete power loss occurs. This type of rotary actuator will deliver controlled torque from near 0 to 15,000 in. · lb with pressures of 5 to 300 psi.

As shown in Fig. 7-18, applications for rotary actuators include conveyor sorting, valve turning, air bending operations, flipover between work stations, positioning for welding, lifting, rotating, and dumping. The symbol for a rotary actuator is shown at the lower right-hand side of Fig. 7-18.

7.4 GEAR MOTORS

Hydraulic motors are actuators that can rotate continuously and as such have the same basic configuration as pumps. However, instead of pushing on the fluid as pumps do, motors are pushed upon by the fluid. In this way, hydraulic motors develop torque and produce continuous rotary motion. Since the casing of a hydraulic motor is pressurized from an outside source, most hydraulic motors have casing drains to protect shaft seals. There are three basic types of hydraulic motors: gear, vane, and piston. Let's first examine the operation and configuration of the gear motor.

A gear motor develops torque due to hydraulic pressure acting on the surfaces of the gear teeth, as illustrated in Fig. 7-19. The direction of rotation of the

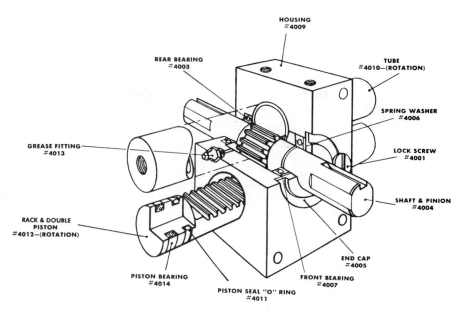

Figure 7-16. Rack and pinion drive rotary actuator. (*Courtesy of Carter Controls, Inc., Lansing, Illinois.*)

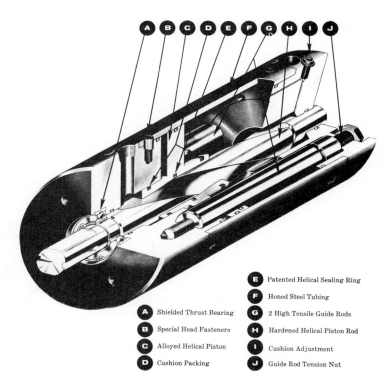

A — Shielded Thrust Bearing
B — Special Head Fasteners
C — Alloyed Helical Piston
D — Cushion Packing
E — Patented Helical Sealing Ring
F — Honed Steel Tubing
G — 2 High Tensile Guide Rods
H — Hardened Helical Piston Rod
I — Cushion Adjustment
J — Guide Rod Tension Nut

Figure 7-17. Rotary actuator with helical piston and rod. (*Courtesy of Carter Controls, Inc., Lansing, Illinois.*)

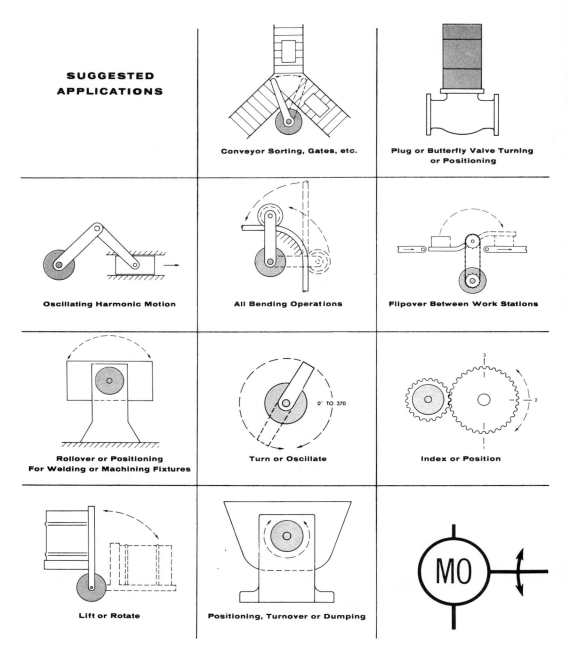

Figure 7-18.   Applications of rotary actuators. (*Courtesy of Carter Controls, Inc., Lansing, Illinois.*)

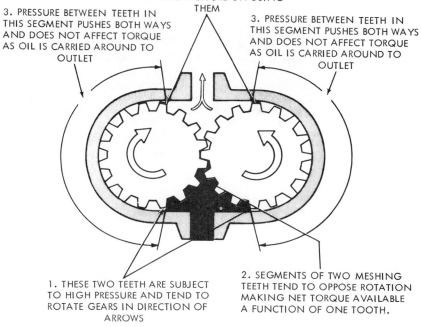

4. THESE TWO TEETH HAVE ONLY
TANK LINE PRESSURE OPPOSING
THEM

3. PRESSURE BETWEEN TEETH IN
THIS SEGMENT PUSHES BOTH WAYS
AND DOES NOT AFFECT TORQUE
AS OIL IS CARRIED AROUND TO
OUTLET

3. PRESSURE BETWEEN TEETH IN
THIS SEGMENT PUSHES BOTH WAYS
AND DOES NOT AFFECT TORQUE
AS OIL IS CARRIED AROUND TO
OUTLET

1. THESE TWO TEETH ARE SUBJECT
TO HIGH PRESSURE AND TEND TO
ROTATE GEARS IN DIRECTION OF
ARROWS

2. SEGMENTS OF TWO MESHING
TEETH TEND TO OPPOSE ROTATION
MAKING NET TORQUE AVAILABLE
A FUNCTION OF ONE TOOTH.

Figure 7-19. Torque development by a gear motor. (*Courtesy of Sperry Vickers, Sperry Rand Corp., Troy, Michigan.*)

Figure 7-20. External gear motor. (*Courtesy of Webster Electric Company, Inc., subsidiary of STA-RITE Industries, Inc., Racine, Wisconsin.*)

Figure 7-21. Screw motor. (*Courtesy of DeLaval, IMO Pump Division, Trenton, New Jersey.*)

motor can be reversed by reversing the direction of flow. As is the case for gear pumps, the volumetric displacement of a gear motor is fixed. The gear motor shown in Fig. 7-19 is not balanced with respect to pressure loads. The high pressure at the inlet, coupled with the low pressure at the outlet, produces a large side load on the shaft and bearings. Actual gear motors balance off this side load by internal passages and ports, which locate identical pressure conditions 180° apart. Gear motors are normally limited to 2000-psi operating pressures and 2400-rpm operating speeds. They are available with a maximum flow capacity of 150 gpm.

The main advantages of a gear motor are its simple design and subsequent low cost. Figure 7-20 shows a cutaway view of an actual gear motor. Also shown is the hydraulic symbol used in hydraulic circuits for representing fixed displacement motors.

Hydraulic motors can also be of the internal gear design. This type can operate at higher pressures and speeds and also has greater displacements than the external gear motor.

As in the case of pumps, screw-type hydraulic motors exist using three meshing screws (a power rotor and two idler rotors). Such a motor is illustrated in Fig. 7-21. The rolling screw set results in extremely quiet operation. Torque is developed by differential pressure acting on the thread area of the screw set. Motor torque is proportional to differential pressure across the screw set. This particular motor can operate at pressures up to 3000 psi and can possess volumetric displacements up to 13.9 in.[3]

7.5 VANE MOTORS

Vane motors develop torque by the hydraulic pressure acting on the exposed surfaces of the vanes, which slide in and out of the rotor connected to the drive shaft (see Fig. 7-22, view A). As the rotor revolves, the vanes follow the surface of

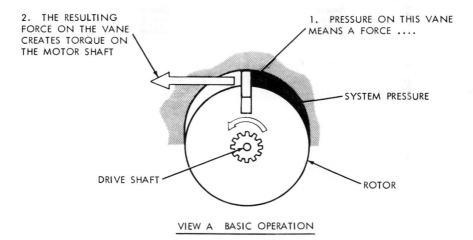

2. THE RESULTING
FORCE ON THE VANE
CREATES TORQUE ON
THE MOTOR SHAFT

1. PRESSURE ON THIS VANE
MEANS A FORCE ....

SYSTEM PRESSURE

DRIVE SHAFT

ROTOR

VIEW A   BASIC OPERATION

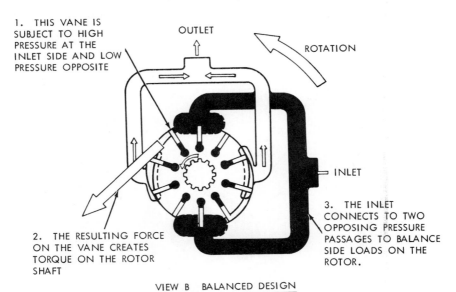

1. THIS VANE IS
SUBJECT TO HIGH
PRESSURE AT THE
INLET SIDE AND LOW
PRESSURE OPPOSITE

OUTLET

ROTATION

INLET

2. THE RESULTING FORCE
ON THE VANE CREATES
TORQUE ON THE ROTOR
SHAFT

3. THE INLET
CONNECTS TO TWO
OPPOSING PRESSURE
PASSAGES TO BALANCE
SIDE LOADS ON THE
ROTOR.

VIEW B   BALANCED DESIGN

Figure 7-22.   Operation of a vane motor. (*Courtesy of Sperry Vickers, Sperry Rand Corp., Troy, Michigan.*)

the cam ring because springs (not shown in Fig. 7-22) are used to force the vanes radially outward. No centrifugal force exists until the rotor starts to revolve. Therefore, the vanes must have some means other than centrifugal force to hold them against the cam ring. Some designs use springs, whereas other types use pressure-loaded vanes. The sliding action of the vanes forms sealed chambers, which carry the fluid from the inlet to the outlet.

Vane motors are universally of the balanced design illustrated in view B of Fig. 7-22. In this design, pressure buildup at either port is directed to two interconnected cavities located 180° apart. The side loads that are created are therefore canceled out. Since vane motors are hydraulically balanced, they are fixed displacement units.

Figure 7-23 shows a design where pivoted rocker arms are attached to the

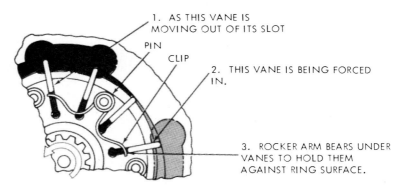

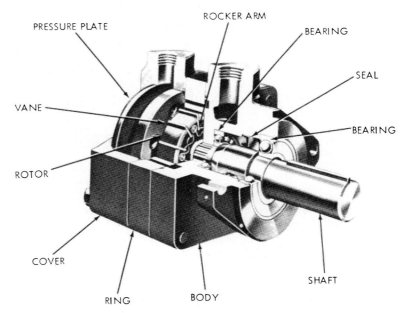

Figure 7-23.   Vane motors with spring-loaded vanes. (*Courtesy of Sperry Vickers, Sperry Rand Corp, Troy, Michigan.*)

rotor and serve as springs to force the vanes outward against the elliptical cam ring. This type of motor is available to operate at pressures up to 2500 psi and at speeds up to 4000 rpm. The maximum flow delivery is 250 gpm.

7.6 PISTON MOTORS

Piston motors can be either fixed or variable displacement units. They generate torque by pressure acting on the ends of pistons reciprocating inside a cylinder block. Figure 7-24 illustrates the in-line design in which the motor driveshaft and

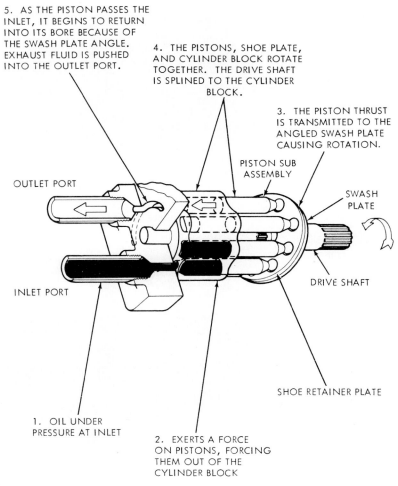

5. AS THE PISTON PASSES THE INLET, IT BEGINS TO RETURN INTO ITS BORE BECAUSE OF THE SWASH PLATE ANGLE. EXHAUST FLUID IS PUSHED INTO THE OUTLET PORT.

4. THE PISTONS, SHOE PLATE, AND CYLINDER BLOCK ROTATE TOGETHER. THE DRIVE SHAFT IS SPLINED TO THE CYLINDER BLOCK.

3. THE PISTON THRUST IS TRANSMITTED TO THE ANGLED SWASH PLATE CAUSING ROTATION.

PISTON SUB ASSEMBLY

OUTLET PORT

SWASH PLATE

INLET PORT

DRIVE SHAFT

SHOE RETAINER PLATE

1. OIL UNDER PRESSURE AT INLET

2. EXERTS A FORCE ON PISTONS, FORCING THEM OUT OF THE CYLINDER BLOCK

Figure 7-24.   In-line piston motor operation. (*Courtesy of Sperry Vickers, Sperry Rand Corp., Troy, Michigan.*)

cylinder block are centered on the same axis. Pressure acting on the ends of the pistons generates a force against an angled swash plate. This causes the cylinder block to rotate with a torque that is proportional to the area of the pistons. The torque is also a function of the swash plate angle. The in-line piston motor is designed either as a fixed or variable displacement unit (see Fig. 7-25). As illustrated in Fig. 7-26, the swash plate angle determines the volumetric displacement.

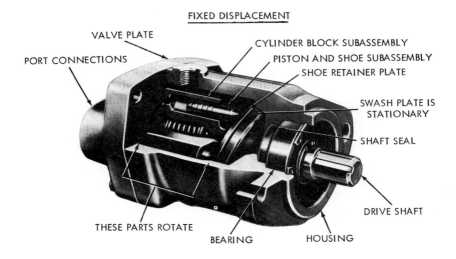

Figure 7-25. Two configurations of in-line piston motors. (*Courtesy of Sperry Vickers, Sperry Rand Corp., Troy, Michigan.*)

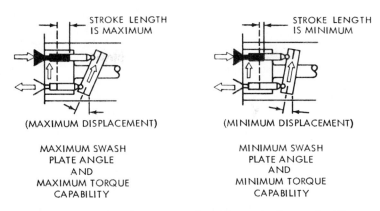

Figure 7-26. Motor displacement varies with swash plate angle. (*Courtesy of Sperry Vickers, Sperry Rand Corp., Troy, Michigan.*)

In variable displacement units, the swash plate is mounted in a swinging yoke. The angle of the swash plate can be altered by various means such as a lever, handwheel, or servo control. If the swash plate angle is increased, the torque capacity is increased, but the drive shaft speed is decreased. Mechanical stops are usually incorporated so that the torque and speed capacities stay within prescribed limits.

A bent-axis piston motor is illustrated in Fig. 7-27. This type of motor also develops torque due to pressure acting on reciprocating pistons. This design, however, has the cylinder block and driveshaft mounted at an angle to each other so that the force is exerted on the driveshaft flange.

Speed and torque depend on the angle between the cylinder block and driveshaft. The larger the angle, the greater the displacement and torque but the smaller the speed. This angle varies from a minimum of $7\frac{1}{2}°$ to a maximum of $30°$. Figure 7-28 shows a fixed displacement, bent-axis motor, whereas Fig. 7-29 illustrates the variable displacement design in which the displacement is varied by a handwheel.

Piston motors are the most efficient of the three basic types and are capable of operating at the highest speeds and pressures. Operating speeds of 12,000 rpm and pressures of 5000 psi can be obtained with piston motors. Large piston motors are capable of delivering flows up to 450 gpm.

A direct drive wheel hub motor is illustrated in Fig. 7-30. This type of motor imparts torque directly to drive wheels of vehicles such as tractors without any intermediate reduction gears. Designed to be mounted directly into a standard 15- or 20-in. wheel rim, these simplified power packages eliminate axles, gear boxes, torque converters, conventional hydrostatic transmissions, and reduction gears.

These wheel motors are of multistroke radial piston design, working against a cam ring. The special design of the cam permits full rated torque from start-up

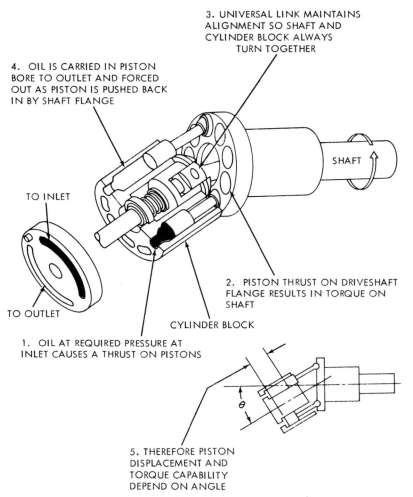

3. UNIVERSAL LINK MAINTAINS
ALIGNMENT SO SHAFT AND
CYLINDER BLOCK ALWAYS
TURN TOGETHER

4. OIL IS CARRIED IN PISTON
BORE TO OUTLET AND FORCED
OUT AS PISTON IS PUSHED BACK
IN BY SHAFT FLANGE

SHAFT

TO INLET

TO OUTLET

2. PISTON THRUST ON DRIVESHAFT
FLANGE RESULTS IN TORQUE ON
SHAFT

CYLINDER BLOCK

1. OIL AT REQUIRED PRESSURE AT
INLET CAUSES A THRUST ON PISTONS

θ

5. THEREFORE PISTON
DISPLACEMENT AND
TORQUE CAPABILITY
DEPEND ON ANGLE

Figure 7-27.   Bent-axis piston motor. (*Courtesy of Sperry Vickers, Sperry Rand Corp., Troy, Michigan.*)

through maximum rpm. In most applications, the inherent dynamic braking is sufficient. A secondary, static braking system is available, which provides fail-safe "holding" where such is required.

Other significant features of this wheel motor are instantaneous reversing through simply changing the direction of the oil flow; two-speed ranges, with full or half displacement; high external loading; operations at 5000 psi; low noise level; free wheeling; and ultrasmooth performance.

The theoretical torque capacity of a hydraulic motor can be determined by the following equation, which is identical to that used for hydraulic actuators:

Figure 7-28. Fixed displacement bent-axis piston motor. (*Courtesy of Sperry Vickers, Sperry Rand Corp., Troy, Michigan.*)

$$T = \frac{P \cdot V_D}{2\pi}$$

$$T = \frac{PV_D}{6.28} \qquad \qquad (7\text{-}11)$$

where does this come from?

Thus, the torque capacity is not only proportional to the pressure but also to the volumetric displacement.

The horsepower (*HP*) output can also be mathematically expressed:

$$HP = \frac{TN}{63,000} = \frac{PV_D N}{395,000} \qquad (7\text{-}12)$$

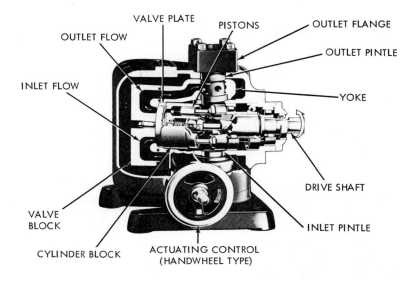

Figure 7-29. Variable displacement bent-axis piston motor with handwheel control. (*Courtesy of Sperry Vickers, Sperry Rand Corp., Troy, Michigan.*)

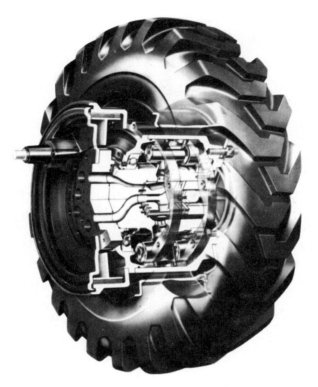

Figure 7-30.   Wheel hub motor. (*Courtesy of Bird-Johnson Co., Walpole, Massachusetts.*)

Also, as is the case for pumps, the following equation gives the relationship between speed, volumetric displacement, and flow:

$$Q = \frac{V_D N}{231} \qquad (7\text{-}13)$$

The use of Eqs. (7-11), (7-12), and (7-13) can be illustrated by an example problem.

EXAMPLE 7-4

A hydraulic motor has a 5-in.^{3} volumetric displacement. If it has a pressure rating of 1000 psi and it receives oil from a 10-gpm pump, find the motor

    a. Speed
    b. Torque capacity
    c. Horsepower capacity

Solution

a. From Eq. (7-13) we solve for motor speed:

$$N = \frac{231Q}{V_D} = \frac{(231)(10)}{5} = 462 \text{ rpm}$$

b. Torque capacity is found using Eq. (7-11):

$$T = \frac{V_D P}{6.28} = \frac{(5)(1000)}{6.28} = 795 \text{ in.} \cdot \text{lb}$$

c. Horsepower output is obtained from Eq. (7-12):

$$HP = \frac{TN}{63,000} = \frac{(795)(462)}{63,000} = 5.83 \text{ hp}$$

7.7 HYDRAULIC MOTOR PERFORMANCE

The performance of any hydraulic motor depends on the seal between the inlet and outlet sides. Internal leakage (slippage) between the inlet and outlet reduces efficiency.

Gear motors typically have an overall efficiency of 70% to 75% as compared to 75% to 85% for vane motors and 85% to 95% for piston motors.

Some systems require that a hydraulic motor start under load. Such systems should include a stall-torque factor when making design calculations. For example, only about 80% of the maximum torque can be expected if the motor is required to start either under load or operate at speeds below 500 rpm.

Hydraulic motor performance is evaluated on the same three efficiency parameters as used for hydraulic pumps. They are defined for motors as follows:

1. *Volumetric efficiency* (η_v):

$$\eta_v = \frac{\text{theoretical flow rate motor should consume}}{\text{actual flow rate consumed by motor}} \times 100 = \frac{Q_T}{Q_A} \times 100 \qquad (7\text{-}14)$$

Observe that volumetric efficiency for a motor is the inverse of that for a pump, because a pump does not produce as much flow as it theoretically should, whereas a motor uses more flow than it theoretically should due to slippage.

Determination of volumetric efficiency requires the calculation of the theoretical flow rate, which is defined for a motor by

$$Q_T = \frac{V_D N}{231} \qquad (7\text{-}15)$$

2. *Mechanical efficiency* (η_m):

$$\eta_m = \frac{\text{actual torque delivered by motor}}{\text{torque motor should theoretically deliver}} \times 100 = \frac{T_A}{T_T} \times 100 \qquad (7\text{-}16)$$

Equations (7-17) and (7-18) allow for the calculation of T_T and T_A, respectively:

$$T_T(\text{in.} \cdot \text{lb}) = \frac{V_D(\text{in.}^3) \cdot P \text{ (psi)}}{6.28} \qquad (7\text{-}17)$$

$$T_A(\text{in.} \cdot \text{lb}) = \frac{\text{actual } HP \text{ delivered by motor} \times 63,000}{N(\text{rpm})} \qquad (7\text{-}18)$$

3. *Overall efficiency* (η_o):

$$\eta_o = \frac{\eta_v \eta_m}{100} \qquad (7\text{-}19)$$

$$\eta_o = \frac{\text{actual } HP \text{ delivered by motor}}{\text{actual } HP \text{ delivered to motor}} \times 100$$

$$= \frac{T_A N / 63,000}{P Q_A / 1714} \times 100 \qquad (7\text{-}20)$$

where P = measured pressure at motor inlet (psi)
N = measured motor speed (rpm)
T_A = measured output torque of motor (in. $\cdot$ lb)
Q_A = actual flow rate consumed by motor (gpm)

Figure 7-31 represents typical performance curves obtained for a 6-in.^{3} variable displacement motor operating at full displacement. The upper graph gives curves of overall and volumetric efficiencies as a function of motor speed (rpm) for pressure levels of 3000 and 5000 psi. The lower graph gives curves of motor input flow (gpm) and motor output torque (in. $\cdot$ lb) as a function of motor speed (rpm) for the same two pressure levels.

EXAMPLE 7-5

A hydraulic motor has a displacement of 10 in.^{3} and operates with a pressure of 1000 psi and a speed of 2000 rpm. If the actual flow rate consumed by the motor is 95 gpm and the actual torque delivered by the motor is 1500 in. $\cdot$ lb, find

    a. η_v
    b. η_m

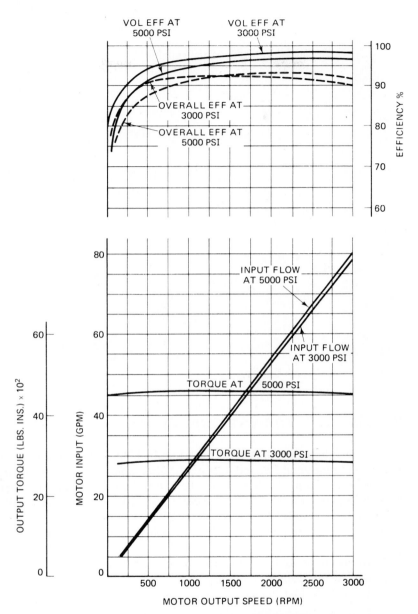

Figure 7-31.   Performance curves for 6-in.³ variable displacement motor. (*Courtesy of Abex Corp., Denison Division, Columbus, Ohio.*)

c. η_o

d. The actual horsepower delivered by the motor

Solution

a. First calculate the theoretical flow rate:

$$Q_T = \frac{V_D N}{231} = \frac{(10)(2000)}{231} = 86.6 \text{ gpm}$$

$$\eta_v = \frac{Q_T}{Q_A} \times 100 = \frac{86.6}{95} \times 100 = 91.1\%$$

b. To find η_m, we need to calculate the theoretical torque:

$$T_T = \frac{V_D P}{6.28} = \frac{(10)(1000)}{6.28} = 1592 \text{ in.} \cdot \text{lb}$$

$$\eta_m = \frac{T_A}{T_T} \times 100 = \frac{1500}{1592} \times 100 = 94.2\%$$

c.

$$\eta_o = \frac{\eta_v \eta_m}{100} = \frac{91.1 \times 94.2}{100} = 85.8\%$$

d.

$$HP = \frac{T_A N}{63,000} = \frac{(1500)(2000)}{63,000} = 47.6 \text{ hp}$$

7.8 HYDROSTATIC TRANSMISSIONS

A system consisting of a hydraulic pump, a hydraulic motor, and appropriate valves and pipes can be used to provide adjustable speed drives for many practical applications. Such a system is called a hydrostatic transmission. There, of course, must be a prime mover such as an electric motor or gasoline engine. Applications in existence include tractors, rollers, front-end loaders, hoes, and lift trucks. Some of the advantages of hydrostatic transmissions are the following:

1. Infinitely variable speed and torque in either direction and over the full speed and torque ranges.
2. Extremely high horsepower-to-weight ratio.
3. Can be stalled without damage.
4. Low inertia of rotating members permits fast starting and stopping with smoothness and precision.
5. Flexibility and simplicity of design.

Figure 7-32 illustrates the internal features of a variable displacement piston pump and a fixed displacement piston motor used in a heavy-duty hydrostatic transmission. Both pump and motor are of the swash plate in-line piston design. This type of hydrostatic transmission is expressly designed for application in the agricultural, construction, materials-handling, garden tractor, recreational vehicle, and industrial markets.

For the transmission of Fig. 7-32, the operator has complete control of the system, with one lever for starting, stopping, forward motion, or reverse motion. Control of the variable displacement pump is the key to controlling the vehicle. Prime mover horsepower is transmitted to the pump. When the operator moves the control lever, the swash plate in the pump is tilted from neutral. When the pump swash plate is tilted, a positive stroke of the pistons occurs. This, in turn, at any given input speed, produces a certain flow from the pump. This flow is transferred through high-pressure lines to the motor. The ratio of the volume of flow from the pump to the displacement of the motor determines the speed at which the motor will run. Moving the control lever to the opposite side of neutral causes the flow through the pump to reverse its direction. This reverses the direction of rotation of the motor. Speed of the output shaft is controlled by adjusting the displacement (flow) of the pump. Load (working pressure) is determined by the external conditions (grade, ground conditions, etc.), and this establishes the demand on the system.

Figure 7-33 illustrates the plumbing of the entire system for the hydrostatic transmission. The shutoff valve is included to facilitate a filter change without a large loss of fluid from the reservoir. The heat exchanger ensures that the maximum continuous oil temperature will not exceed 180°F.

EXAMPLE 7-6

A hydrostatic transmission, operating at 1000-psi pressure, has the following characteristics:

Pump	*Motor*
$V_D = 5$ in.^{3}	$V_D = ?$
$\eta_v = 82\%$	$\eta_v = 92\%$
$\eta_m = 88\%$	$\eta_m = 90\%$
$N = 500$ rpm	$N = 400$ rpm

Find the

a. Displacement of the motor
b. Motor output torque

FIXED DISPLACEMENT MOTOR

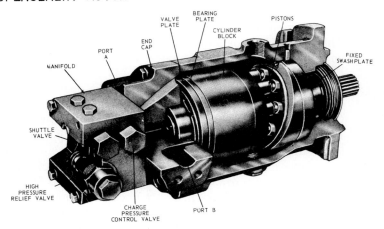

VARIABLE DISPLACEMENT PUMP

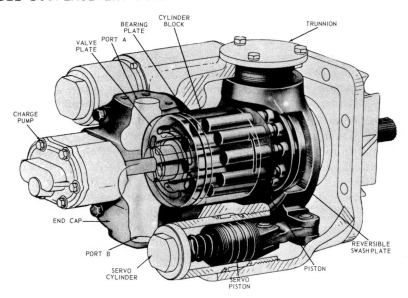

Figure 7-32. Pump and motor of a heavy-duty hydrostatic transmission. (*Courtesy of Sundstrand Hydro-Transmission Division, Sundstrand Corp., Ames, Iowa.*)

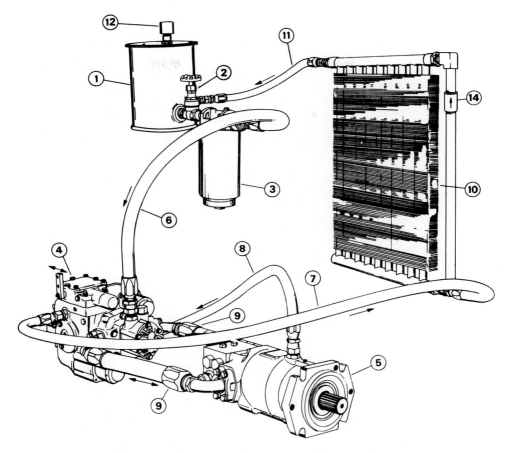

1. Reservoir
2. Shut-off Valve
3. Filter
4. Variable Displacement
   Pump
5. Fixed Displacement Motor
6. Inlet Line
7. Pump Case Drain Line
8. Motor Case Drain Line
9. High Pressure Lines
10. Heat Exchanger
11. Reservoir Return Line
12. Reservoir Fill Cap or
    Breather
14. Heat Exchanger
    By-pass Valve

Figure 7-33.   Plumbing installation of hydrostatic transmission system. (*Courtesy of Sundstrand Hydro-Transmission Division, Sundstrand Corp., Ames, Iowa.*)

Solution

a. pump theoretical flow rate $= \dfrac{\text{displacement of pump} \times \text{pump speed}}{231}$

$$= \frac{(5)(500)}{231} = 10.8 \text{ gpm}$$

pump actual flow rate $=$ pump theoretical flow rate

$\times$ pump volumetric efficiency

$$= (10.8)(0.82) = 8.86 \text{ gpm}$$

motor theoretical flow rate $=$ pump actual flow rate

$\times$ motor volumetric efficiency

$$= (8.86)(0.92) = 8.15 \text{ gpm}$$

motor displacement $= \dfrac{\text{motor theoretical flow rate} \times 231}{\text{motor speed}}$

$$= \frac{(8.15)(231)}{400} = 4.71 \text{ in.}^3$$

b.　　　HP delivered to motor $= \dfrac{\text{system pressure} \times \text{actual flow rate to motor}}{1714}$

$$= \frac{(1000)(8.86)}{1714} = 5.17 \text{ hp}$$

HP delivered by motor $= 5.17 \times 0.92 \times 0.90 = 4.28$

torque delivered by motor $= \dfrac{HP \text{ delivered by motor} \times 63{,}000}{\text{motor speed}}$

$$= \frac{4.28 \times 63{,}000}{400} = 674 \text{ in.} \cdot \text{lb}$$

7.9 *ELECTROHYDRAULIC STEPPING MOTORS*

An electrohydraulic stepper (or stepping) motor (EHSM) is a device that uses a small electrical stepper motor to control the huge power available from a hydraulic motor. The electrohydraulic stepper motor consists of three components: electrical stepper motor, hydraulic servo valve, and hydraulic motor (see Fig. 7-34). These three independent components, when integrated in a particular fashion, provide for the hydraulic motor to accurately follow the electrical stepper but with

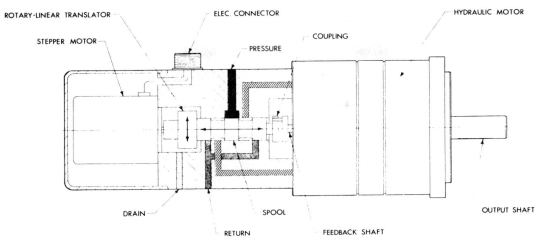

Figure 7-34.   Schematic drawing of electrohydraulic stepper motor (EHSM). (*Courtesy of Motion Products, Minneapolis, Minnesota.*)

a torque output that is several hundred times greater than the capabilities of the electrical stepper.

The electric stepper motor rotates a precise, fixed amount per each electrical pulse received. This motor is directly coupled to the rotary linear translator of the servo valve. The output torque of the electric motor must be capable of overcoming the flow forces in the servo valve. The flow forces in the servo valve are directly proportional to the rate of flow through the valve. Figure 7-35(a) gives the axial force on the valve spool for various flows. The torque required to operate the rotary linear translator against this axial force is dependent on the flow gain in the servo valve. Figure 7-35(b) gives the torque required for the operation of the servo

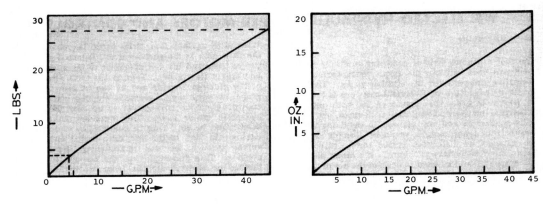

Figure 7-35.   Characteristics of electrohydraulic stepper motor (EHSM). (*Courtesy of Motion Products, Minneapolis, Minnesota.*)

valve at various flow rates. This is the torque demanded of the electric stepper, and as can be seen from the curve, a very small electric stepper can produce enormous output horsepower to be available at the hydraulic motor shaft. The electric motor is run submerged in oil. This is achieved by diverting the casing drain flow of the hydraulic motor and servo valve through the housing surrounding the electric stepper. This technique has the advantage of protecting the electric stepper from overheating due to the ability of the oil to act as a very efficient heat sink.

The mechanical input servo valve performs the following two functions:

1. The hydraulic motor comes to rest only when there is no positional error between the electric stepper and the hydraulic motor. Hence, the servo valve makes the hydraulic motor reproduce exactly the position of the electric stepper. A closed-loop circuit exists between the electric stepper and the hydraulic motor, through the servo valve. The hydraulic motor exactly reproduces the position of the electrical stepper, ignoring any minor mechanical imperfections.

2. The amount of oil flow through the servo valve is such that it tends to cancel out the lag between the electric motor and the hydraulic motor. Thus, the speed and direction of rotation of the hydraulic motor would always try to reproduce the motion of the electric motor. The flow through the servo valve is directly proportional to the lag between the electric and hydraulic motors. Thus, as the lag between the electric stepper and the hydraulic motor increases, this causes additional flow to pass through the servo valve, which, in turn, speeds up the hydraulic motor. The hydraulic motor is faithfully trying to reproduce the position of the electric stepper, and the servo valve provides the feedback through mechanical linkage.

The hydraulic motor is the most important component of the EHSM package. The performance characteristics of the hydraulic motor determine the performance that can be achieved from the whole package. The hydraulic motor has very desirable characteristics for operation under these conditions. Among these are high starting torque, excellent reliability, low-speed operating capability, and high overall efficiency. The motor is a fixed clearance, axial rolling vane motor and has a balanced rotor design, as shown in Fig. 7-36.

Figure 7-37(a) shows an actual electrohydraulic stepper motor and Figure 7-37(b) shows an electronic control system called a preset indexer. This electronic control system performs the counting function, acceleration-deceleration control, and the general control interface between the motor, machine, and operator. The output shaft speed of the hydraulic motor is directly proportional to the input pulse rate to the electric stepper motor. This rate is controlled by the preset indexer. The position of the output shaft is directly related to the number of input pulses received by the electric stepper motor.

Electrohydraulic stepper motors are typically used for precision control of position and speed. They are available with displacements from 0.4 to 7 in.^{3}

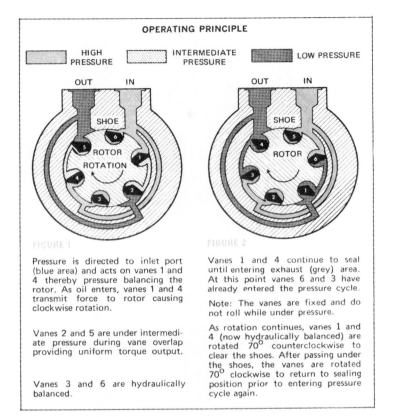

DESIGN ADVANTAGES

HIGH STARTING TORQUE

Low internal friction allows a starting torque actually greater than running torque. Motor will hold full torque at stall without overheating.

RELIABILITY

Low internal friction also reduces wear, providing long, trouble free operation. With proper installation and care, many thousands of hours of life can be expected.

LOW SPEED

Combination of fixed clearance and hydraulically balanced design features reduce stick-slip to minimum, for steady rotation at near stall speeds.

GENERAL NOTES

1. Moving parts do not touch metal to metal but maintain a small, fixed clearance.
2. Rotor is hydraulically balanced throughout complete revolution.
3. Vanes are rotated only when hydraulically balanced.
4. Rotation is reversible with no changes to the motor.

HIGH EFFICIENCIES

Volumetric	96-98%
Torque (Stall)	99%
(Running)	95%
Overall	90%

OPERATING PRINCIPLE

HIGH PRESSURE   INTERMEDIATE PRESSURE   LOW PRESSURE

FIGURE 1

Pressure is directed to inlet port (blue area) and acts on vanes 1 and 4 thereby pressure balancing the rotor. As oil enters, vanes 1 and 4 transmit force to rotor causing clockwise rotation.

Vanes 2 and 5 are under intermediate pressure during vane overlap providing uniform torque output.

Vanes 3 and 6 are hydraulically balanced.

FIGURE 2

Vanes 1 and 4 continue to seal until entering exhaust (grey) area. At this point vanes 6 and 3 have already entered the pressure cycle.

Note: The vanes are fixed and do not roll while under pressure.

As rotation continues, vanes 1 and 4 (now hydraulically balanced) are rotated 70° counterclockwise to clear the shoes. After passing under the shoes, the vanes are rotated 70° clockwise to return to sealing position prior to entering pressure cycle again.

Figure 7-36.   Working principle of hydraulic motor. (*Courtesy of Motion Products, Minneapolis, Minnesota.*)

Horsepower capabilities run from 3.5 to 35 hp. Typical applications include textile drives, paper mills, roll feeds, automatic storage systems, machine tools, conveyor drives, hoists, and elevators.

In Figure 7-38 we see a cutaway view of an electrohydraulic stepping motor. This particular motor produces a rotation of 1.8°/pulse and can accept pulses at a rate of 2500 pulses/sec.

7.10 LOW-SPEED, HIGH-TORQUE MOTORS

In many applications, there is a need for low speed accompanied by high-torque output. There are special hydraulic motors available specifically designed to meet these conditions. Figure 7-39 shows a cutaway view of such a motor, which

(a)

(b)

Figure 7-37. Electrohydraulic stepper motor with electronic control system. (*Courtesy of Motion Products, Minneapolis, Minnesota.*)

contains five cylinders. It is designed to operate at speeds as low as 3 rpm with no need for cumbersome, power-wasting speed reducers. This type of motor can generate up to 192,000 in. · lb of torque in continuous operation and can be reversed and stalled without damage.

Figure 7-38. Cutaway of an electrohydraulic stepping motor. (*Courtesy of Bird-Johnson Co., Walpole, Massachusetts.*)

The secret to the operation of this type of motor (called Staffa by its manufacturer) is illustrated in Fig. 7-40, along with specifications for this motor, which is applicable to the B series of 5, 7, or 10 cylinders. With its machined tolerances so closely held, and because it is completely balanced in a film of oil, a Staffa motor has an overall efficiency as high as 98%. Typical applications for this type of motor include machine tools, vibratory rollers, swing drive of log loaders, and underground coal mining machinery.

Figure 7-39. Low-speed, high-torque motor. (*Courtesy of Brown & Sharpe Mfg. Co., Manchester, Michigan.*)

MODEL	MAX. PRESSURE (PSI)		DISPLACE-MENT		TORQUE (FT-LBS)		SPEED	HORSE-POWER
	Cont.	Inter-mittent	Cu.-In.	Gal.	Max. Cont.	Starting	Max. RPM	Max. Contin.
B10	3000	3500	11.5	.050	440	410	500	33
B30	3000	3500	27	.117	1,020	1,050	450	50
B45	3000	3500	45	.195	1,700	1,700	400	80
B80	3000	4000	82	.355	3,150	3,680	300	130
B125	3000	4000	125	.541	4,670	5,300	220	100
B150	3000	3500	151	.653	5,850	6,240	220	120
B200	3000	3500	188	.812	7,350	8,500	175	125
B270	3000	3500	263	1.138	10,250	11,600	125	170
B400	3000	3500	415	1.796	16,000	18,400	100	225

All models have 5 cylinders, except B270, which has 7, and B400, which has 10. All models have standard continuous pressure ratings of 3,000 PSI, but this and other specs can vary with modifications available within the range of a single model number.

HYDRAULICALLY BALANCED VALVE

The rotary valve has large inlet (high pressure) and discharge (low pressure) slots to direct oil to and from the cylinders. To offset the side thrust that these unequal pressure areas would cause, the Staffa design provides a total counterbalancing effect: Oil from each large slot is allowed access to two small slots on the opposite side, one at each end. This keeps the valve centered with no frictional or efficiency loss.

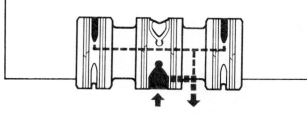

BALANCED PISTON ASSEMBLY

Staffa motors have greater starting torque and wear less than all other motors because there is *no metal-to-metal* contact between the piston assembly and the crankshaft. Hydraulic oil is metered down through the assembly to a pocket under the connecting rod, where it bleeds out under the shoe. The presence of this oil creates an upward hydraulic force equal to the main downward force. This results in the creation of a hydrostatic "float" for the piston assembly.

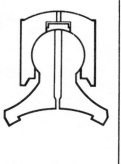

Figure 7-40. Operational features of low-speed, high-torque motor. (*Courtesy of Brown & Sharpe Mfg. Co., Manchester, Michigan.*)

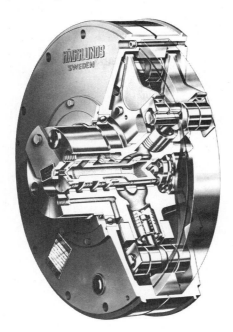

Figure 7-41.   Radial piston low-speed, high-torque motor. (*Courtesy of Bird-Johnson Co., Walpole, Massachusetts.*)

In Fig. 7-41 we see a photograph showing a cutaway view of a second type of low-speed, high-torque hydraulic motor, which is capable of delivering torques up to 1,108,000 in. · lb. This motor is of radial piston design and offers high power in a narrow profile. The hub remains stationary as pairs of opposing pistons turn the rotating casing. Each pair of pistons is connected to rollers, which ride a specially designed cam ring. A rotary valve distributes oil to pairs of pistons in sequence, and they, in turn, produce a smooth, nonpulsating rotation, with torque being constant throughout the full 360°, at all speeds and all operating pressures (see Fig. 7-42).

Direction of rotation is determined by feeding oil to one or the other of two alternative ports. By fitting a special valve to the motor, two speed ranges are possible using the same rate of inlet flow. The speed is doubled, whereas the output torque is halved in the higher range.

A key feature of the motors is their ability to develop full operational torque when starting and operating at very low speeds. Full torque at fractional rpm can be attained.

Special roller bearings prevent side loading of the piston; hydraulic and dynamic balance reduces the thrust on the heavy-duty main bearings. This design permits much higher external loading than other concepts. Freewheeling, in which the housing rotates freely, may be accomplished by retracting the pistons, thus eliminating pressured oil circulation through the rotary valve.

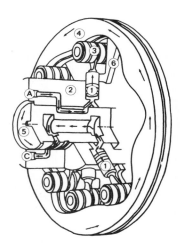

The pistons (1) are contained in a stationary cylinder housing (2) and are connected to rollers (3) which bear on the cam ring (4). The rotary valve (5) distributes oil from the inlet (A) to each cylinder in sequence. The oil pressure displaces the pistons outward, driving the rollers over the cam ring, causing it and the outer casing of the motor to rotate. Fixed side guides (6) absorb reaction forces on the rollers, thus unloading the pistons from any side (tangential) forces, and eliminating this source of cylinder wear. Oil escapes from the motor through outlet (C). An even number of opposed pistons assures hydraulic balance and unloads the main bearing from any piston forces.

Connection of an integral two-speed (4-way) valve to the motor provides a choice of two speed ranges. In the higher speed range, the torque is halved. In this arrangement two oil inlet lines are provided, each of which communicates with half the number of pistons.

Free wheeling can be accomplished, by pressurizing the case and retracting the piston/roller assemblies, permitting the housing to rotate freely without any circulation of hydraulic fluid.

Figure 7-42. Operating principles of low-speed, high-torque motor. (*Courtesy of Bird-Johnson Co., Walpole, Massachusetts.*)

7.11 HYDRAULIC MOTOR PERFORMANCE RATING IN METRIC UNITS

As in the case for pumps, performance data for hydraulic motors are measured and specified in metric units as well as English units. Figure 7-43 shows actual performance and dimensional data for Vickers M2 series high performance motors designed for use in mobil equipment. The performance of these motors has been proven via rugged earth-moving, agricultural, mining, material-handling applications, and with hydrostatic transmissions.

Examples 7-7, 7-8, and 7-9 show how to analyze the performance of hydraulic motors and hydrostatic transmissions using metric units.

EXAMPLE 7-7

A hydraulic motor has an 82 cm³ (0.082 liter) volumetric displacement. If it has a pressure rating of 70 bars and it receives oil from a 0.0006 m³/s (0.60 ℓ/s or 36.0 ℓ/min) pump, find the motor

    a. speed

    b. torque capacity

    c. power capacity

M2 Series High Performance Motors

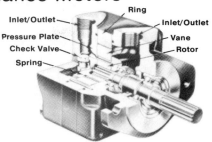

Characteristics. These motors are an economical, efficient, and compact means of converting hydraulic energy to mechanical energy. They can be stalled under load without damage when protected by a relief valve.

Hydraulic Balance. Two internal areas are diametrically opposed as are two outlet areas. This construction eliminates bearing loads resulting from pressure, a major cause of shaft bearing wear in designs without this feature.

Efficiency. Optimum running clearances and hydraulic balance assure sustained high efficiency over the life of the motor.

Smooth Operation. Inertia of rotating parts is low . . . parts are symmetrical, providing dynamic balance and freedom from vibration.

- Speeds to 2800 RPM
- Pressures to 2000 PSI (138 bar)
- Choice of shafts, port positions and mountings

Model Series	Torque lb. in. (Nm) 100 psi (7 bar)	Displacement cu. in./rev.	Max. Torque lb. in. (Nm) at Max. Pressure	Max. Speeds and Pressures	DIMENSIONS: INCHES (MMS)			Approx. Weight lbs. (kgs)
					●Length	Width	Height	
M2-200*	25 (34) 35 (47)	1.51 2.16	410 (556) 500 (678)	2200 rpm @ 2000 psi (138 bar) 1800 rpm @ 1750 psi (121 bar)	5.9 (149.9)	4.2 (106.7)	4.8 (121.9)	21 (10)
M2U	18 (24) 25 (34) 35 (47)	1.32 1.55 2.29	340 (461) 410 (556) 600 (813)	2800 rpm @ 2000 psi (138 bar) 2500 rpm @ 2000 psi (138 bar) 2000 rpm @ 2000 psi (138 bar)	5.1 (129.5) 5.4 (137.2) 5.6 (142.2)	4.1 (104.1)	4.9 (124.5)	17 (8)

* Shaft rotation is reversible in this series.   ● From face of mounting flange to end of motor cover.

Figure 7-43. Performance and dimensional English/metric data for Vickers M2 series high performance motors. (*Courtesy of Vickers Inc., Troy, Michigan.*)

Solution

a. From Eq. (7-13) we solve for the motor speed.

$$N = \frac{Q}{V_D} = \frac{0.0006 \text{ m}^3/\text{s}}{0.000082 \text{ m}^3/\text{rev}} = 7.32 \text{ rev/s} = 439 \text{ rpm}$$

b. Torque capacity is found using Eq. (7-11).

$$T = \frac{V_D P}{6.28} = \frac{(0.000082 \text{ m}^3)(70 \times 10^5 \text{N/m}^2)}{6.28} = 91.4 \text{ N} \cdot \text{m}$$

c. Power output is obtained as follows:

$$\text{Power} = TN = (91.4\text{N} \cdot \text{m})(7.32 \times 2\pi \text{ radians/s})$$

$$= 4200 \text{ watts} = 4.20 \text{ kW}$$

Example 7-8

A hydraulic motor has a displacement of 164 cm³ and operates with a pressure of 70 bars and a speed of 2000 rpm. If the actual flow rate consumed by the motor is 0.006

m³/s and the actual torque delivered by the motor is 170 N · m, find

    a. η_v
    b. η_m
    c. η_o
    d. The actual kW delivered by the motor

Solution

    a. First calculate the theoretical flow rate.

$$Q = V_D N = (0.000164 \text{ m}^3/\text{rev}) \left(\frac{2000}{60} \text{ rev/s}\right) = 0.00547 \text{ m}^3/\text{s}$$

$$\eta_v = \frac{Q_T}{Q_A} \times 100 = \frac{0.00547}{0.006} \times 100 = 91.2\%$$

    b. To find η_m, we need to calculate the theoretical torque.

$$T_T = \frac{V_D P}{6.28} = \frac{(0.000164)(70 \times 10^5)}{6.28} = 182.8 \text{ N} \cdot \text{m}$$

$$\eta_m = \frac{T_A}{T_T} \times 100 = \frac{170}{182.8} \times 100 = 93.0\%$$

    c.

$$\eta_o = \frac{\eta_v \eta_m}{100} = \frac{(91.2)(93.0)}{100} = 84.8\%$$

    d.   Power $= T_A N = (170) \left(2000 \times \frac{2\pi}{60}\right) = 35,600$ watts $= 35.6$ kW

EXAMPLE 7-9

A hydrostatic transmission, operating at 70 bars pressure, has the following characteristics:

	Pump	Motor
V_D	$= 82 \text{ cm}^3$	$= ?$
η_v	$= 82\%$	$= 92\%$
η_m	$= 88\%$	$= 90\%$
N	$= 500 \text{ rpm}$	$= 400 \text{ rpm}$

Find the

a. Displacement of the motor
b. Motor output torque

Solution

a. Pump theoretical flow rate = displacement of pump × pump speed

$$= (0.000082) \left(\frac{500}{60} \right) = 0.000683 \text{ m}^3/\text{s}$$

Pump actual flow rate = pump theoretical flow rate × pump volumetric efficiency

$$= (0.000683)(0.82) = 0.000560 \text{ m}^3/\text{s}$$

Motor theoretical flow rate = pump actual flow rate

$$\times \text{ motor volumetric efficiency}$$

$$= (0.000560)(0.92) = 0.000515 \text{ m}^3/\text{s}$$

$$\text{Motor displacement} = \frac{\text{motor theoretical flow rate}}{\text{motor speed}}$$

$$= \frac{0.000515}{400/60} = 0.0000773 \text{ m}^3 = 77.3 \text{ cm}^3$$

b. Power delivered to motor = system pressure × actual flow rate to motor

$$= (70 \times 10^5)(0.000560) = 3920 \text{ watts}$$

Power delivered by motor = (3920)(0.92)(0.90) = 3246 watts

$$\text{Torque delivered by motor} = \frac{\text{power delivered by motor}}{\text{motor speed}}$$

$$= \frac{3246}{400 \times 2\pi/60} = 77.5 \text{ N} \cdot \text{m}$$

EXERCISES

7-1. What is the difference between a single-acting and a double-acting hydraulic cylinder?

7-2. Name four different types of hydraulic cylinder mountings.

7-3. A pump supplies oil at 25 gpm to a $1\frac{1}{2}$-in.-diameter double-acting hydraulic cylinder. If the load is 1200 lb (extending and retracting) and the rod diameter is $\frac{3}{4}$ in., find the

(a) Hydraulic pressure during the extending stroke

(b) Piston velocity during the extending stroke

(c) Cylinder horsepower during the extending stroke

(d) Hydraulic pressure during the retracting stroke

(e) Piston velocity during the retracting stroke

(f) Cylinder horsepower during the retracting stroke

7-4. What is a cylinder cushion? What is its purpose?

7-5. A pump delivers oil at a rate of 20 gpm to the blank end of a 2-in.-diameter hydraulic cylinder as shown in Fig. 7-8. The piston contains a ¾-in.-diameter cushion plunger that is 1 in. long. The cylinder drives a 1000-lb load, which slides on a flat horizontal surface having a coefficient of friction equal to 0.15. The pump pressure relief valve setting equals 500 psi. Find the maximum pressure developed by the cushion.

7-6. What is a double-rod cylinder? When would it normally be used?

7-7. What is a telescoping rod cylinder? When would it normally be used?

7-8. What is a limited rotation hydraulic actuator? How does it differ from a hydraulic motor?

7-9. A rotary actuator has the following physical data:

  1. Outer radius of rotor = 0.4 in.

  2. Outer radius of vane = 1.25 in.

  3. Width of vane = 0.75 in.

    If the torque load is 750 in. · lb, what pressure must be developed to overcome the load?

7-10. Why are gear motors hydraulically balanced?

7-11. What are the main advantages of gear motors?

7-12. Why are vane motors fixed displacement units?

7-13. Name one way in which vane motors differ from vane pumps.

7-14. Can a piston pump be used as a piston motor?

7-15. A hydraulic motor has a 6-in.³ volumetric displacement. If it has a pressure rating of 2000 psi and receives oil from a 15 gpm pump, find the motor

  (a) Speed

  (b) Torque capacity

  (c) Horsepower capacity

7-16. For a hydraulic motor, define volumetric, mechanical, and overall efficiency.

7-17. Why does a hydraulic motor use more flow than it theoretically should?

7-18. What is a hydrostatic transmission? Name four advantages it typically possesses.

7-19. A hydraulic motor has a displacement of 8 in.³ and operates with a pressure of 1500 psi and a speed of 2000 rpm. If the actual flow rate consumed by the motor is 75 gpm and the actual torque delivered by the motor is 1800 in. · lb, find

  (a) η_v

  (b) η_m

  (c) η_o

  (d) The horsepower delivered by the motor

7-20. A hydrostatic transmission operating at 1500-psi pressure has the following characteristics:

Pump	Motor
$V_D = 6$ in.^{3}	$V_D = ?$
$\eta_v = 85\%$	$\eta_v = 94\%$
$\eta_m = 90\%$	$\eta_m = 92\%$
$N = 1000$ rpm	$N = 600$ rpm

Find the

(a) Displacement of the motor

(b) Motor output torque

7-21. What is an electrohydraulic stepping motor, and how does it work?

7-22. Explain why, theoretically, the torque output from a fixed displacement hydraulic motor operating at constant pressure is the same regardless of changes in speed.

7-23. Why does the rod of a double-acting cylinder retract at a greater velocity than it extends for the same input flow rate?

7-24. How are single-acting cylinders retracted?

7-25. A pump supplies oil at 0.0016 m³/s to a 40 mm diameter double-acting hydraulic cylinder. If the load is 5000 N (extending and retracting) and the rod diameter is 20 mm find the

a. Hydraulic pressure during the extending stroke

b. Piston velocity during the extending stroke

c. Cylinder kW power during the extending stroke

d. Hydraulic pressure during the retracting stroke

e. Piston velocity during the retracting stroke

f. Cylinder kW power during the retracting stroke

7-26. A rotary actuator has the following physical data:

1. Outer radius of rotor = 10 mm.

2. Outer radius of vane = 32 mm.

3. Width of vane = 20 mm.

If the torque load is 85 N · m, what pressure must be developed to overcome the load?

7-27. A hydraulic motor has a 100 cm³ volumetric displacement. If it has a pressure rating of 140 bars and receives oil from a 0.001 m³/s pump, find the motor

a. Speed

b. Torque capacity

c. kW power capacity

7-28. A hydraulic motor has a displacement of 130 cm³ and operates with a pressure of 105 bars and a speed of 2000 rpm. If the actual flow rate consumed by the motor is 0.005 m³/s and the actual torque delivered by the motor is 200 N · m, find

    a. η_v

    b. η_m

    c. η_o

    d. kW power delivered by the motor

7-29. A hydrostatic transmission operating at 105 bars pressure has the following characteristics:

Pump	Motor
$V_D = 100 \text{ cm}^3$	$V_D = \, ?$
$\eta_v = 85\%$	$\eta_v = 94\%$
$\eta_m = 90\%$	$\eta_m = 92\%$
$N = 1000$ rpm	$N = 600$ rpm

Find the

    a. Displacement of the motor

    b. Motor output torque

7-30. A hydraulic motor has a volumetric efficiency of 90% and operates at a speed of 1750 rpm and a pressure of 1000 psi. If the actual flow rate consumed by the motor is 75 gpm and the actual torque delivered by the motor is 1300 in. $\cdot$ lb, find the overall efficiency of the motor.

7-31. The torque output from a fixed displacement hydraulic motor operating at constant pressure is the same regardless of changes in speed—true or false? Explain your answer.

7-32. What determines the speed of a hydraulic actuator?

7-33. What determines the pump size required in a hydraulic system?

7-34. Determine the value of constants C_1 and C_2 in the following equations for the speed of a hydraulic cylinder.

$$v(\text{in./min}) = \frac{C_1 Q(\text{gpm})}{A(\text{in.}^2)} \qquad v(m/s) = \frac{C_2 Q(m^3/s)}{A(m^2)}$$

7-35. Define the displacement and torque ratings of a hydraulic motor.

7-36. Explain how the vanes are held in contact with the cam ring in high performance vane motors.

7-37. How is torque developed in an in-line-type piston motor?

7-38. If a hydraulic motor is pressure compensated, what is the effect of an increase in the working load?

7-39. What type of hydraulic motor is generally most efficient?

7-40. Knowing the displacement and speed of a hydraulic motor, how do you calculate the gpm flowing through it?

7-41. A hydraulic cylinder has a rod diameter equal to one-half the piston diameter. Determine the difference in load-carrying capacity between extension and retraction if the pressure is constant.

7-42. For the cylinder in Exercise 7-41, what would happen if the pressure were applied to both sides of the cylinder at the same time?

7-43. The pressure rating of the components in a hydraulic system is 1000 psi. The system contains a hydraulic motor to turn a 10-in. radius drum at 30 rpm to lift a 1000-lb weight W as shown in Fig. 7-44. Determine the flow rate in units of gpm and the horsepower of the motor.

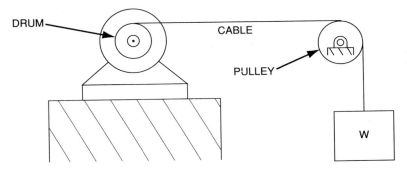

Figure 7-44.   System in Exercise 7-43.

7-44. The system of Exercise 7-43, as shown in Fig. 7-44, has the following data using metric units:

Pressure = 1×10^5 kPa
Drum radius = 0.3 m
Motor speed = 30 rpm
Weight of load = 4000 N

Determine the flow rate in units of m^3/s and the power of the motor in kW.

7-45. A hydraulic system contains a pump that discharges oil at 2000 psi and 100 gpm to a hydraulic motor as shown in Fig. 7-45. The pressure at the motor inlet is 1800 psi due to a pressure drop in the line. If oil leaves the motor at 200 psi, determine the power delivered by the motor.

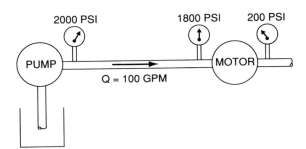

Figure 7-45.   System in Exercise 7-45.

7-46. If the pipeline between the pump and motor in Exercise 7-45 is horizontal and of constant diameter, what is the cause of the 200 psi pressure drop?

7-47. What theoretical torque could a 4-HP hydraulic motor deliver at a rated speed of 1750 rpm?

7-48. In Exercise 7-47, the pressure remains constant at 2000 psi.

a. What effect would doubling the speed have on torque?

b. What effect would halving the speed have on torque?

7-49. A gear motor has an overall efficiency of 84% at a pressure drop of 3000 psi across its ports and when the ratio of flow rate to speed is 0.075 gpm/rpm. Determine the torque and displacement of the motor.

8

Control Components in Hydraulic Systems

8.1 INTRODUCTION

One of the most important considerations in any fluid power system is control. If control components are not properly selected, the entire system will not function as required. Fluid power is controlled primarily through the use of control devices called valves. The selection of these control devices not only involves the type but also the size, the actuating technique, and remote-control capability. There are three basic types of control devices: (1) directional control valves, (2) pressure control valves, and (3) flow control valves.

Directional control valves determine the path through which a fluid traverses within a given circuit. For example, they establish the direction of motion of a hydraulic cylinder or motor. This control of the fluid path is accomplished primarily by check valves, shuttle valves, and two-way, three-way, and four-way directional control valves.

Pressure control valves protect the system against overpressure, which may occur due to a gradual buildup as fluid demand decreases or due to a sudden surge as valves open or close. The gradual buildup of pressure is controlled by pressure relief, pressure reducing, sequence, unloading, and counterbalance valves. Of course, pressure-compensated pumps can also be used. Pressure surges can produce an instantaneous increase in pressure as much as four times the normal system pressure. Shock absorbers are hydraulic devices designed to smooth out pressure surges and to dampen out hydraulic shock.

In addition, fluid flow rate must be controlled in various lines of a hydraulic circuit. For example, the control of actuator speeds depends on flow rates. This type of control is accomplished through the use of flow control valves. A variable displacement pump can also be used to control actuator speed unless the system contains several actuators, each of which must operate at different speeds. In such a case separate flow control valves are required. Noncompensated flow control valves are used where precise speed control is not required since flow rate varies with pressure drop across a flow control valve. Pressure-compensated flow control valves automatically adjust to changes in pressure drop to produce a constant flow rate.

It is important to know the primary function and operation of the various types of control components. This type of knowledge is not only required for a good functioning system, but it also leads to the discovery of innovative ways to improve a fluid power system for a given application. This is one of the biggest challenges facing the fluid power system designer. Circuit applications of each of the control devices discussed in this chapter are presented in Chapter 9.

8.2 DIRECTIONAL CONTROL VALVES

As the name implies, directional control valves are used to control the direction of flow in a hydraulic circuit. The simplest type is a check valve (see Fig. 8-1), which is in reality a one-way directional control valve. It is a one-way valve because it permits free flow in one direction and prevents any flow in the opposite direction.

Figure 8-2 shows the internal operation of a check valve. As shown, a light spring holds the poppet in the closed position. In the free-flow direction, the fluid pressure overcomes the spring force at about 5 psi. If flow is attempted in the opposite direction, the fluid pressure pushes the poppet (along with the spring force) in the closed position. Therefore, no flow is permitted. The higher the pressure, the greater will be the force pushing the poppet against its seat. Thus, increased pressure will not result in any tendency to allow flow in the no-flow direction.

Figure 8-3 shows the free-flow direction implied when using the symbolic

Figure 8-1.   Check valve. (*Courtesy of Sperry Vickers, Sperry Rand Corp., Troy, Michigan.*)

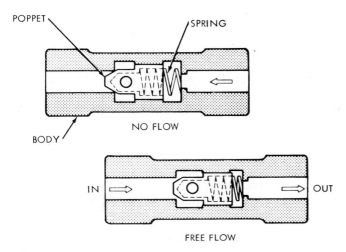

Figure 8-2. Operation of check valve. (*Courtesy of Sperry Vickers, Sperry Rand Corp., Troy, Michigan.*)

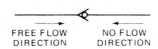

Figure 8-3. Symbol for a check valve with its free-flow direction defined.

representation of a check valve. This symbol, which clearly shows the function of a check valve, will be used when drawing total hydraulic circuits. Note that a check valve is analogous to a diode in electric circuits.

A second type of check valve is the pilot-operated check valve, shown in Fig. 8-4 with its symbol. This type of check valve always permits free flow in one

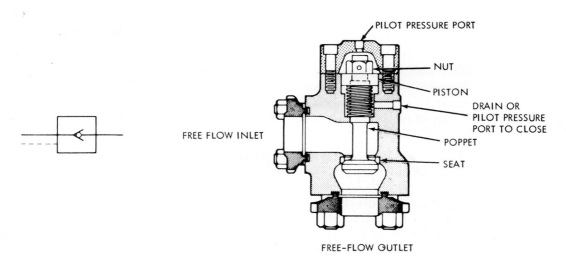

Figure 8-4. Pilot-operated check valve. (*Courtesy of Sperry Vickers, Sperry Rand Corp., Troy, Michigan.*)

direction but permits flow in the normally blocked opposite direction only if pilot pressure is applied at the pilot pressure port of the valve. In the design of Fig. 8-4, the check valve poppet has the pilot piston attached to the threaded poppet stem by a nut. The light spring holds the poppet seated in a no-flow condition by pushing against the pilot piston. The purpose of the separate drain port is to prevent oil from creating a pressure buildup on the bottom of the piston. The dashed line (which is part of the symbol shown in Fig. 8-4) represents the pilot pressure line connected to the pilot pressure port of the valve. Pilot check valves are frequently used for locking hydraulic cylinders in position.

Additional types of directional control valves are the two-way and four-way valves used to direct inlet flow to either of two outlet ports. As illustrated in Fig. 8-5, flow entering at the pump port *P* (this is the port that is connected to the pump discharge line) can be directed to either the outlet port *A* or *B*. Most directional control valves use a sliding spool to change the path of flow through the valve. For a given position of the spool, a unique flow path configuration exists within the valve. Directional control valves are designed to operate with either two positions of the spool or three positions of the spool. The flow path configuration for each unique spool position is shown symbolically by a rectangle, sometimes called an envelope. Therefore, the valves of Fig. 8-5 are two-position valves (the graphical symbols show two side-by-side rectangles for each valve).

The following is a description of each valve of Fig. 8-5:

1. *Two-way valve:* Notice that the flow can go through the value in *two* unique ways depending on the spool position.
   a. *Spool position 1:* Flow can go from *P* to *B* as shown by the straight-through line and arrowhead. Ports *A* and *T* are blocked as shown.
   b. *Spool position 2:* Flow can go from *P* to *A*. Notice that the ports are labeled for only one envelope so that the reader must mentally identify the ports.
2. *Four-way valve:* Observe that the flow can go through the valve in *four* unique ways depending on the spool position.
   a. *Spool position 1:* Flow can go from *P* to *A* and *B* to *T*.
   b. *Spool position 2:* Flow can go from *A* to *T* and *P* to *B*.

Two-way valves can be used to direct pump flow to either of two different parts of a circuit. Four-way valves are typically used to control double-acting hydraulic cylinders. The spool of a directional control valve can be positioned manually, mechanically, by using pilot pressure, or by using electrical solenoids.

A physical understanding of the flow path configuration of two-way and four-way valves is provided by Figs. 8-6 and 8-7, respectively. Notice that a spool is a cylindrical member that has large-diameter lands machined to slide in a very close-fitting bore of the valve body. The radial clearance is usually less than 0.001 in. The grooves between the lands provide the flow passages between ports.

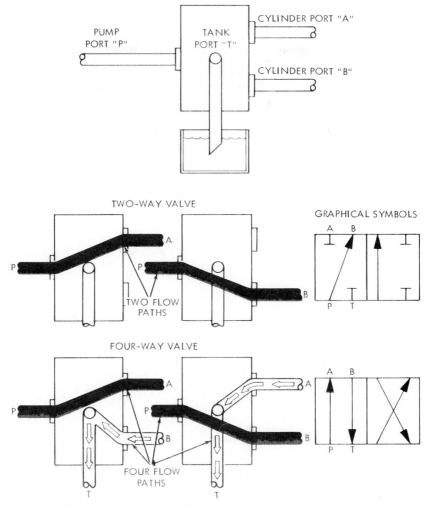

Figure 8-5. Two-way and four-way directional control valves. (*Courtesy of Sperry Vickers, Sperry Rand Corp., Troy, Michigan.*)

Observe that the graphical symbol shows only one tank port *T* even though the physical design may have two. The tank port is the port of the valve that is piped back to the hydraulic oil tank. Therefore, each tank port provides the same function. Recall that the graphical symbol is only concerned with the function of a component and not its internal design.

Figure 8-8 is a photograph of a cutaway of a four-way valve. Notice that it is manual-actuated (see hand lever). Since the spool is spring-loaded at both ends, it is a spring-centered, three-position, directional control valve. Thus, when the

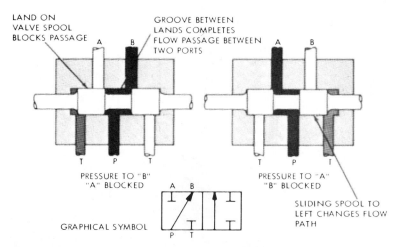

Figure 8-6. Spool positions inside two-way valve. (*Courtesy of Sperry Vickers, Sperry Rand Corp., Troy, Michigan.*)

valve is unactuated (no hand force on lever), the valve will assume its center position due to the balancing opposing spring forces.

In Fig. 8-9 we see the symbolic representation of the four-way valve of Fig. 8-8. Notice that the ports are labeled on the center envelope, which represents the flow path configuration in the spring-centered position of the spool. Also observe the spring and lever actuation symbols used at the ends of the right and left envelopes. These imply a spring-centered, manually actuated valve. It should be

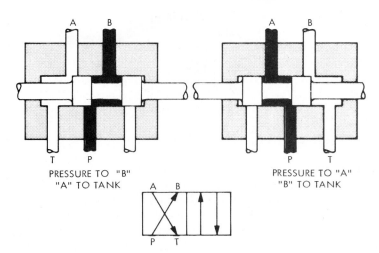

Figure 8-7. Spool positions inside four-way valve. (*Courtesy of Sperry Vickers, Sperry Rand Corp., Troy, Michigan.*)

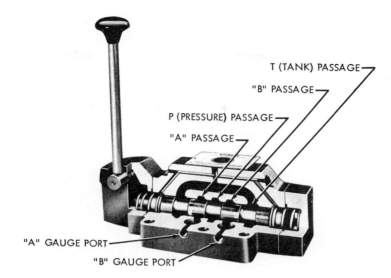

Figure 8-8.   Manually actuated, spring-centered, three-position, four-way valve. (*Courtesy of Sperry Vickers, Sperry Rand Corp., Troy, Michigan.*)

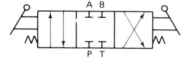

Figure 8-9.   Symbol for a four-way, spring-centered, three-position, manually actuated valve.

noted that a three-position valve is used when it is necessary to stop or hold a hydraulic actuator at some intermediate position within its entire stroke range.

In Fig. 8-10 we see a two-position, four-way valve that is spring offset. In this case the lever shifts the spool, and the spring returns the spool to its original position when the lever is released. There are only two unique operating positions, as indicated by the graphical symbol. Notice that the ports are labeled at the envelope representing the neutral (spring offset or return) or unactuated position of the spool.

The directional control valves of Figs. 8-8 and 8-10 are manually actuated by the use of a lever. Figure 8-11 shows a two-position, four-way spring offset valve that is mechanically rather than manually actuated. This is depicted in the cutaway view, with the spool end containing a roller that is typically actuated by a cam-type mechanism. Notice that the basic graphical symbol is the same but that actuation is depicted as being mechanical rather than manual.

Directional control valves can also be shifted by applying air pressure against a piston at either end of the valve spool. Such a design is illustrated by the cutaway view of Fig. 8-12. As shown, springs (located at both ends of the spool) push against centering washers to center the spool when no air is applied. When air is introduced through the left end passage, its pressure pushes against the

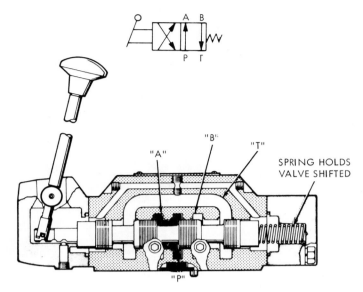

Figure 8-10.   Two-position, spring-offset, four-way valve. (*Courtesy of Sperry Vickers, Sperry Rand Corp., Troy, Michigan.*)

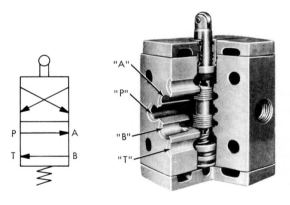

Figure 8-11.   Mechanically actuated four-way valve. (*Courtesy of Sperry Vickers, Sperry Rand Corp., Troy, Michigan.*)

piston to shift the spool to the right. Removal of this left end air supply and introduction of air through the right end passage causes the spool to shift to the left. Therefore, this is a four-way, three-position, spring-centered, air pilot-actuated directional control valve. It is graphically represented in Fig. 8-13, which gives a complete story of its operation. Once again the dashed lines represent pilot pressure lines.

A very common way to actuate a spool valve is by using a solenoid, illustrated in Fig. 8-14. As shown, when the electric coil (solenoid) is energized, it creates a magnetic force that pulls the armature into the coil. This causes the armature to push on the push rod to move the spool of the valve.

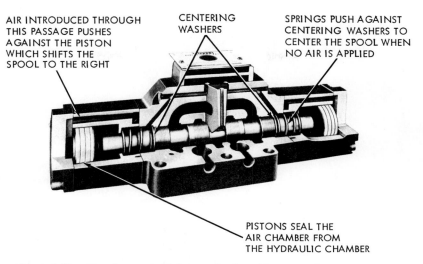

AIR INTRODUCED THROUGH
THIS PASSAGE PUSHES
AGAINST THE PISTON
WHICH SHIFTS THE
SPOOL TO THE RIGHT

CENTERING
WASHERS

SPRINGS PUSH AGAINST
CENTERING WASHERS TO
CENTER THE SPOOL WHEN
NO AIR IS APPLIED

PISTONS SEAL THE
AIR CHAMBER FROM
THE HYDRAULIC CHAMBER

Figure 8-12.   Air pilot-actuated four-way valve. (*Courtesy of Sperry Vickers, Sperry Rand Corp., Troy, Michigan.*)

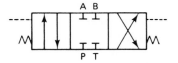

Figure 8-13.   Graphic symbol for a pilot-actuated, four-way, three-position, spring-centered directional control valve.

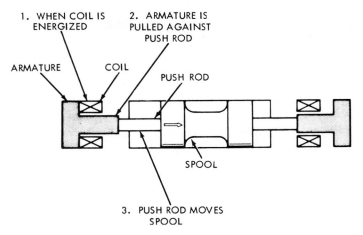

1. WHEN COIL IS
ENERGIZED

2. ARMATURE IS
PULLED AGAINST
PUSH ROD

ARMATURE   COIL

PUSH ROD

SPOOL

3. PUSH ROD MOVES
SPOOL

Figure 8-14.   Operation of solenoid to shift spool of valve. (*Courtesy of Sperry Vickers, Sperry Rand Corp., Troy, Michigan.*)

Figure 8-15. Actual solenoid-actuated directional control valve. (*Courtesy of Continental Hydraulics, Division of Continental Machines Inc., Savage, Minnesota.*)

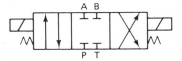

Figure 8-16. Graphic symbol for a solenoid-actuated, four-way, three-position, spring-centered directional control valve.

Figure 8-15 is a cutaway view of an actual solenoid-actuated directional control valve. This valve has a flow capacity of 12 gpm and a maximum operating pressure of 3500 psi. It has a wet armature solenoid, which means that the plunger or armature of the solenoid moves in a tube that is open to the tank cavity of the valve. The fluid around the armature serves to cool it and cushion its stroke without appreciably affecting response time. There are no seals around this armature to wear or restrict its movement. This allows all the power developed by the solenoid to be transmitted to the valve spool without having to overcome seal friction. Impact loads, which frequently cause premature solenoid failure, are eliminated with this construction. This valve has a solenoid at each end of the spool. Specifically, it is a solenoid-actuated, four-way, three-position, spring-centered directional control valve and is represented by the graphical symbol in Fig. 8-16. Notice the symbol used to represent the solenoid at both ends of the spool.

Figure 8-17 shows a single solenoid-actuated four-way, two-position, spring-offset directional control valve. Its graphical symbol is given in Fig. 8-18.

In Fig. 8-19 we see a solenoid-controlled, pilot-operated directional control valve. Notice that the pilot valve is actually mounted on top of the main valve body. The upper pilot stage spool (which is solenoid-actuated) controls the pilot pressure, which can be directed to either end of the main stage spool. This 35-gpm, 3000-psi valve is of the four-way, three-position, spring-centered configuration and has a manual override to shift the pilot stage mechanically when trouble-shooting.

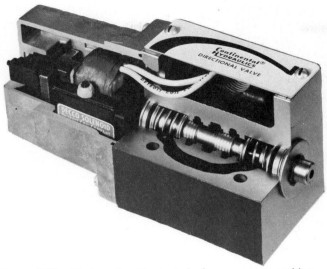

Figure 8-17. Single solenoid-actuated, four-way, two-position, spring-offset directional control valve. (*Courtesy of Continental Hydraulics, Division of Continental Machines Inc., Savage, Minnesota.*)

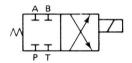

Figure 8-18. Graphic symbol for valve of Figure 8-17.

Solenoids are commonly used to actuate small spool valves. This is illustrated in Fig. 8-20, which compares the size of a micro miniature directional control valve to the size of a U.S silver dollar. This three-way valve is specially designed to be used with small cylinders and other similar devices requiring low flow rates.

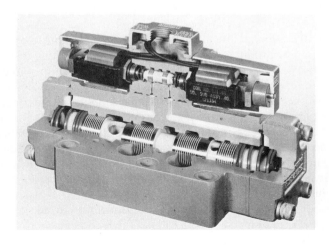

Figure 8-19. Solenoid-controlled, pilot-operated, directional control valve. (*Courtesy of Continental Hydraulics, Division of Continental Machines Inc., Savage, Minnesota.*)

Figure 8-20. Micro miniature solenoid valve. (*Courtesy of Skinner Precision Industries, Inc., New Britain, Connecticut.*)

Most three-position valves have a variety of possible flow path configurations. Each four-way valve has identical flow path configurations in the actuated positions but different spring-centered flow paths. This is illustrated in Fig. 8-21.

Notice that the open-center type connects all ports together. In this design the pump flow can return directly back to the tank at essentially atmospheric pressure. At the same time, the actuator (cylinder or motor) can be moved freely by applying an external force.

The closed-center design has all ports blocked, as is the case for the valve of Figs. 8-8 and 8-12. In this way the pump flow can be used for other parts of the circuit. At the same time, the actuator connected to ports *A* and *B* is hydraulically locked. This means it cannot be moved by the application of an external force.

The tandem design also results in a locked actuator. However, it also unloads the pump at essentially atmospheric pressure. For example, the closed-center design forces the pump to produce flow at the high-pressure setting of the pressure relief valve. This not only wastes pump design horsepower but promotes wear and shortens pump life, especially if operation in the center position occurs for long periods of time. Another factor is that the wasted horsepower shows up as heat, which raises the temperature of the oil. This promotes oil oxidation, which increases the acidity of the oil. Such an oil tends to corrode the critical metallic parts not only of the pump but also of the actuators and valves. Also affected is the viscosity of the oil. Higher temperature lowers the viscosity, which in turn increases leakage and reduces the lubricity of the oil. To keep the temperature at a safe level, an expensive oil cooler may be required.

Although most directional control valves are of the spool design, other types are in use. One such design is the rotary four-way valve, which consists simply of a rotor closely fitted in a valve body, as shown in Fig. 8-22. Passages in the rotor

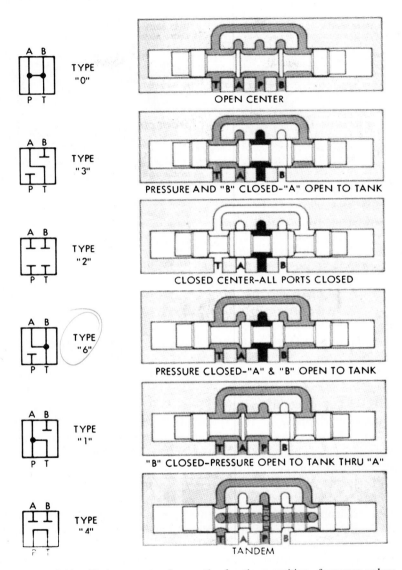

Figure 8-21. Various center flow paths for three-position, four-way valves. (*Courtesy of Sperry Vickers, Sperry Rand Corp., Troy, Michigan.*)

connect or block off the ports in the valve body to provide the four flow paths as illustrated. The design shown is a three-position valve in which the centered position has all four ports blocked. Notice that the graphical symbol indicates the flow path configuration through the valve for each of its three positions. The

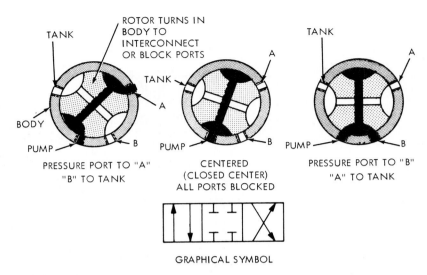

Figure 8-22. Rotary four-way valve. (*Courtesy of Sperry Vickers, Sperry Rand Corp., Troy, Michigan.*)

symbolic representation of a valve is not affected by the physical construction of the valve. Rotary valves are usually actuated manually or mechanically.

The operating features of an actual rotary valve are illustrated in Figs. 8-23 and 8-24. This design (called a shear-seal valve by its manufacturer) contains lapped metal-to-metal sealing surfaces, which form a virtually leakproof seal. The gradual overlapping of the round flow passages produces a smooth shearing action, which results in a low handle load and no sudden surges. There is no external leakage because of a static seal on a rotating shaft (not reciprocating and not under pressure). The high-pressure regions are confined to flow passages. This type of valve can take higher velocities and more flow than a spool valve of the same pipe size. For example, a $\frac{1}{2}$-in., 2000-psi valve will experience only a 14-psi pressure drop for a 14-gpm flow rate, which corresponds to a velocity of 20 ft/sec. Although the design shown is manually operated, solenoid, and air pilot-actuated models are available. These valves can be obtained in a variety of three-way and four-way, two- and three-position flow path configurations. Certain models operate with working pressures as high as 10,000 psi.

A shuttle valve is another type of directional control valve. It permits a system to operate from either of two fluid power sources. One application is for safety in the event that the main pump can no longer provide hydraulic power to operate emergency devices. The shuttle valve will shift to allow fluid to flow from a secondary backup pump. As shown in Fig. 8-25, a shuttle valve consists of a floating piston that can be shuttled to one side or the other of the valve depending on which side of the piston has the greater pressure. Shuttle valves may be spring-

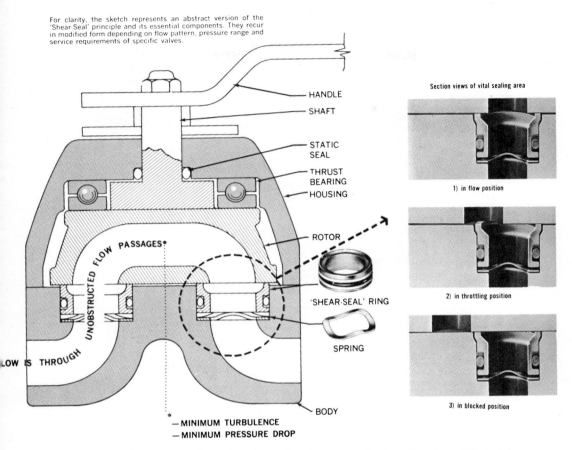

For clarity, the sketch represents an abstract version of the
'Shear-Seal' principle and its essential components. They recur
in modified form depending on flow pattern, pressure range and
service requirements of specific valves.

HANDLE

SHAFT

STATIC
SEAL

THRUST
BEARING

HOUSING

ROTOR

'SHEAR-SEAL' RING

SPRING

BODY

UNOBSTRUCTED FLOW PASSAGES*

LOW IS THROUGH

*—MINIMUM TURBULENCE
—MINIMUM PRESSURE DROP

Section views of vital sealing area

1) in flow position

2) in throttling position

3) in blocked position

Figure 8-23.   Shear-flow rotary directional control valve. (*Courtesy of DeLaval*
Turbine, Inc., Barksdale Controls Division, Los Angeles, California.)

loaded (biased as shown in Fig. 8-25) in one direction to favor one of the supply
sources or unbiased so that the direction of flow through the valve is determined
by circuit conditions. A shuttle valve is essentially a direct-acting double check
valve with a cross-bleed, as shown by the graphical symbol of Fig. 8-25. As shown
by the double arrows on the graphical symbol, reverse flow is permitted.

8.3 PRESSURE CONTROL VALVES

The most widely used type of pressure control valve is the pressure relief valve,
since it is found in practically every hydraulic system. It is a normally closed valve
whose function is to limit the pressure to a specified maximum value by diverting

NO CREEPING CYLINDERS

Lapped metal to metal sealing surfaces form a virtually leakproof seal, which actually improves with use through continued lapping action.

Note un-retouched mirror-like sealing surfaces of this rotor after 4 years of water service. (5% soluble oil is recommended minimum for normal applications.)

COMPLETE CONTROL

ON — OFF — THROTTLING TO ANY DEGREE.

The gradual overlapping of round flow passages produces a smooth shearing action — low handle load — no sudden surges.

DIRT CANNOT SCORE SEALS

because FLOW IS THROUGH 'SHEAR-SEAL' RINGS rather than across sealing surfaces. Sealing surfaces remain in constant intimate contact and thus wipe away any dirt (like a window squeegee).

LONG MAINTENANCE-FREE SERVICE

because a SPRING COMPENSATES FOR THE WEAR as 'Shear-Seals' and rotor continue to lap-in with use. Sealing qualities do not diminish even after this much wear.

NO EXTERNAL LEAKAGE

because of a static seal on a rotating shaft (not reciprocating and not under high pressure); HIGH PRESSURE CONFINED TO FLOW PASSAGES. Eliminates safety hazard of slippery floors, and danger of fire.

APPEARANCE AND VERSATILITY

Your machine gains in sales appeal with the neat, finished look of Barksdale Valves.

A CHOICE OF PORTING (in-line, straight or manifold), at no extra cost on the 'O.E.M.' line, offers the installation which is most attractive and least expensive for **your** application. PANEL MOUNTING PROVISION IS STANDARD.

LOW MAINTENANCE COST

because 3 'SHEAR-SEALS' AND 'O' RINGS GENERALLY MAKE A NEW VALVE. Without removing valve from line, all parts are easily accessible.

Figure 8-24.   Operating features of the shear-flow rotary valve. (*Courtesy of DeLaval Turbine, Inc., Barksdale Controls Division, Los Angeles, California.*)

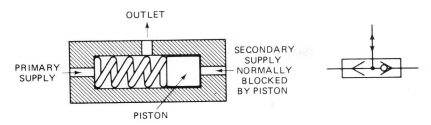

Figure 8-25.   Shuttle valve (schematic and graphic symbol.)

pump flow back to the tank. Figure 8-26 illustrates the operation of a simple relief valve. A poppet is held seated inside the valve by a heavy spring. When the system pressure reaches a high enough value, the poppet is forced off its seat. This permits flow through the outlet to the tank as long as this high pressure level is maintained. Notice the external adjusting screw, which varies the spring force and, thus, the pressure at which the valve begins to open (cracking pressure).

It should be noted that the poppet must open sufficiently to allow full pump flow. The pressure that exists at full pump flow can be substantially greater than the cracking pressure. This is shown in Fig. 8-27, where system pressure is plotted

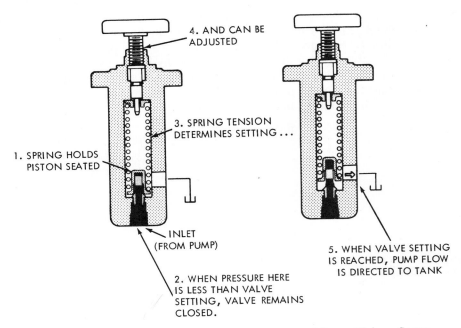

Figure 8-26.   Simple pressure relief valve. (*Courtesy of Sperry Vickers, Sperry Rand Corp., Troy, Michigan.*)

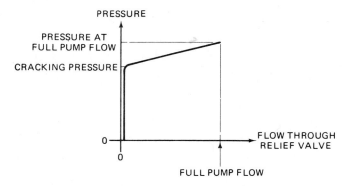

Figure 8-27.   Pressure versus flow curve for simple relief valve.

versus flow through the relief valve. The pressure at full pump flow is the pressure level that is specified when referring to the pressure setting of the relief valve. It is the maximum pressure level permitted by the relief valve.

The symbolic representation of a simple relief valve is given in Fig. 8-28. As can be seen, the symbol shows the function of a relief valve but not its internal configuration.

Figure 8-29 shows a partial hydraulic circuit containing a pump and pressure relief valve, which are drawn symbolically. If the hydraulic system (not shown) does not accept any flow, then all the pump flow must return back to the tank via the relief valve. The pressure relief valve provides protection against any overloads experienced by the actuators in the hydraulic system. Of course, a relief valve is not needed if a pressure-compensated vane pump is used. Obviously one important function of a pressure relief valve is to limit the force or torque produced by hydraulic cylinders and motors.

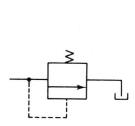

Figure 8-28.   Symbol for simple relief valve.

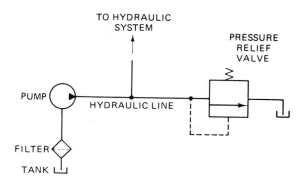

Figure 8-29.   Symbolic representation of partial hydraulic circuit.

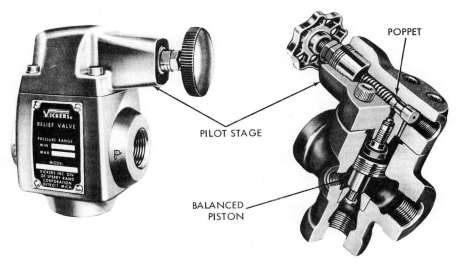

POPPET

PILOT STAGE

BALANCED PISTON

Figure 8-30. External and cutaway views of an actual compound relief valve. (*Courtesy of Sperry Vickers, Sperry Rand Corp., Troy, Michigan.*)

A compound pressure relief valve (see Fig. 8-30 for external and cutaway views of an actual design) is one that operates in two stages. As shown in Fig. 8-30, the pilot stage is located in the upper valve body and contains a pressure-limiting poppet that is held against a seat by an adjustable spring. The lower body contains the port connections. Diversion of the full pump flow is accomplished by the balanced piston in the lower body.

The operation is as follows (refer to Fig. 8-31): In normal operation, the balanced piston is in hydraulic balance. Pressure at the inlet port acts under the piston and also on its top because an orifice is drilled through the large land. For pressures less than the valve setting, the piston is held on its seat by a light spring. As soon as pressure reaches the setting of the adjustable spring, the poppet is forced off its seat. This limits the pressure in the upper chamber. The restricted flow through the orifice and into the upper chamber results in an increase in pressure in the lower chamber. This causes an unbalance in hydraulic forces, which tends to raise the piston off its seat. When the pressure difference between the upper and lower chambers reaches approximately 20 psi, the large piston lifts off its seat to permit flow directly to the tank. If the flow increases through the valve, the piston lifts farther off its seat. However, this compresses only the light spring, and hence very little override occurs. Compound relief valves may be remotely operated by using the outlet port from the chamber above the piston. For example, this chamber can be vented to tank via a solenoid directional control valve. When this valve vents the pressure relief valve to the tank, the 20-psi pressure in the bottom chamber overcomes the light spring and unloads the pump to the tank.

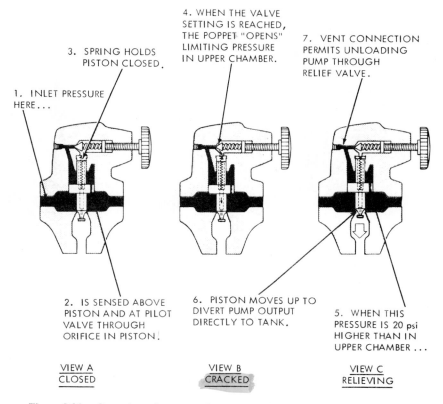

3. SPRING HOLDS
PISTON CLOSED.

4. WHEN THE VALVE
SETTING IS REACHED,
THE POPPET "OPENS"
LIMITING PRESSURE
IN UPPER CHAMBER.

7. VENT CONNECTION
PERMITS UNLOADING
PUMP THROUGH
RELIEF VALVE.

1. INLET PRESSURE
HERE . . .

2. IS SENSED ABOVE
PISTON AND AT PILOT
VALVE THROUGH
ORIFICE IN PISTON.

6. PISTON MOVES UP TO
DIVERT PUMP OUTPUT
DIRECTLY TO TANK.

5. WHEN THIS
PRESSURE IS 20 psi
HIGHER THAN IN
UPPER CHAMBER . . .

VIEW A
CLOSED

VIEW B
CRACKED

VIEW C
RELIEVING

Figure 8-31. Operation of compound pressure relief valve. (*Courtesy of Sperry Vickers, Sperry Rand Corp., Troy, Michigan.*)

Figure 8-32 is a photograph of a compound pressure relief valve that has this remote operation capability. This particular model has its own built-in solenoid-actuated two-way vent valve, which is located between the cap and body of the main valve. Manual override of the solenoid return spring is a standard feature. The pressure relief valve is vented when the solenoid is de-energized and devented when energized. This relief valve has a maximum flow capacity of 53 gpm and can be adjusted to limit system pressures up to 5000 psi. Clockwise tightening of the hex locknut prevents accidental setting changes by use of the knurled knob.

A second type of pressure control valve is the pressure-reducing valve. This type of valve (which is normally open) is used to maintain reduced pressures in specified locations of hydraulic systems. It is actuated by downstream pressure and tends to close as this pressure reaches the valve setting. Figure 8-33 illustrates the operation of a pressure-reducing valve that uses a spring-loaded spool to

Figure 8-32.   Pressure relief valve with integral solenoid-actuated, two-way vent valve. (*Courtesy of Abex Corp., Denison Division, Columbus, Ohio.*)

control the downstream pressure. If downstream pressure is below the valve setting, fluid will flow freely from the inlet to the outlet. Notice that there is an internal passageway from the outlet, which transmits outlet pressure to the spool end opposite the spring. When the outlet (downstream) pressure increases to the valve setting, the spool moves to the right to partially block the outlet port, as shown in view B. Just enough flow is passed to the outlet to maintain its preset pressure level. If the valve closes completely, leakage past the spool could cause downstream pressure to build up above the valve setting. This is prevented from occurring because a continuous bleed to the tank is permitted via a separate drain line to the tank. Figure 8-33 also provides the graphical symbol for a pressure-reducing valve. Observe that the symbol shows that the spring cavity has a drain to the tank.

An additional pressure control device is the unloading valve. This valve is used to permit a pump to build up to an adjustable pressure setting and then allow it to discharge to the tank at essentially zero pressure as long as pilot pressure is maintained on the valve from a remote source. Hence, the pump has essentially no load and is therefore developing a minimum amount of horsepower. This is the case in spite of the fact that the pump is delivering a full pump flow because the pressure is practically zero. This is not the same with a pressure relief valve because the pump is delivering full pump flow at the pressure relief valve setting and thus is operating at maximum horsepower conditions. Figure 8-34 shows a schematic of an unloading valve used to unload the pump connected to port A when the pressure at port X is maintained at the value that satisfies the valve setting. The high-flow poppet is controlled by the spring-loaded ball and the pressure applied to port X. Flow entering at port A is blocked by the poppet at low pressures. The pressure signal from A passes through the orifice in the main poppet to the topside area and on to the ball. There is no flow through these sections of the valve until the pressure rises to the maximum permitted by the

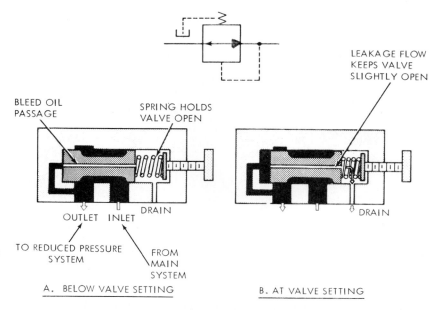

Figure 8-33. Operation of a pressure-reducing valve. (*Courtesy of Sperry Vickers, Sperry Rand Corp., Troy, Michigan.*)

adjustably set spring-loaded ball. When that occurs, the poppet lifts and flow goes from port *A* to port *B*, which is typically connected to the tank. The pressure signal to port *X* (sustained by another part of the system) acts against the solid control piston and forces the ball farther off the seat. This causes the topside pressure on the main poppet to go to a very low value and allows flow from *A* to *B* with a very low pressure drop as long as signal pressure at *X* is maintained. The ball reseats, and the main poppet closes with a snap action when the pressure at *X* falls to approximately 90% of the maximum pressure setting of the spring-loaded ball. Also included in Fig. 8-34 is the graphical symbol of an unloading valve. Figure 8-35 shows a photograph of the actual unloading valve.

EXAMPLE 8-1

A pressure relief valve has a pressure setting of 1000 psi. Compute the horsepower loss across this valve if it returns all the flow back to the tank from a 20-gpm pump.

Solution

$$HP = \frac{PQ}{1714} = \frac{(1000)(20)}{1714} = 11.7 \text{ hp}$$

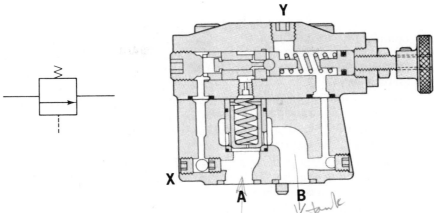

Figure 8-34.   Schematic of unloading valve. (*Courtesy of Abex Corp., Denison Division, Columbus, Ohio.*)

EXAMPLE 8-2

An unloading valve is used to unload the pump of Example 8-1. If the pump discharge pressure (during unloading) equals 25 psi, how much hydraulic horsepower is being wasted?

Solution

$$HP = \frac{PQ}{1714} = \frac{(25)(20)}{1714} = 0.29 \text{ hp}$$

Figure 8-35.   Photograph of unloading valve. (*Courtesy of Abex Corp., Denison Division, Columbus, Ohio.*)

Still another pressure control device is the sequence valve, which is designed to cause a hydraulic system to operate in a pressure sequence. After the components connected to port *A* (see Fig. 8-36) have reached the adjusted pressure of the sequence valve, it passes fluid through port *B* to do additional work in a different portion of the system. The high-flow poppet of the sequence valve is controlled by the spring-loaded cone. Flow entering at port *A* is blocked by the poppet at low pressures. The pressure signal at *A* passes through orifices to the topside of the poppet and to the cone. There is no flow through these sections until the pressure rises at *A* to the maximum permitted by the adjustably set spring-loaded cone. When the pressure at *A* reaches that value, the main poppet lifts, passing flow to port *B*. It maintains the adjusted pressure at port *A* until the pressure at *B* rises to the same value. A small pilot flow (about ¼ gpm) goes through the control piston and past the pilot cone to the external drain at this time. When the pressure at *B* rises to the pressure at *A*, the control piston seats and prevents further pilot flow loss. The main poppet opens fully and allows the pressure at *A* and *B* to rise to higher values together. Flow may go either way at this time. The spring cavity of the control cone drains externally from port *Y*, generally to the tank. This sequence valve may be remotely controlled from vent port *X*. Figure 8-36 also includes the graphical symbol for a sequence valve.

A final pressure control valve to be presented here is the counterbalance valve. The purpose of a counterbalance valve is to maintain control of a vertical cylinder to prevent it from descending due to gravity. As shown in Fig. 8-37, the primary port of this valve is connected to the bottom of the cylinder, and the secondary port is connected to a directional control valve (DCV). The pressure

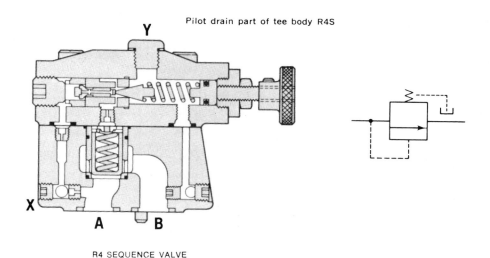

Figure 8-36. Schematic of sequence valve. (*Courtesy of Abex Corp., Denison Division, Columbus, Ohio.*)

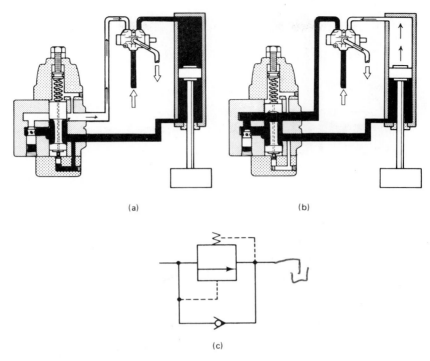

(a)   (b)

(c)

Figure 8-37.   Application of counterbalance valve. (*Courtesy of Sperry Vickers, Sperry Rand Corp., Troy, Michigan.*)

setting of the counterbalance valve is somewhat higher than is necessary to prevent the cylinder load from falling. As shown in Fig. 8-37(a), when pump flow is directed (via the DCV) to the top of the cylinder, the cylinder piston is pushed downward. This causes pressure at the primary port to increase to raise the spool. This opens a flow path for discharge through the secondary port to the DCV and back to the tank. When raising the cylinder [see Fig. 8-37(b)], an integral check valve opens to allow free flow for retracting the cylinder. Figure 8-37(c) gives the graphical symbol for a counterbalance valve.

8.4 *FLOW CONTROL VALVES*

Flow control valves are used to regulate the speed of hydraulic cylinders and motors controlling the flow rate to these actuators. They may be as simple as a fixed orifice or an adjustable needle valve. Needle valves are designed to give fine control of flow in small-diameter piping. As illustrated in Fig. 8-38, their name is

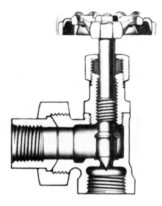

Figure 8-38. Needle valve. (*Courtesy of Crane Co., Chicago, Illinois.*)

derived from their sharp, pointed conical disk and matching seat. The graphical symbol for a needle valve (which is a variable orifice) is also given in Fig. 8-38.

Figure 8-39 shows a flow control valve that is easy to read and adjust. The stem has several color rings, which, in conjunction with a numbered knob, permits reading of a given valve opening as shown. Charts are available that allow quick determination of the controlled flow rate for given valve settings and pressure drops. A locknut prevents unwanted changes in flow.

There are two basic types of flow control valves: nonpressure-compensated and pressure-compensated. The nonpressure-compensated type is used where

Figure 8-39.   Easy read and adjust flow control valve. (*Courtesy of Deltrol Corp., Bellwood, Illinois.*)

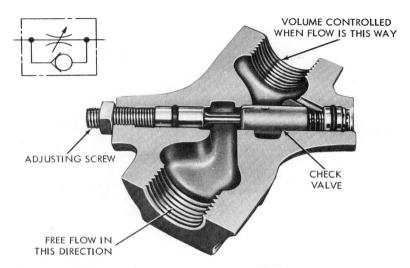

Figure 8-40.   Noncompensated flow control valve. (*Courtesy of Sperry Vickers, Sperry Rand Corp., Troy, Michigan.*)

system pressures are relatively constant and motoring speeds are not too critical. They work on the principle that the flow through an orifice will be constant if the pressure drop remains constant. Figure 8-40 gives a cutaway view of a nonpressure-compensated flow control valve and its graphical symbol. The design shown also includes a check valve, which permits free flow in the direction opposite to the flow control direction.

If the load on an actuator changes significantly, system pressure will change appreciably. Thus, the flow rate through a nonpressure-compensated valve will change for the same flow rate setting. Figure 8-41 illustrates the operation of a pressure-compensated valve. This design incorporates a hydrostat that maintains a constant 20-psi differential across the throttle, which is an orifice whose area can be adjusted by an external knob setting. The orifice area setting determines the flow rate to be controlled. The hydrostat is held normally open by a light spring. However, it starts to close as inlet pressure increases and overcomes the light spring force. This closes the opening through the hydrostat and thereby blocks off all flow in excess of the throttle setting. As a result, the only oil that will pass through the valve is that amount which 20 psi can force through the throttle. Flow exceeding this amount can be used by other parts of the circuit or return to the tank via the pressure relief valve. Also included in Fig. 8-41 is the graphical symbol for a pressure-compensated flow control valve.

In Fig. 8-42 we have a see-through model of an actual pressure-compensated flow control valve, which has a pressure rating of 3000 psi. Pressure compensation will maintain preset flow within 1–5% depending on the basic flow rate as long as there is 150-psi pressure differential between the inlet and outlet ports. The dial is

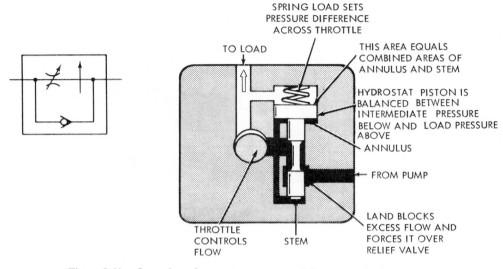

SPRING LOAD SETS
PRESSURE DIFFERENCE
ACROSS THROTTLE

TO LOAD

THIS AREA EQUALS
COMBINED AREAS OF
ANNULUS AND STEM

HYDROSTAT PISTON IS
BALANCED BETWEEN
INTERMEDIATE PRESSURE
BELOW AND LOAD PRESSURE
ABOVE

ANNULUS

FROM PUMP

LAND BLOCKS
EXCESS FLOW AND
FORCES IT OVER
RELIEF VALVE

THROTTLE
CONTROLS
FLOW

STEM

Figure 8-41. Operation of pressure-compensated flow control valve. (*Courtesy of Sperry Vickers, Sperry Rand Corp., Troy, Michigan.*)

calibrated for easy and repeatable flow settings. Adjustments over the complete valve capacity of 12 gpm are obtained within a 270° arc. A dial key lock prevents tampering with valve settings. A sharp-edged orifice design means that the valve is immune to temperature or fluid viscosity changes.

Figure 8-42. Pressure-compensated flow control valve. (*Courtesy of Continental Hydraulics, Division of Continental Machines Inc., Savage, Minnesota.*)

8.5 SERVO VALVES

A servo valve is a directional control valve that has infinitely variable positioning capability. Thus, it can control not only the direction of fluid flow but also the amount. Servo valves are coupled with feedback-sensing devices, which allow for the very accurate control of position, velocity, and acceleration of an actuator.

Figure 8-43 shows the mechanical-type servo valve, which is essentially a force amplifier used for positioning control. In this design, a small input force shifts the spool of the servo valve to the right by a specified amount. The oil flows through port P_1, retracting the hydraulic cylinder to the right. The action of the feedback link shifts the sliding sleeve to the right until it blocks off the flow to the hydraulic cylinder. Thus, a given input motion produces a specific and controlled amount of output motion. Such a system where the output is fed back to modify the input is called a closed-loop system. One of the most common applications of this type of mechanical-hydraulic servo valve is the hydraulic power steering system of automobiles and other transportation vehicles.

In recent years, the electrohydraulic servo valve has arrived on the industrial scene. Typical electrohydraulic servo valves use an electrical torque motor, a double-nozzle pilot stage, and a sliding spool second stage. Figure 8-44 gives a cutaway view of an actual electrohydraulic servo valve. This servo valve is an electrically controlled, proportional metering valve suitable for a variety of mobile vehicles and industrial control applications such as earth-moving vehicles, articulated arm devices, cargo-handling cranes, lift trucks, logging equipment, farm machinery, steel mill controls, utility construction, fire trucks, and servicing vehicles.

The construction and operational features of an electrohydraulic servo valve can be seen by referring to the schematic drawing of Fig. 8-45. The torque motor includes coils, pole pieces, magnets, and an armature. The armature is supported for limited movement by a flexure tube. The flexure tube also provides a fluid seal

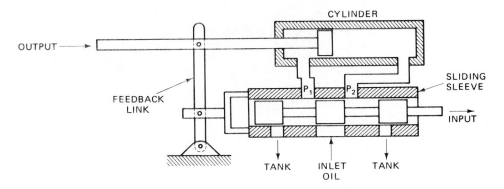

Figure 8-43.   Mechanical-hydraulic servo valve.

Figure 8-44. Electrohydraulic servo valve. (*Courtesy of Moog Inc., Industrial Division, East Aurora, New York.*)

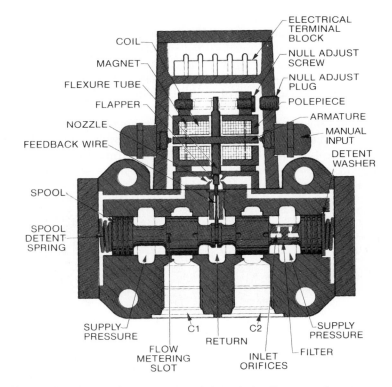

Figure 8-45. Schematic cross section of electrohydraulic servo valve. (*Courtesy of Moog Inc., Industrial Division, East Aurora, New York.*)

between the hydraulic and electromagnetic portions of the valve. The flapper attaches to the center of the armature and extends down, inside the flexure tube. A nozzle is located on each side of the flapper so that flapper motion varies the nozzle openings. Pressurized hydraulic fluid is supplied to each nozzle through an inlet orifice located in the end of the spool. This pilot stage flow is filtered by a 40-micron screen that is wrapped around the shank of the spool. Differential pressure between the ends of the spool is varied by flapper motion between the nozzles.

The four-way valve spool directs flow from supply to either control port $C1$ or $C2$ in an amount proportional to spool displacement. The spool contains flow metering slots in the control lands that are uncovered by spool motion. Spool movement deflects a feedback wire that applies a torque to the armature/flapper. Spool detent springs are provided to center the spool whenever hydraulic driving pressures are absent. Electrical current in the torque motor coils causes either clockwise or counterclockwise torque on the armature. This torque displaces the flapper between the two nozzles. The differential nozzle flow moves the spool to either the right or left. The spool continues to move until the feedback torque counteracts the electromagnetic torque. At this point the armature/flapper is returned to center, so the spool stops and remains displaced until the electrical input changes to a new level. Therefore, valve spool position is proportional to the electrical signal. The actual flow from the valve to the load will depend on the load pressure.

Rated flow is achieved with either a plus or minus 100% electrical signal. The amount of rated flow now depends on the valve pressure drop.

The overall operation of an electrohydraulic system is as follows: The electrohydraulic servo valve operates from an electrical signal to its torque motor, which positions the spool of a directional control valve. The signal to the torque motor comes from an electrical device such as a potentiometer (see Fig. 8-46). The signal from the potentiometer is electrically amplified to drive the torque motor of the servo valve. The hydraulic flow output of the servo valve powers an actuator, which in turn drives the load. The velocity or position of the load is fed back in electrical form to the input of the servo valve via a feedback device such as a

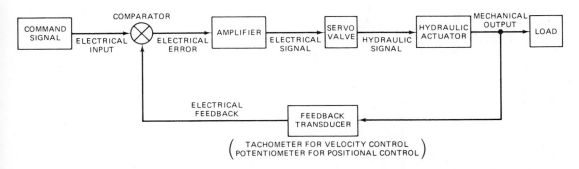

Figure 8-46. Block diagram of an electrohydraulic servo system (closed-loop).

tachometer generator or potentiometer. Because the loop is closed by this action, this type of system is commonly referred to as a closed-loop system. This feedback signal is compared to the command input signal, and the difference between the two signals is sent to the torque motor as an error signal. This produces a correction in the velocity or position of the load until it matches up with the desired value. At this point the error signal to the torque motor becomes zero, and no additional changes are made to the load until the original command input signal is changed as desired. Electrohydraulic servo systems (closed-loop systems) can provide very precise control of the position, velocity, or acceleration of a load.

8.6 CARTRIDGE VALVES

Market pressures and worldwide competition make the need for more efficient and economical hydraulic systems greater than ever. Integrated hydraulic circuits offer a proven way to achieve these improvements. Integrated hydraulic circuits are compact hydraulic systems formed by integrating various cartridge valves and other components into a single, machined, ported manifold block.

A cartridge valve is designed to be assembled into a cavity of a ported manifold block (alone or along with other cartridge valves and hydraulic components) in order to perform the valve's intended function. (See Fig. 8-47 for cutaway views of several threaded cartridge valves.) The cartridge valve is assembled

Figure 8-47. Cutaway views of threaded cartridge valves. (*Courtesy of Parker Hannifin Corp., Elyria, Ohio.*)

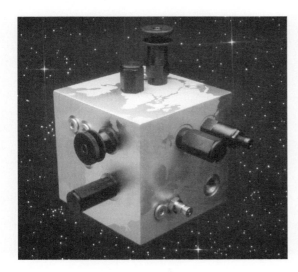

Figure 8-48.   Manifold block containing cartridge valves. (*Courtesy of Parker Hannifin Corp., Elyria, Ohio.*)

into the manifold block either by screw threads (threaded design) or by a bolted cover (slip-in design). Figure 8-48 shows a manifold block containing a number of cartridge valves and other hydraulic components. The world map was etched on the outside surfaces of the manifold block to reflect one's entering the "world" of integrated hydraulic circuits.

The use of cartridge valves in ported manifold blocks provides a number of advantages over discrete, conventional, ported valves mounted at various locations in pipelines of hydraulic systems. The advantages include the following:

1. Reduced number of fittings to connect hydraulic lines between various components in a system.
2. Reduced oil leakages and contamination due to fewer fittings.
3. Lower system installation time and costs.
4. Reduced service time since faulty cartridge valves can be easily changed without disconnecting fittings.
5. Smaller space requirements of overall system.

A variety of different valve functions can be provided using cartridge valves. These include directional control, pressure relief, pressure reduction, unloading, counterbalance, and sequence and flow control capability. Figure 8-49 shows five different solenoid operated directional control cartridge valves from left to right as follows:

1. 2-way, spool-type, N.C. or N.O.
2. 2-position, 3-way, spool-type

Figure 8-49.   Solenoid-operated directional control cartridge valves. (*Courtesy of Parker Hannifin Corp., Elyria, Ohio.*)

3. 2 position, 4-way, spool-type
4. 2-way, poppet-type, N.C. or N.O.
5. 3-position, 4-way, spool-type

Figures 8-50 through 8-52 show a cartridge pressure-relief valve, check valve, and solenoid operated flow control (proportional) valve, respectively. Internal me-

Figure 8-50.   Cartridge pressure relief valve. (*Courtesy of Parker Hannifin Corp., Elyria, Ohio.*)

Figure 8-51. Cartridge check valve. (*Courtesy of Parker Hannifin Corp., Elyria, Ohio.*)

Figure 8-52. Cartridge solenoid operated flow control valve. (*Courtesy of Parker Hannifin Corp., Elyria, Ohio.*)

chanical design and fluid flow operating features of a pressure relief valve, check valve, solenoid directional control valve, and flow control (proportional) valve are shown in Figs. 8-53 through 8-56 respectively.

Integrated hydraulics technology can provide easier installation and servicing, greater reliability, reduced leakage, expanded design flexibility, and lighter, neater hydraulic packages for a variety of hydraulic applications. Figure 8-57 shows several manifold blocks superimposed on a symbolic hydraulic circuit diagram to represent a complete integrated hydraulic system.

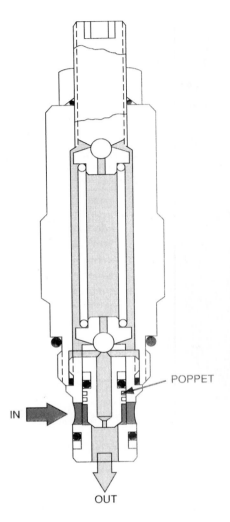

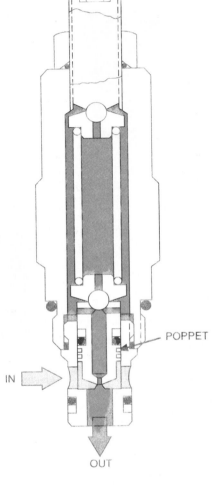

No Flow Condition

- System pressure is lower than relief valve setting.

- Poppet is seated, held in position by spring force.

- Flow is blocked at Inlet.

Throttled Flow Condition

- System pressure has reached relief valve setting.

- Pressure has moved Poppet away from Seat, allowing flow to pass through valve.

- Valve is throttling flow to maintain relief pressure at Inlet.

Figure 8-53.   Cartridge pressure relief valve. (*Courtesy of Parker Hannifin Corp., Elyria, Ohio.*)

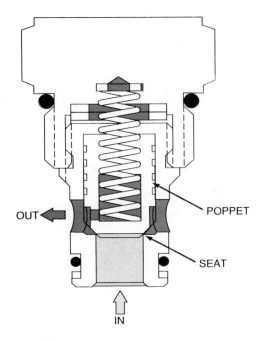

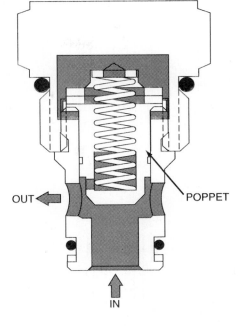

No Flow Condition

- Poppet is held on Seat by spring and pressure forces.

- Flow is blocked.

Full Flow Condition

- Pressure on Inlet of valve creates force against Poppet.

- Reverse flow through valve is blocked by Poppet.

- Pressure force is greater than spring force lifting Poppet.

- Flow passes through valve.

Figure 8-54.   Cartridge check valve. (*Courtesy of Parker Hannifin Corp., Elyria, Ohio.*)

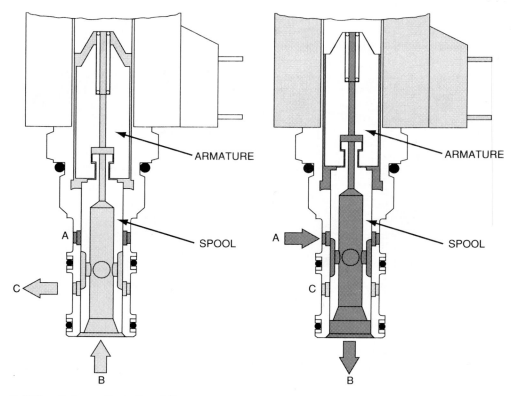

Full Flow Between Ports C and B

- Armature, Plunger and Spool are held down by spring force.

- Plunger Spring extended but under tension.

- Flow is blocked at Port A.

- Fluid flows through valve (either direction).

- Hollow spool center is part of flow path.

Full Flow Between Ports A and C

- Armature, Plunger and Spool are held upward by electromagnetic force of Solenoid.

- Spring is compressed.

- Flow is blocked at Port C.

- Fluid flows through valve (either direction).

- Hollow spool center is part of flow path.

Figure 8-55. Cartridge solenoid directional control valve. (*Courtesy of Parker Hannifin Corp., Elyria, Ohio.*)

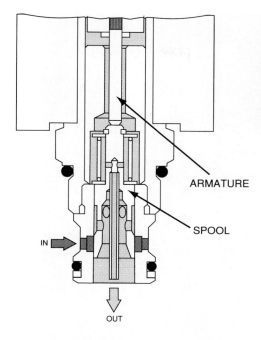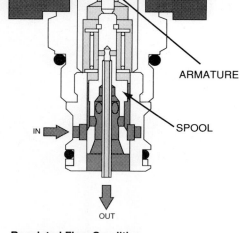

No Flow Condition

- Spool and Armature are held in neutral position by spring force.

- Flow is blocked in either direction.

Regulated Flow Condition

- Spool and Armature are held partially down by electromagnetic forces of Solenoid.

- Partial flow passes from Inlet to Outlet.

- Increasing or decreasing percent of current to the Solenoid changes controlled flow.

- Hollow Spool center is part of flow path.

Figure 8-56.   Cartridge flow control valve. (*Courtesy of Parker Hannifin Corp., Elyria, Ohio.*)

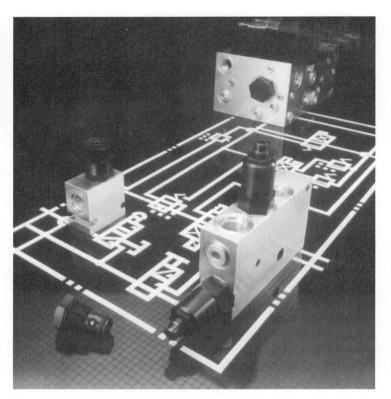

Figure 8-57.   Integrated hydraulic circuit. (*Courtesy of Parker Hannifin Corp., Elyria, Ohio.*)

8.7 HYDRAULIC FUSES

Figure 8-58(a) shows a schematic drawing of a hydraulic fuse, which is analogous to an electric fuse. It prevents hydraulic pressure from exceeding an allowable value in order to protect circuit components from damage. When the hydraulic pressure exceeds a design value, the thin metal disk ruptures to relieve the pressure as oil is drained back to the oil tank. After rupture, a new metal disk must be inserted before operation can be started again. Hydraulic fuses are used mainly with pressure-compensated pumps for fail-safe overload protection in case the compensator control on the pump fails to operate. Figure 8-58(b) shows the symbolic representation of a partial circuit consisting of a pressure-compensated pump and a hydraulic fuse. A hydraulic fuse is analogous to an electrical fuse because they both are one-shot devices. On the other hand, a pressure relief valve is analogous to an electrical circuit breaker because they both are resettable devices.

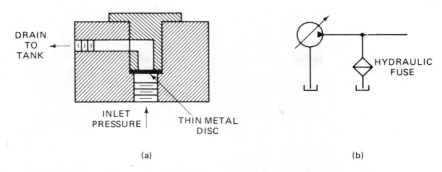

Figure 8-58.   Hydraulic fuse (schematic and graphic symbol.)

8.8 PRESSURE AND TEMPERATURE SWITCHES

A pressure switch is an instrument that automatically senses a change in pressure and opens or closes an electrical switching element when a predetermined pressure point is reached. A pressure-sensing element is the portion of the pressure switch that moves due to a change in pressure. There are four types of sensing elements, which produce the following four different models:

1. *Diaphragm* (see Fig. 8-59): This model can operate from vacuum pressures up to 150 psi. It has a weld-sealed diaphragm direct-acting on a snap-action switch.
2. *Bourdon tube* (see Fig. 8-60): This model can operate with pressures varying from 50 to 18,000 psi. It has a weld-sealed Bourdon tube acting on a snap-action switch.

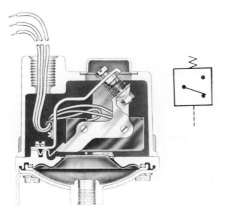

Figure 8-59.   Diaphragm pressure switch. (*Courtesy of Barksdale Controls, Los Angeles, California.*)

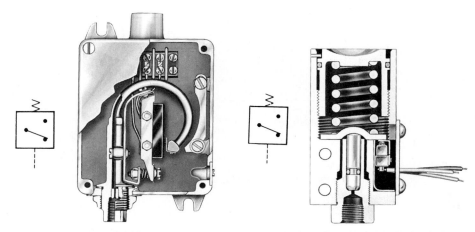

Figure 8-60. Bourdon tube pressure
switch. (*Courtesy of Barksdale Controls,
Los Angeles, California.*)

Figure 8-61. Sealed piston pressure
switch. (*Courtesy of Barksdale Controls,
Los Angeles, California.*)

3. *Sealed piston* (see Fig. 8-61): This model can operate with pressures from 15
to 12,000 psi. It has an O-ring sealed piston direct-acting on a snap-action
switch.

4. *Dia-seal piston* (see Fig. 8-62): This model can operate with 0.5- to 1600-psi
pressures. It has a dia-seal piston direct-acting on a snap-action switch. This
design combines diaphragm accuracy with piston long life and high-proof
pressure tolerance.

The electrical switching element in a pressure switch opens or closes an
electrical circuit in response to the actuating force it receives from the pressure-
sensing element. The designs shown use single-pole, double-throw snap-action
switches for maximum reliability.

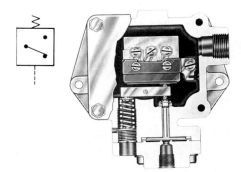

Figure 8-62. Dia-Seal Piston pressure
switch. (*Courtesy of Barksdale Controls,
Los Angeles, California.*)

There are two types of switching elements: normally open (N.O.) and normally closed (N.C.). A normally open switch is one in which no current can flow through the switching element until the switch is actuated. In a normally open switch, a plunger pin is held down by a snap-action leaf spring, and force must be applied to the plunger pin to close the circuit. In a normally closed switch, current flows through the switching element until the switch is actuated. In a normally closed switch, a plunger pin is held down by a snap-action leaf spring, and force must be applied to the plunger pin to open the circuit.

Pressure switches have three electrical terminals identified: C (common), N.C. (normally closed), and N.O. (normally open). When wiring in a switch, only two terminals are used. The common terminal is always used, plus a second terminal, either N.C. or N.O. depending on whether the switch is to operate as a normally open or normally closed switch.

A temperature switch is an instrument that automatically senses a change in temperature and opens or closes an electrical switching element when a predetermined temperature point is reached. Figure 8-63 shows a temperature switch that has a rated accuracy of $\pm 1°F$ maximum. This particular temperature switch incorporates a compensating device to cancel out the adverse effects of ambient fluctuations. At the top end is an adjustment screw to change the actuation point. The threads near the bottom end are used to mount locally on the hydraulic system where the temperature is to be measured. As in the case of pressure switches, temperature switches can be wired either normally open or normally closed. Applications of pressure and temperature switches are presented in Chapter 12.

Figure 8-63.   Temperature switch. (*Courtesy of Barksdale Controls, Los Angeles, California.*)

8.9 SHOCK ABSORBERS

A shock absorber is a device that brings a moving load to a gentle rest through the use of metered hydraulic fluid. Figure 8-64 shows a shock absorber that can provide a uniform gentle deceleration of any moving load from 25 to 25,000 lb or where the velocity and weight combination equals 3300 in. · lb. Heavy-duty units are available with load capacities of over 11 million in. · lb. and strokes up to 20 in.

The construction and operation of the shock absorber of Fig. 8-64 is described as follows:

These shock absorbers are filled completely with oil. Therefore, they may be mounted in any position or at any angle. The spring-return units are entirely self-contained, extremely compact types that require no external hoses, valves, or fittings. In this spring-returned type a built-in cellular accumulator accommodates oil displaced by the piston rod as the rod moves inward. See Fig. 8-65 for a cutaway view. Since the shock absorber is always filled with oil, there are no air pockets to cause spongy or erratic action.

These shock absorbers are multiple-orifice hydraulic devices. The orifices may be fixed in size as they are in the standard and heavy-duty types or of adjustable size as they are in the adjustable types. When a moving load strikes the bumper of the shock absorber, it sets the rod and piston in motion. The moving piston pushes oil through a series of holes from an inner high-pressure chamber to an outer low-pressure chamber.

The resistance to the oil flow caused by the holes (restrictions) creates a pressure that acts against the piston to oppose the moving load. Holes are spaced geometrically according to a proven formula that produces constant pressure on the side of the piston opposite the load (constant resisting force) from the beginning to nearly the end of the stroke. The piston progressively shuts off these

Figure 8-64. Shock absorber. (*Courtesy of E.G.D. Inc., Glenview, Illinois.*)

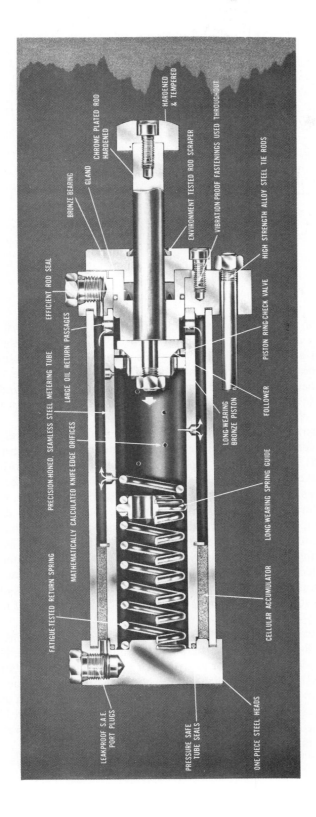

HARDENED & TEMPERED

CHROME PLATED ROD HARDENED

GLAND

BRONZE BEARING

ENVIRONMENT TESTED ROD SCRAPER

VIBRATION-PROOF FASTENINGS USED THROUGHOUT

HIGH STRENGTH ALLOY STEEL TIE RODS

EFFICIENT ROD SEAL

PRECISION-HONED, SEAMLESS STEEL METERING TUBE

LARGE OIL RETURN PASSAGES

MATHEMATICALLY CALCULATED KNIFE-EDGE ORIFICES

PISTON RING-CHECK VALVE

FOLLOWER

LONG-WEARING BRONZE PISTON

FATIGUE-TESTED RETURN SPRING

LONG-WEARING SPRING GUIDE

CELLULAR ACCUMULATOR

LEAKPROOF S.A.E. PORT PLUGS

PRESSURE-SAFE TUBE SEALS

ONE-PIECE STEEL HEADS

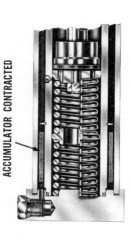

ACCUMULATOR CONTRACTED

Figure 8-65. Cutaway view of shock absorber. (*Courtesy of E.G.D. Inc., Glenview, Illinois.*)

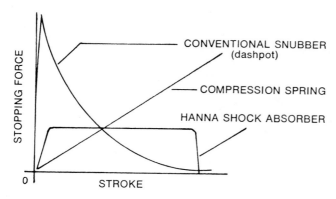

CONVENTIONAL SNUBBER
(dashpot)

COMPRESSION SPRING

HANNA SHOCK ABSORBER

STOPPING FORCE

0

STROKE

Figure 8-66. Shock absorbers create a uniform stopping force. (*Courtesy of E.G.D. Inc., Glenview, Illinois.*)

orifices as the piston and rod move inward. Therefore, the total orifice area continually decreases and the load decelerates uniformly. At the end of the stroke, the load comes to a gentle rest and pressure drops to zero gage pressure. This results in uniform deceleration and gentle stopping, with no bounce back. In bringing a moving load to a stop, the shock absorber converts work and kinetic energy into heat, which is dissipated to the surroundings or through a heat exchanger.

Figure 8-66 illustrates various methods of decelerating the same weight from the same velocity over the same distance. The area under each curve represents the energy absorbed. The snubber or dashpot produces a high peak force at the beginning of the stroke; then the resistance is sharply reduced during the remainder of the stopping distance. The snubber, being a single-orifice device, produces a nonuniform deceleration, and the initial peak force can produce damaging stresses on the moving load and structural frame. Compression springs have a low initial stopping force and build up to a peak at the end of the stroke. The springs store the energy during compression only to return it later, causing bounce back. The rising force deflection curve requires a longer stroke to stay below a given maximum deceleration force. Liquid springs rely on the slight compressibility of the hydraulic fluid to stop a load. The reaction of a liquid spring is similar to that of a mechanical spring.

One application for shock absorbers is energy dissipation of moving cranes, as illustrated in Fig. 8-67. Shock absorbers prevent bounce back of the bridge or trolley and thus provide protection for the operator, crane, and building structure.

Perhaps the most common application of shock absorbers is for the suspension systems of automobiles. Figure 8-68 is a schematic showing design details of such a shock absorber, whereas Fig. 8-69 provides a cutaway view of an actual design. As shown, the coil spring compression and the replenishing valve provide for smooth valve action and positive compression control.

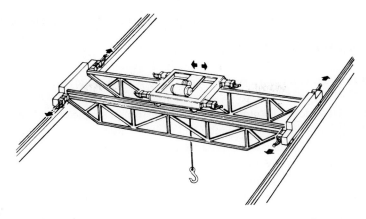

Figure 8-67. Crane application of shock absorbers. (*Courtesy of E.G.D. Inc., Glenview, Illinois.*)

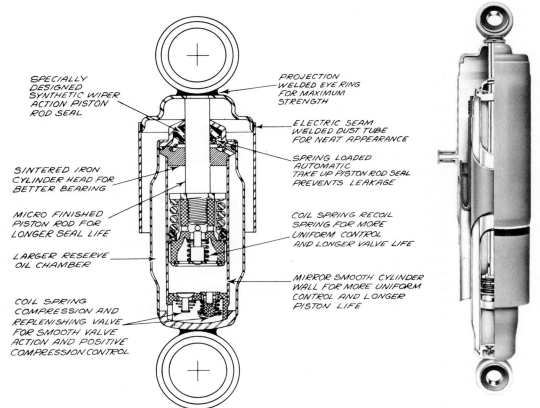

SPECIALLY
DESIGNED
SYNTHETIC WIPER
ACTION PISTON
ROD SEAL

SINTERED IRON
CYLINDER HEAD FOR
BETTER BEARING

MICRO FINISHED
PISTON ROD FOR
LONGER SEAL LIFE

LARGER RESERVE
OIL CHAMBER

COIL SPRING
COMPRESSION AND
REPLENISHING VALVE
FOR SMOOTH VALVE
ACTION AND POSITIVE
COMPRESSION CONTROL

PROJECTION
WELDED EYE RING
FOR MAXIMUM
STRENGTH

ELECTRIC SEAM
WELDED DUST TUBE
FOR NEAT APPEARANCE

SPRING LOADED
AUTOMATIC
TAKE UP PISTON ROD SEAL
PREVENTS LEAKAGE

COIL SPRING RECOIL
SPRING FOR MORE
UNIFORM CONTROL
AND LONGER VALVE LIFE

MIRROR SMOOTH CYLINDER
WALL FOR MORE UNIFORM
CONTROL AND LONGER
PISTON LIFE

Figure 8-68. (Left) Schematic of automotive shock absorber. (*Courtesy of Texaco Inc., New York, New York.*)

Figure 8-69.
(Right) Cutaway
view of automotive
shock absorber.
(*Courtesy of Texaco Inc., New York, New York.*)

EXERCISES

8-1. What is the purpose of a directional control valve?

8-2. What is a check valve, and what does it accomplish?

8-3. How does a pilot check valve differ from a simple check valve?

8-4. What is a four-way directional control valve?

8-5. What is a four-way, spring-centered, three-position valve?

8-6. Name three ways in which directional control valves may be actuated.

8-7. What is a solenoid, and how does it work?

8-8. What is the difference between an open-center and closed-center type of directional control valve?

8-9. What is a shuttle valve? Name one application.

8-10. What is the purpose of a pressure relief valve?

8-11. What is a pressure-reducing valve. What is its purpose?

8-12. What does an unloading valve accomplish?

8-13. What is a sequence valve? What is its purpose?

8-14. Name one application of a counterbalance valve.

8-15. What is the purpose of a flow control valve?

8-16. What is a pressure-compensated flow control valve?

8-17. What is a servo valve? How does it work?

8-18. What is the difference between a mechanical-hydraulic and an electrohydraulic servo valve?

8-19. Explain what is meant by a closed-loop system by drawing a block diagram.

8-20. What is a hydraulic fuse? What electrical device is it analogous to?

8-21. What is a pressure switch? Name four types.

8-22. What is a temperature switch?

8-23. What is the difference between a normally closed and normally open electric switch?

8-24. What is the purpose of a shock absorber? Name two applications.

8-25. A pressure relief valve has a pressure setting of 2000 psi. Compute the horsepower loss across this valve if it returns all the flow back to the tank from a 25-gpm pump.

8-26. An unloading valve is used to unload the pump of Exercise 8-25. If the pump discharge pressure during unloading equals 30 psi, how much hydraulic horsepower is being wasted?

8-27. Explain how the pilot-operated check valve shown in Fig. 8-4 works.

8-28. Name one application for a pilot-operated check valve.

8-29. Explain how the four-way directional control valve of Fig. 8-5 operates.

8-30. Name two advantages of the shear-seal rotary directional control valve as compared to the spool type.

8-31. How do a simple pressure relief valve and a compound relief valve differ in operation?

8-32. How does an unloading valve differ from a sequence valve in mechanical construction?

8-33. Explain the operational features of the pressure-compensated flow control valve of Fig. 8-41.

8-34. Describe how the Bourdon tube pressure switch of Fig. 8-60 operates.

8-35. Discuss the construction and operating features of the shock absorber of Fig. 8-64.

8-36. A pressure relief valve has a pressure setting of 140 bars. Compute the kW power loss across this valve if it returns all the flow back to the tank from a 0.0016 m³/s pump.

8-37. An unloading valve is used to unload the pump of Exercise 8-36. If the pump discharge pressure during unloading equals 2 bars, how much hydraulic kW power is being wasted?

8-38. How many positions does a spring-offset valve have?

8-39. How many positions does a spring-centered valve have?

8-40. What are rotary valves and how do they operate?

8-41. How are solenoids most often used in valves?

8-42. Name two ways of regulating flow to a hydraulic actuator.

8-43. What is cracking pressure?

8-44. Where are the ports of a relief valve connected?

8-45. Name the three basic functions of valves.

8-46. What is the difference in function between a pressure-relief valve and a hydraulic fuse?

8-47. Relative to directional control valves, distinguish between the terms *position, way,* and *port.*

8-48. What is a cartridge valve?

8-49. What is the difference between slip-in and screw-type cartridge valves?

8-50. Name five benefits of using cartridge valves.

8-51. Name five different valve functions that can be provided using cartridge valves.

8-52. Relative to the use of cartridge valves, what are integrated hydraulic circuits?

9

Hydraulic Circuit Design and Analysis

9.1 INTRODUCTION

The material presented in previous chapters dealt with basic fundamentals and system components. In this chapter we discuss basic hydraulic circuits. A hydraulic circuit is a group of components such as pumps, actuators, control valves, and conductors so arranged that they will perform a useful task. When analyzing or designing a hydraulic circuit, the following three important considerations must be taken into account:

1. Safety of operation
2. Performance of desired function
3. Efficiency of operation

It is very important for the fluid power technician or designer to have a working knowledge of components and how they operate in a circuit. Hydraulic circuits are developed through the use of graphical symbols for all components. Before hydraulic circuits can be understood, it is necessary to know these fluid power symbols. Figure 9-1 gives a table of symbols that conform to the American National Standards Institute (ANSI) specifications. Many of these symbols are presented in previous chapters, and ANSI symbols are used throughout this book. Although complete memorization of basic symbols is not necessary, Fig. 9-1 should be studied so that the symbols become familiar. The discussions that follow will cover circuits that represent basic hydraulic technology.

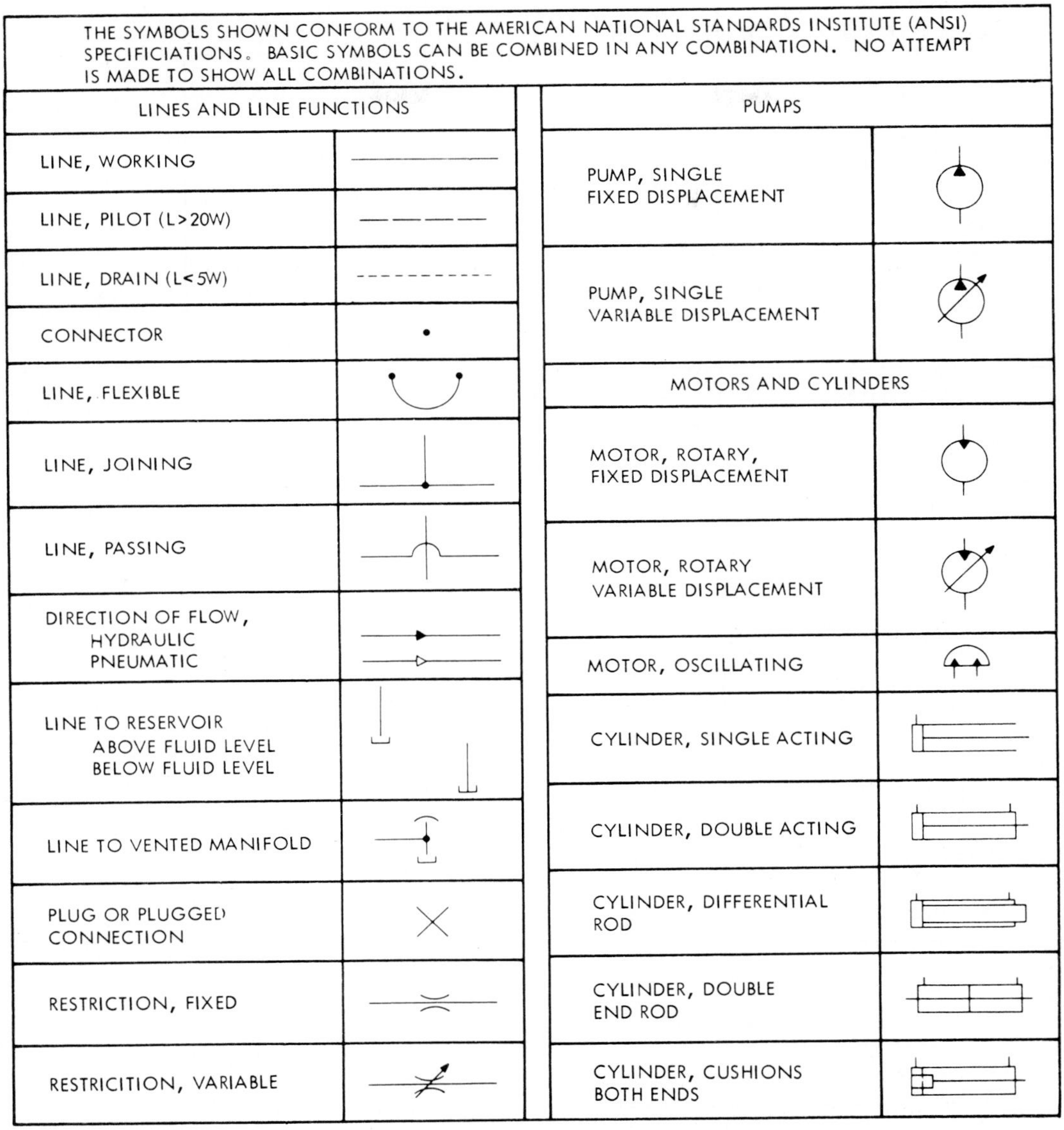

| THE SYMBOLS SHOWN CONFORM TO THE AMERICAN NATIONAL STANDARDS INSTITUTE (ANSI) SPECIFICIATIONS. BASIC SYMBOLS CAN BE COMBINED IN ANY COMBINATION. NO ATTEMPT IS MADE TO SHOW ALL COMBINATIONS. |||

LINES AND LINE FUNCTIONS		PUMPS	
LINE, WORKING		PUMP, SINGLE FIXED DISPLACEMENT	
LINE, PILOT (L>20W)			
LINE, DRAIN (L<5W)		PUMP, SINGLE VARIABLE DISPLACEMENT	
CONNECTOR			
LINE, FLEXIBLE		MOTORS AND CYLINDERS	
LINE, JOINING		MOTOR, ROTARY, FIXED DISPLACEMENT	
LINE, PASSING		MOTOR, ROTARY VARIABLE DISPLACEMENT	
DIRECTION OF FLOW, HYDRAULIC PNEUMATIC		MOTOR, OSCILLATING	
LINE TO RESERVOIR ABOVE FLUID LEVEL BELOW FLUID LEVEL		CYLINDER, SINGLE ACTING	
LINE TO VENTED MANIFOLD		CYLINDER, DOUBLE ACTING	
PLUG OR PLUGGED CONNECTION		CYLINDER, DIFFERENTIAL ROD	
RESTRICTION, FIXED		CYLINDER, DOUBLE END ROD	
RESTRICITION, VARIABLE		CYLINDER, CUSHIONS BOTH ENDS	

Figure 9-1. ANSI symbols of hydraulic components.

MISCELLANEOUS UNITS		BASIC VALVE SYMBOLS (CONT.)	
DIRECTION OF ROTATION (ARROW IN FRONT OF SHAFT)		VALVE, SINGLE FLOW PATH, NORMALLY OPEN	
COMPONENT ENCLOSURE		VALVE, MAXIMUM PRESSURE (RELIEF)	
RESERVOIR, VENTED		BASIC VALVE SYMBOL, MULTIPLE FLOW PATHS	
RESERVOIR, PRESSURIZED		FLOW PATHS BLOCKED IN CENTER POSITION	
PRESSURE GAGE		MULTIPLE FLOW PATHS (ARROW SHOWS FLOW DIRECTION)	
TEMPERATURE GAGE		VALVE EXAMPLES	
FLOW METER (FLOW RATE)		UNLOADING VALVE, INTERNAL DRAIN, REMOTELY OPERATED	
ELECTRIC MOTOR		DECELERATION VALVE, NORMALLY OPEN	
ACCUMULATOR, SPRING LOADED		SEQUENCE VALVE, DIRECTLY OPERATED, EXTERNALLY DRAINED	
ACCUMULATOR, GAS CHARGED		PRESSURE REDUCING VALVE	
FILTER OR STRAINER		COUNTER BALANCE VALVE WITH INTEGRAL CHECK	
HEATER			
COOLER		TEMPERATURE AND PRESSURE COMPENSATED FLOW CONTROL WITH INTEGRAL CHECK	
TEMPERATURE CONTROLLER			
INTENSIFIER		DIRECTIONAL VALVE, TWO POSITION, THREE CONNECTION	
PRESSURE SWITCH			
BASIC VALVE SYMBOLS		DIRECTIONAL VALVE, THREE POSITION, FOUR CONNECTION	
CHECK VALVE			
MANUAL SHUT OFF VALVE		VALVE, INFINITE POSITIONING (INDICATED BY HORIZONTAL BARS)	
BASIC VALVE ENVELOPE			
VALVE, SINGLE FLOW PATH, NORMALLY CLOSED			

Figure 9-1. Con't.

METHODS OF OPERATION		METHODS OF OPERATION	
PRESSURE COMPENSATOR		LEVER	
DETENT		PILOT PRESSURE	
MANUAL		SOLENOID	
MECHANICAL		SOLENOID CONTROLLED, PILOT PRESSURE OPERATED	
PEDAL OR TREADLE		SPRING	
PUSH BUTTON		SERVO	

Figure 9-1. Con't.

9.2 CONTROL OF A SINGLE-ACTING HYDRAULIC CYLINDER

Figure 9-2 shows how a two-position, three-way, manually actuated, spring offset directional control valve (DCV) can be used to control the operation of a single-acting cylinder. In the spring offset mode, full pump flow goes to the tank via the pressure relief valve. The spring in the rod end of the cylinder retracts the piston as oil from the blank end drains back to the tank. When the valve is manually actuated into its left envelope flow path configuration, pump flow extends the cylinder. At full extension, pump flow goes through the relief valve. Deactivation of the DCV allows the cylinder to retract as the DCV shifts into its spring offset mode.

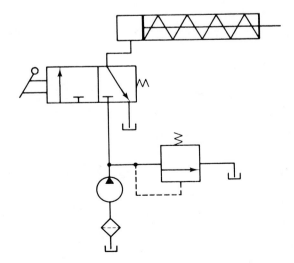

Figure 9-2. Control of single-acting hydraulic cylinder.

9.3 CONTROL OF A DOUBLE-ACTING HYDRAULIC CYLINDER

Figure 9-3 gives a circuit used to control a double-acting hydraulic cylinder. The operation is described as follows:

1. When the four-way valve is in its spring-centered position (tandem design), the cylinder is hydraulically locked. Also the pump is unloaded back to the tank at essentially atmospheric pressure.

2. When the four-way valve is actuated into the flow path configuration of the left envelope, the cylinder is extended against its load force F_{load} as oil flows from port P through port A. Also, oil in the rod end of the cylinder is free to flow back to the tank via the four-way valve from port B through port T. Note that the cylinder could not extend if this oil were not allowed to leave the rod end of the cylinder.

3. When the four-way valve is deactivated, the spring-centered envelope prevails, and the cylinder is once again hydraulically locked.

4. When the four-way valve is actuated into the right envelope configuration, the cylinder retracts as oil flows from port P through port B. Oil in the blank end is returned to the tank via the flow path from port A to port T.

5. At the ends of the stroke, there is no system demand for oil. Thus, the pump flow goes through the relief valve at its pressure-level setting unless the

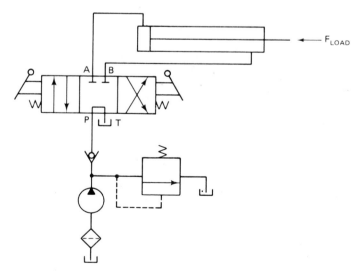

Figure 9-3.   Control of a double-acting hydraulic cylinder.

four-way valve is deactivated. In any event, the system is protected from any cylinder overloads.

6. The check valve prevents the load (if it becomes excessive) from retracting the cylinder while it is being extended using the left envelope flow path configuration.

9.4 *REGENERATIVE CIRCUIT*

Figure 9-4 shows a regenerative circuit that is used to speed up the extending speed of a double-acting hydraulic cylinder. Notice that the pipelines to both ends of the hydraulic cylinder are connected in parallel and that one of the ports of the four-way valve is blocked. The operation of the cylinder during the retraction stroke is the same as that of a regular double-acting cylinder. Fluid flows through the DCV via the right envelope during retraction. In this mode, fluid from the pump bypasses the DCV and enters the rod end of the cylinder. Fluid in the blank end drains back to the tank through the DCV as the cylinder retracts.

When the DCV is shifted into its left envelope configuration, the cylinder extends. The speed of extension is greater than that for a regular double-acting cylinder because flow from the rod end (Q_R) regenerates with the pump flow (Q_P) to provide a total flow rate (Q_T), which is greater than the pump flow rate to the blank end of the cylinder.

The equation for the extending speed can be obtained as follows (refer to Fig. 9-4): The total flow rate entering the blank end of the cylinder equals the

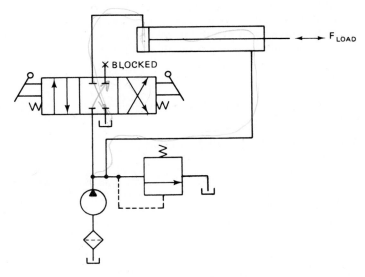

Figure 9-4. Regenerative circuit.

pump flow rate plus the regenerative flow rate coming from the rod end of the cylinder:

$$Q_T = Q_P + Q_R$$

Solving for the pump flow, we have

$$Q_P = Q_T - Q_R$$

We know that the total flow rate equals the piston area multiplied by the extending speed of the piston ($v_{P_{ext}}$). Similarly, the regenerative flow rate equals the difference of the piston and rod areas ($A_P - A_r$) multiplied by the extending speed of the piston. Substituting these two relationships into the preceding equation yields

$$Q_P = A_P v_{P_{ext}} - (A_P - A_r)v_{P_{ext}}$$

Solving for the extending speed of the piston, we have

$$v_{P_{ext}} = \frac{Q_P}{A_r} \qquad (9\text{-}1)$$

From Eq. (9-1), we see that the extending speed equals the pump flow divided by the area of the rod. Thus, a small rod area (which produces a large regenerative flow) provides a large extending speed. In fact the extending speed can be greater than the retracting speed if the rod area is made small enough. Let's find the ratio of extending and retracting speeds to determine under what conditions the extending and retracting speeds are equal. We know that the retracting speed ($v_{P_{ret}}$) equals the pump flow divided by the difference of the piston and rod areas:

$$v_{P_{ret}} = \frac{Q_P}{A_P - A_r} \qquad (9\text{-}2)$$

Dividing Eq. (9-1) by Eq. (9-2), we have

$$\frac{v_{P_{ext}}}{v_{P_{ret}}} = \frac{Q_P/A_r}{Q_P/(A_P - A_r)} = \frac{A_P - A_r}{A_r}$$

Upon further simplification we obtain the desired equation:

$$\frac{v_{P_{ext}}}{v_{P_{ret}}} = \frac{A_P}{A_r} - 1 \qquad (9\text{-}3)$$

From Eq. (9-3), we see that when the piston area equals two times the rod area, the extension and retraction speeds are equal. In general, the greater the ratio of piston area to rod area, the greater the ratio of extending speed to retracting speed.

It should be kept in mind that the load-carrying capacity of a regenerative cylinder during extension is less than that obtained from a regular double-acting cylinder. The load-carrying capacity (F_{load}) for a regenerative cylinder equals the pressure times the piston rod area rather than the pressure times piston area. This is due to the same system pressure acting on both sides of the piston during the extending stroke of the regenerative cylinder. This is in accordance with Pascal's law.

$$F_{load} = PA_r \qquad (9\text{-}4)$$

Thus, we are not obtaining more power from the regenerative cylinder because the extending speed is increased at the expense of load-carrying capacity.

EXAMPLE 9-1

A double-acting cylinder is hooked up in the regenerative circuit of Fig. 9-4. The relief valve setting is 1000 psi. The piston area is 25 in.^{2}, and the rod area is 7 in.^{2} If the pump flow is 20 gpm, find the cylinder speed and load-carrying capacity for the

    a. Extending stroke
    b. Retracting stroke

Solution

    a. $\quad v_{P_{ext}} = \dfrac{Q_P}{A_r} = \dfrac{(20 \text{ gpm})(231 \text{ in.}^3/1 \text{ gal})(1 \text{ min}/60 \text{ sec})}{7 \text{ in.}^2} = 11.0 \text{ in./sec}$

    $\quad F_{load_{ext}} = PA_r = 1000 \text{ lb/in.}^2 \times 7 \text{ in.}^2 = 7000 \text{ lb}$

    b. $\quad v_{P_{ret}} = \dfrac{Q_P}{A_P - A_r} = \dfrac{20 \times \dfrac{231}{60}}{25 - 7} = 4.28 \text{ in./sec}$

    $\quad F_{load_{ret}} = P(A_P - A_r) = 1000 \text{ lb/in.}^2 \times (25 - 7) \text{ in.}^2 = 18{,}000 \text{ lb}$

Figure 9-5 shows an application using a four-way valve having a spring-centered design with a closed tank port and a pressure port open to outlet ports *A* and *B*.

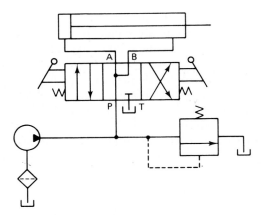

Figure 9-5. Drilling machine application.

The application is for a drilling machine, where the following operations take place:

1. The spring-centered position gives rapid spindle advance (extension).
2. The left envelope mode gives slow feed (extension) when the drill starts to cut into the workpiece.
3. The right envelope mode retracts the piston.

Why does the spring-centered position give rapid extension of the cylinder (drill spindle)? The reason is simple. Oil from the rod end regenerates with the pump flow going to the blank end. This effectively increases pump flow to the blank end of the cylinder during the spring-centered mode of operation. Once again we have a regenerative cylinder. It should be noted that the cylinder used in a regenerative circuit is actually a regular double-acting cylinder. What makes it a regenerative cylinder is the way it is hooked up in the circuit. The blank and rod ends are connected in parallel during the extending stroke of a regenerative cylinder. The retraction mode is the same as a regular double-acting cylinder.

9.5 PUMP-UNLOADING CIRCUIT

In Fig. 9-6 we see a circuit using an unloading valve to unload a pump. The unloading valve opens when the cylinder reaches the end of its extension stroke because the check valve keeps high-pressure oil in the pilot line of the unloading valve. When the DCV is shifted to retract the cylinder, the motion of the piston reduces the pressure in the pilot line of the unloading valve. This resets the unloading valve until the cylinder is fully retracted, at which point the unloading valve unloads the pump. Thus, the unloading valve unloads the pump at the ends

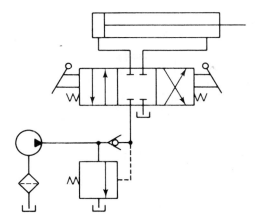

Figure 9-6.   Pump-unloading circuit.

of the extending and retraction strokes as well as in the spring-centered position of the DCV.

9.6 DOUBLE-PUMP HYDRAULIC SYSTEM

Figure 9-7 shows a circuit that uses a high-pressure, low-flow pump in conjunction with a low-pressure, high-flow pump. A typical application is a punch press in which the hydraulic ram must extend rapidly over a large distance with very low pressure but high flow requirements. However, during the short motion portion when the punching operation occurs, the pressure requirements are high due to the punching load. Since the cylinder travel is small during the punching operation, the flow-rate requirements are also low.

The circuit shown eliminates the necessity of having a very expensive high-pressure, high-flow pump. When the punching operation begins, the increased pressure opens the unloading valve to unload the low-pressure pump. The purpose of the relief valve is to protect the high-pressure pump from overpressure at the end of the cylinder stroke. The check valve protects the low-pressure pump from high pressure, which occurs during the punching operation, at the ends of the cylinder stroke, and when the DCV is in its spring-centered mode.

9.7 PRESSURE INTENSIFIER CIRCUIT

One way to eliminate the high-pressure, low-flow pump in the punch press application of Fig. 9-7 is to use a pressure intensifier. This is done in the circuit of Fig. 9-8, which also includes a pilot check valve and sequence valve. Very high pressures can be supplied by a pressure intensifier operating on a low-pressure pump. The intensifier should be installed near the cylinder to keep the high-pressure lines

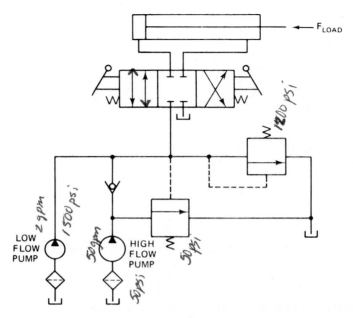

Figure 9-7. Double-pump hydraulic system.

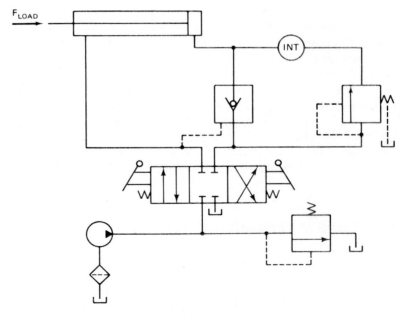

Figure 9-8. Pressure intensifier circuit.

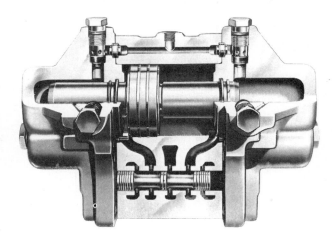

Figure 9-9.   Automatic pressure intensifier. (*Courtesy of Rexnord Inc.,
Hydraulic Components Division, Racine, Wisconsin.*)

as short as possible. An automatic-type pressure intensifier similar to that illustrated in Fig. 9-9 is utilized. When the pressure in the cylinder reaches the sequence valve pressure setting, the intensifier starts to operate. The high-pressure
output of the intensifier closes the pilot check valve and pressurizes the blank end
of the cylinder to perform the punching operation. A pilot check valve is used
instead of a regular check valve to permit retraction of the cylinder.

9.8 COUNTERBALANCE VALVE APPLICATION

Figure 9-10 illustrates the use of a counterbalance or back-pressure valve to keep
a vertically-mounted cylinder in the upward position while the pump is idling. The
counterbalance valve is set to open at slightly above the pressure required to hold
the piston up. This permits the cylinder to be forced downward when pressure is
applied on the top. The open-center directional control valve unloads the pump.
The DCV is a solenoid-actuated, spring-centered valve with an open-center flow
path configuration.

9.9 HYDRAULIC CYLINDER SEQUENCING CIRCUIT

As stated earlier, a sequence valve causes operations in a hydraulic circuit to
behave sequentially. Figure 9-11 is an example where two sequence valves are
used to control the sequence of operations of two double-acting cylinders. When
the DCV is shifted into its left envelope mode, the left cylinder extends completely, and then the right cylinder extends. If the DCV is then shifted into its right
envelope mode, the right cylinder retracts fully, and then the left cylinder retracts.

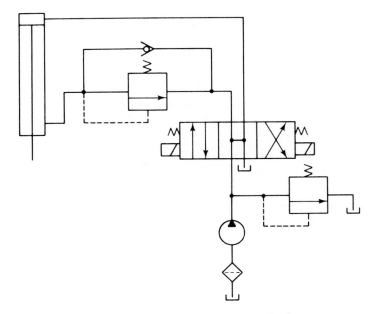

Figure 9-10. Counterbalance valve application.

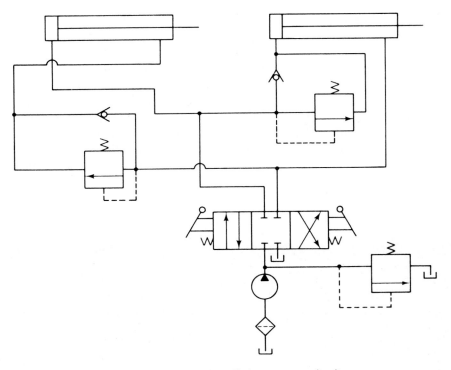

Figure 9-11. Hydraulic cylinder sequence circuit.

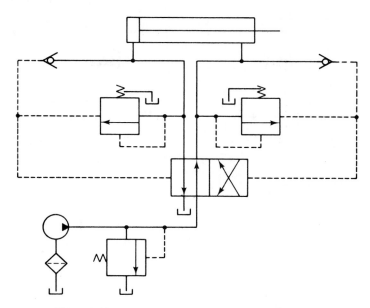

Figure 9-12.   Automatic cylinder reciprocating system.

This sequence of cylinder operation is controlled by the sequence valves. The spring-centered position of the DCV locks both cylinders in place.

One application of this circuit is a production operation. For example, the left cylinder could extend and clamp a workpiece via a power vise jaw. Then the right cylinder extends to drive a spindle to drill a hole in the workpiece. The right cylinder then retracts the drill spindle, and then the left cylinder retracts to release the workpiece for removal. Obviously these machining operations must occur in the proper sequence as established by the sequence valves in the circuit.

9.10 AUTOMATIC CYLINDER RECIPROCATING SYSTEM

Figure 9-12 is a circuit that produces continuous reciprocation of a hydraulic cylinder. This is accomplished by using two sequence valves, each of which senses a stroke completion by the corresponding buildup of pressure. Each check valve and corresponding pilot line prevents shifting of the four-way valve until the particular stroke of the cylinder has been completed. The check valves are needed to allow pilot oil to leave either end of the DCV while pilot pressure is applied to the opposite end. This permits the spool of the DCV to shift as required.

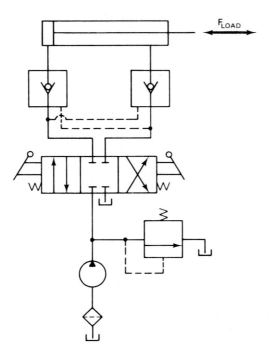

Figure 9-13.   Locked cylinder using pilot check valves.

9.11 LOCKED CYLINDER USING PILOT CHECK VALVES

In many cylinder applications, it is necessary to lock the cylinder so that its piston cannot be moved due to an external force acting on the piston rod. One method for locking a cylinder in this fashion is by using pilot check valves, as shown in Fig. 9-13. The cylinder can be extended and retracted as normally done by the action of the directional control valve. If regular check valves were used, the cylinder could not be extended or retracted by the action of the DCV. An external force, acting on the piston rod, will not move the piston in either direction because reverse flow through either pilot check valve is not permitted under these conditions.

9.12 CYLINDER SYNCHRONIZING CIRCUIT

Figure 9-14 is a very interesting circuit, which seems to show how two identical cylinders can be synchronized by piping them in parallel. However, even if the two cylinders are identical, it would be necessary for the loads on the cylinders to

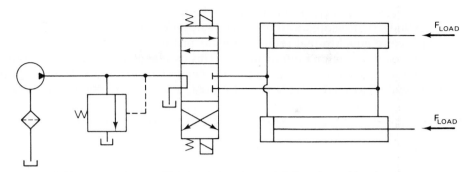

Figure 9-14.   Cylinders hooked in parallel will not operate in synchronization.

be identical in order for them to extend in exact synchronization. If the loads are not exactly identical (as is always the case), the cylinder with the smaller load would extend first because it would move at a lower pressure level. After this cylinder has fully completed its stroke, the system pressure will increase to the higher level required to extend the cylinder with the greater load. It should be pointed out that no two cylinders are really identical. For example, differences in packing friction will vary from cylinder to cylinder. This alone would prevent cylinder synchronization for the circuit of Fig. 9-14.

The circuit of Fig. 9-15 shows a simple way to synchronize two cylinders. Fluid from the pump is delivered to the blank end of cylinder 1, and fluid from the rod end of cylinder 1 is delivered to the blank end of cylinder 2. Fluid returns to the tank from the rod end of cylinder 2 via the DCV. Thus, the cylinders are hooked in series. For the two cylinders to be synchronized, the piston area of

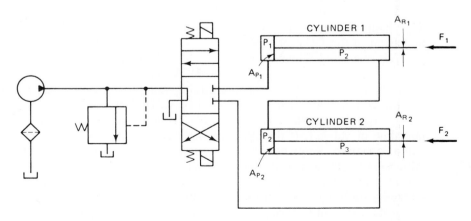

Figure 9-15.   Cylinders hooked in series will operate in synchronization.

cylinder 2 must equal the difference between the areas of the piston and rod for cylinder 1. It should also be noted that the pump must be capable of delivering a pressure equal to that required for the piston of cylinder 1 by itself to overcome the loads acting on both cylinders. This can be shown as follows, noting that the pressures are equal at the blank end of cylinder 2 and the rod end of cylinder 1 per Pascal's law (refer to Fig. 9-15 for area, load, and pressure identifications):

$$P_1 A_{P_1} - P_2(A_{P_1} - A_{R_1}) = F_1$$

and

$$P_2 A_{P_2} - P_3(A_{P_2} - A_{R_2}) = F_2$$

Adding both equations and noting that $A_{P_2} = A_{P_1} - A_{R_1}$ and that $P_3 = 0$ (due to the drain line to the tank), we obtain the desired result:

$$P_1 A_{P_1} = F_1 + F_2 \qquad\qquad (9\text{-}5)$$

9.13 FAIL-SAFE CIRCUITS

Fail-safe circuits are those designed to prevent injury to the operator or damage to equipment. In general they prevent the system from accidentally falling on an operator, and they also prevent overloading of the system. Figure 9-16 shows a fail-safe circuit that prevents the cylinder from accidentally falling in the event a hydraulic line ruptures or a person inadvertently operates the manual override on the pilot-actuated directional control valve when the pump is not operating. To lower the cylinder, pilot pressure from the blank end of the piston must pilot-open the check valve at the rod end to allow oil to return through the DCV to the tank. This happens when the push-button valve is actuated to permit pilot pressure actuation of the DCV or when the DCV is directly manually actuated while the pump is operating. The pilot-operated DCV allows free flow in the opposite direction to retract the cylinder when this DCV returns to its spring offset mode.

Figure 9-17 shows a fail-safe circuit that provides overload protection for system components. Directional control valve 1 is controlled by push-button three-way valve 2. When overload valve 3 is in its spring offset mode, it drains the pilot line of valve 1. If the cylinder experiences excessive resistance during the extension stroke, sequence valve 4 pilot-actuates overload valve 3. This drains the pilot line of valve 1, causing it to return to its spring offset mode. If a person then operates push-button valve 2, nothing will happen unless overload valve 3 is manually shifted into its blocked port configuration. Thus, the system components are protected against excessive pressure due to an excessive cylinder load during its extension stroke.

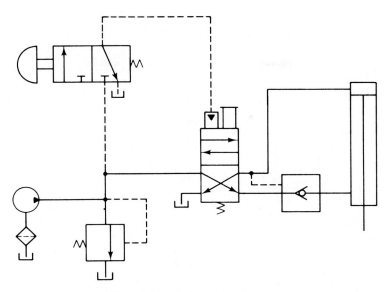

Figure 9-16. Fail-safe circuit.

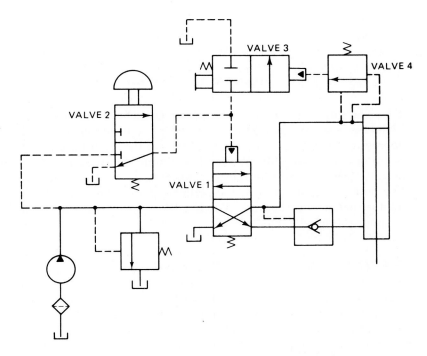

Figure 9-17. Fail-safe circuit with overload protection.

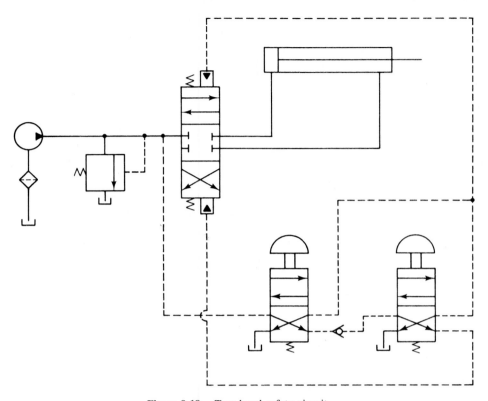

Figure 9-18. Two-hand safety circuit.

The safety circuit of Fig. 9-18 is designed to protect an operator from injury. For the circuit to function (extend and retract the cylinder), the operator must depress both manually actuated valves via the push buttons. Furthermore, the operator cannot circumvent this safety feature by tying down one of the buttons, because it is necessary to release both buttons to retract the cylinder. When the two buttons are depressed, the main three-position directional control valve is pilot-actuated to extend the cylinder. When both push buttons are released, the cylinder retracts.

9.14 SPEED CONTROL OF A HYDRAULIC MOTOR

Figure 9-19 shows a circuit where speed control of a hydraulic motor is accomplished using a pressure-compensated flow control valve.

The operation is as follows:

1. In the spring-centered position of the tandem four-way valve, the motor is hydraulically locked.

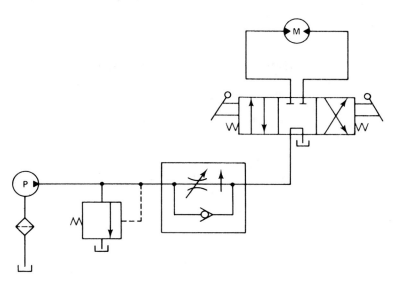

Figure 9-19. Speed control of hydraulic motor using flow control valve.

2. When the four-way valve is actuated into the left envelope, the motor rotates in one direction. Its speed can be varied by adjusting the setting of the throttle of the flow control valve. In this way the speed can be infinitely varied as the excess oil goes through the pressure relief valve.

3. When the four-way valve is deactivated, the motor stops suddenly and becomes locked.

4. When the right envelope of the four-way valve is in operation, the motor turns in the opposite direction. The pressure relief valve provides overload protection if, for example, the motor experiences an excessive torque load.

9.15 HYDRAULIC MOTOR BRAKING SYSTEM

When using a hydraulic motor in a fluid power system, consideration should be given to the type of loading that the motor will experience. A hydraulic motor may be driving a machine having a large inertia. This would create a flywheel effect on the motor, and stopping the flow of fluid to the motor would cause it to act as a pump. In a situation such as this, the circuit should be designed to provide fluid to the motor while it is pumping to prevent it from pulling in air. In addition, provisions should be made for the discharge fluid from the motor to be returned to the tank either unrestricted or through a relief valve. This would stop the motor rapidly but without damage to the system. Figure 9-20 is a motor circuit that possesses these desirable features for either direction of motor rotation.

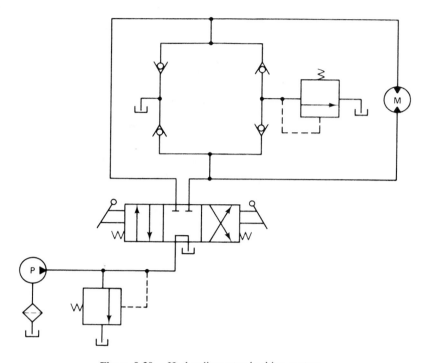

Figure 9-20. Hydraulic motor braking system.

9.16 HYDROSTATIC TRANSMISSION SYSTEM

Figures 9-19 and 9-20 are actually hydrostatic transmissions. They are called open-circuit drives because the pump draws its fluid from a reservoir. Its output is then directed to a hydraulic motor and discharged from the motor back into the reservoir. In a closed-circuit drive, exhaust oil from the motor is returned directly to the pump inlet. Figure 9-21 gives a circuit of a closed-circuit drive that allows for only one direction of motor rotation. The motor speed is varied by changing the pump displacement. The torque capacity of the motor can be adjusted by the pressure setting of the relief valve. Makeup oil to replenish leakage from the closed loop flows into the low-pressure side of the circuit through a line from the reservoir.

Many hydrostatic transmissions are reversible closed-circuit drives that use a variable displacement reversible pump. This allows the motor to be driven in either direction and at infinitely variable speeds depending on the position of the pump displacement control. Figure 9-22 shows a circuit of such a system using a fixed displacement hydraulic motor. Internal leakage losses are made up by a replenishing pump, which keeps a positive pressure on the low-pressure side of

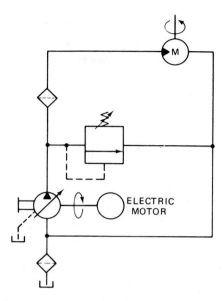

Figure 9-21. Closed circuit-one-direction hydrostatic transmission.

the system. There are two check and two relief valves to accommodate the two directions of flow and motor rotation.

Closed-circuit drives are available as completely integrated units with all the controls and valving enclosed in a single, compact housing. Figure 9-23 shows

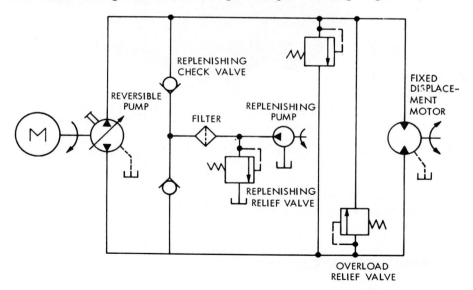

Figure 9-22. Closed circuit-reversible direction hydrostatic transmission. (*Courtesy of Sperry Vickers, Sperry Rand Corp., Troy, Michigan.*)

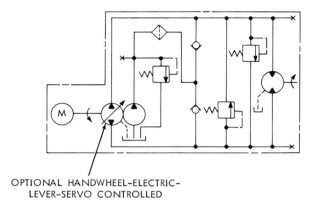

OPTIONAL HANDWHEEL–ELECTRIC–
LEVER–SERVO CONTROLLED

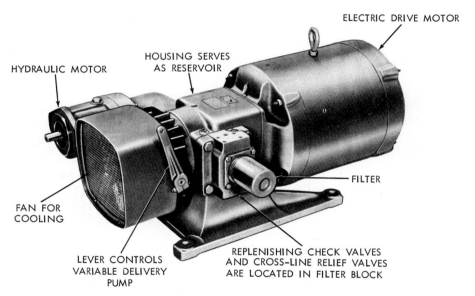

Figure 9-23.   Packaged unit hydrostatic drive-reversible direction. (*Courtesy of Sperry Vickers, Sperry Rand Corp., Troy, Michigan.*)

such a system, which is driven by an electric motor. Notice the lever controls for varying the pump displacement. This unit is not only rugged and compact but also is easy to install.

9.17 *AIR-OVER-OIL CIRCUIT*

Sometimes circuits using both air and oil are utilized to obtain the advantages of each medium. Figure 9-24 shows a counterbalance system, which is an air-over-oil circuit. Compressed air flows through a filter, regulator, lubricator unit (FRL) and

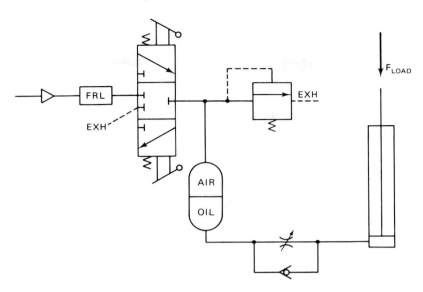

Figure 9-24.   Air-over-oil circuit.

into a surge tank via a directional control valve (upper flow path configuration). Thus, the surge tank is pressurized by compressed air. This pushes oil out the bottom of the surge tank and to the hydraulic cylinder through a check valve and orifice hooked in parallel. This extends the cylinder to lift a load. When the directional control valve is shifted into its lower flow path mode, the cylinder retracts at a controlled rate. This happens because the variable orifice provides a controlled return flow of oil as air leaves the surge tank and exhausts into the atmosphere via the directional control valve. The load can be stopped at any intermediate position by the spring-centered position of the directional control valve. This system eliminates the need for a costly hydraulic pump and tank unit.

9.18 AIR-OVER-OIL INTENSIFIER SYSTEM

In Fig. 9-25 we see an air-over-oil circuit, which drives a cylinder over a large distance at low pressure and then over a small distance at high pressure. Shop air can be used to extend and retract the cylinder during the low-pressure portion of the cycle. The system operates as follows: Valve 1 extends and retracts the cylinder using shop air at approximately 80 psi. Valve 2 applies air pressure to the top end of the hydraulic intensifier. This produces high hydraulic pressure at the bottom end of the intensifier. Actuation of valve 1 directs air to the approach tank. This forces oil at 80 psi through the bottom of the intensifier to the blank end of the cylinder. When the cylinder experiences its load (such as the punching operation in a punch press), valve 2 is actuated, which sends shop air to the top end of the

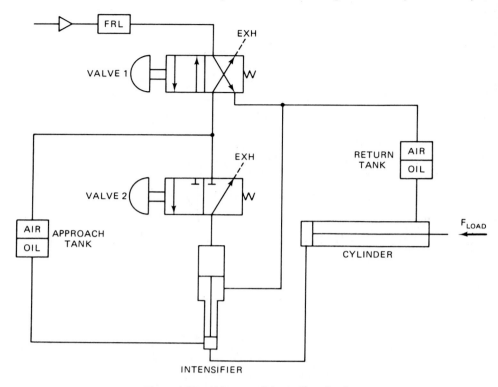

Figure 9-25. Air-over-oil intensifier circuit.

intensifier. The high-pressure oil cannot return to the approach tank because this port is blocked off by the downward motion of the intensifier piston. Thus, the cylinder receives high-pressure oil at the blank end to overcome the load. When valve 2 is released, the shop air is blocked, and the top end of the intensifier is vented to the atmosphere. This terminates the high-pressure portion of the cycle. When valve 1 is released, the air in the approach tank is vented, and shop air is directed to the return tank. This delivers oil at shop pressure to the rod end of the cylinder, causing it to retract. Oil enters the bottom end of the intensifier and flows back to the approach tank. This completes the entire cycle. Figure 9-26 shows an air-oil intensifier and its graphical symbol. This type of intensifier is capable of producing output hydraulic pressures up to 3000 psi.

9.19 ACCUMULATORS AND ACCUMULATOR CIRCUITS

A hydraulic accumulator is a device that stores the potential energy of an incompressible fluid held under pressure by an external source against some dynamic force. This dynamic force can come from three different sources: gravity, me-

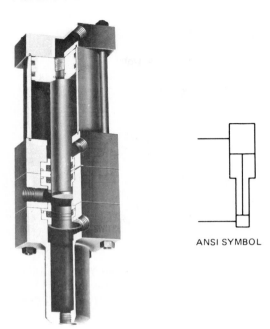

ANSI SYMBOL

Figure 9-26. Cutaway view of an air-oil pressure intensifier. (*Courtesy of the S-P Manufacturing Corp., Cleveland, Ohio.*)

chanical springs, and compressed gases. The stored potential energy in the accumulator is a quick secondary source of fluid power capable of doing useful work as required by the system.

There are three basic types of accumulators used in hydraulic systems. They are identified as follows:

1. Weight-loaded or gravity type
2. Spring-loaded type
3. Gas-loaded type

The weight-loaded type is historically the oldest. This type consists of a vertical, heavy-wall steel cylinder, which incorporates a piston with packings to prevent leakage. A dead weight is attached to the top of the piston (see Fig. 9-27). The force of gravity of the dead weight provides the potential energy in the accumulator. This type of accumulator creates a constant fluid pressure throughout the full volume output of the unit regardless of the rate and quantity of output. In the other types of accumulators, the fluid output pressure decreases as a function of the volume output of the accumulator. The main disadvantage of this type of accumulator is its extremely large size and heavy weight, which makes it unsuitable for mobile equipment. In this section we present the various types of accumulators and several of their common applications. The sizing of gas-loaded

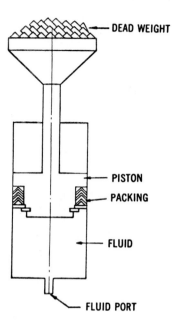

Figure 9-27. Weight-loaded accumulator. (*Courtesy of Greer Olaer Products Division/Greer Hydraulics Inc., Los Angeles, California.*)

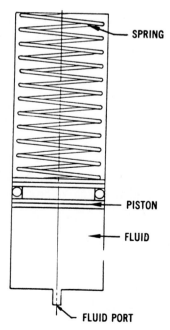

Figure 9-28. Spring-type accumulator. (*Courtesy of Greer Olaer Products Division/Greer Hydraulics Inc., Los Angeles, California.*)

accumulators for given applications is covered in Chapter 10, after Boyle's law of gases is discussed.

A spring-loaded accumulator is similar to the weight-loaded type except that the piston is preloaded with a spring, as illustrated in Fig. 9-28. The spring is the source of energy that acts against the piston, forcing the fluid into the hydraulic system. The pressure generated by this type of accumulator depends on the size and preloading of the spring. In addition, the pressure exerted on the fluid is not a constant. The spring-loaded accumulator typically delivers a relatively small volume of oil at low pressures. Thus, they tend to be heavy and large for high-pressure, large-volume systems. This type of accumulator should not be used for applications requiring high cycle rates because the spring will fatigue and lose its elasticity. The result is an inoperative accumulator.

Gas-loaded accumulators (frequently called hydropneumatic accumulators) have been found to be more practical than the weight- and spring-loaded types. The gas-loaded type operates in accordance with Boyle's law of gases, which states that for a constant temperature process, the pressure of a gas varies inversely with its volume. Thus, for example, the gas volume of the accumulator would be cut in half if the pressure were doubled. The compressibility of gases accounts for the storage of potential energy. This energy forces the oil out of the accumulator when the gas expands due to the reduction of system pressure when, for example, an actuator rapidly moves a load.

Gas-loaded accumulators fall into two main categories:

1. Nonseparator type
2. Separator type

The nonseparator type consists of a fully enclosed shell containing an oil port on the bottom and a gas charging valve on the top (see Fig. 9-29). The gas is

GAS VALVE

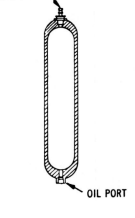

OIL PORT

Figure 9-29. Nonseparator-type accumulator. (*Courtesy of Greer Olaer Products Division/Greer Hydraulics Inc., Los Angeles, California.*)

confined in the top and the oil at the bottom of the shell. There is no physical separator between the gas and oil, and thus the gas pushes directly on the oil. The main advantage of this type is its ability to handle large volumes of oil. The main disadvantage is absorption of the gas in the oil due to the lack of a separator. This type must be installed vertically to keep the gas confined at the top of the shell. This type is not recommended for use with high-speed pumps because the entrapped gas in the oil could cause cavitation and damage to the pump. Absorption of gas in the oil also makes the oil compressible, resulting in spongy operation of the hydraulic actuators.

The commonly accepted design of gas-loaded accumulators is the separator type. In this type there is a physical barrier between the gas and the oil. This barrier effectively utilizes the compressibility of the gas. The three major classifications of the separator accumulator are

1. Piston type
2. Diaphragm type
3. Bladder type

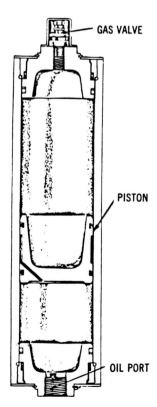

Figure 9-30.   Piston-type accumulator. (*Courtesy of Greer Olaer Products Division/Greer Hydraulics Inc., Los Angeles, California.*)

The piston type consists of a cylinder containing a freely floating piston with proper seals, as illustrated in Fig. 9-30. The piston serves as the barrier between the gas and oil. A threaded lock ring provides a safety feature, which prevents the operator from disassembling the unit while it is precharged. The main disadvantages of the piston type are that they are expensive to manufacture and have practical size limitations. Piston and seal friction may also be a problem in low-pressure systems. Also, appreciable leakage tends to occur over a long period of time, requiring frequent precharging. Piston accumulators should not be used as pressure pulsation dampeners or shock absorbers because of the inertia of the piston and the friction of the seals. The principal advantage of the piston accumulator is its ability to handle very high or low temperature system fluids through the utilization of compatible O-ring seals.

Figure 9-31 shows the internal construction features of a piston-type accumulator that has a safety seal feature. This unique design concept permits the end cap O-rings to lose their pressure seal through a limited degree of housing deformation should pressures exceed safe operating limits and before fracture can occur. A specially designed end cap with a split-ring locking arrangement prevents the pressure within the accumulator from dislodging the end caps.

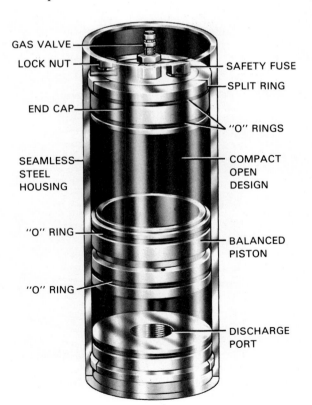

GAS VALVE
LOCK NUT
END CAP
SEAMLESS STEEL HOUSING
"O" RING
"O" RING

SAFETY FUSE
SPLIT RING
"O" RINGS
COMPACT OPEN DESIGN
BALANCED PISTON
DISCHARGE PORT

Figure 9-31. Piston accumulator with safety seal split ring. (*Courtesy of American Bosch, Springfield, Massachusetts.*)

In Fig. 9-32 we see an application of this safety seal split-ring accumulator. As shown, a bank of 19 accumulators is used as the power source of a rotary hydraulic motor system for the emergency closing of valves on natural gas lines.

The diaphragm-type accumulator consists of a diaphragm, secured in the shell, which serves as an elastic barrier between the oil and gas (see Fig. 9-33). A shutoff button, which is secured at the base of the diaphragm, covers the inlet of the line connection when the diaphragm is fully stretched. This prevents the diaphragm from being pressed into the opening during the precharge period. On the gas side, the screw plug allows control of the charge pressure and charging of the accumulator by means of a charging and testing device, which is shown in Fig. 9-34.

Figure 9-35 illustrates the operation of a diaphragm-type accumulator. The hydraulic pump delivers oil into the accumulator and deforms the diaphragm. As the pressure increases, the volume of gas decreases, thus storing hydraulic energy. In the reverse case, where additional oil is required in the circuit, it comes from the accumulator as the pressure drops in the system by a corresponding

Figure 9-32. Application of safety seal split-ring accumulator. (*Courtesy of American Bosch, Springfield, Massachusetts.*)

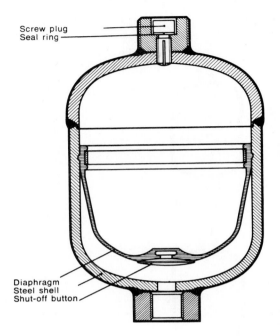

Screw plug
Seal ring

Diaphragm
Steel shell
Shut-off button

Figure 9-33.   Diaphragm-type accumulator. (*Courtesy of Robert Bosch Corp., Broadview, Illinois.*)

amount. The primary advantage of this type of accumulator is its small weight-to-volume ratio, which makes it suitable almost exclusively for airborne applications.

A bladder-type accumulator contains an elastic barrier (bladder) between the oil and gas, as illustrated in Fig. 9-36. The bladder is fitted in the accumulator by means of a vulcanized gas-valve element and can be installed or removed through the shell opening at the poppet valve. The poppet valve closes the inlet when the

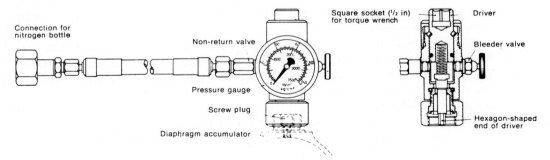

Connection for
nitrogen bottle

Non-return valve

Square socket (½ in)
for torque wrench

Driver

Bleeder valve

Pressure gauge

Screw plug

Diaphragm accumulator

Hexagon-shaped
end of driver

Figure 9-34.   Charging and testing device for diaphragm-type accumulators. (*Courtesy of Robert Bosch, Broadview, Illinois.*)

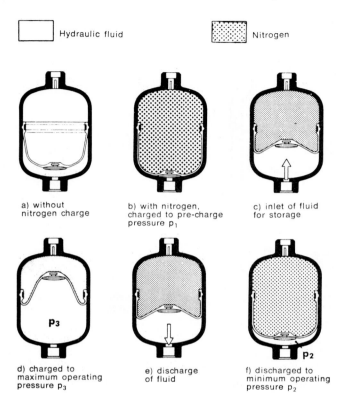

Hydraulic fluid   Nitrogen

a) without
nitrogen charge

b) with nitrogen,
charged to pre-charge
pressure p_1

c) inlet of fluid
for storage

d) charged to
maximum operating
pressure p_3

e) discharge
of fluid

f) discharged to
minimum operating
pressure p_2

Figure 9-35.   Operation of a diaphragm-type accumulator. (*Courtesy of
Robert Bosch Corp., Broadview, Illinois.*)

accumulator bladder is fully expanded. This prevents the bladder from being
pressed into the opening. A shock-absorbing device protects the valve against
accidental shocks during quick opening. The greatest advantage of this type of
accumulator is the positive sealing between the gas and oil chambers. The lightweight bladder provides quick pressure response for pressure regulating, pump
pulsation, and shock-dampening applications.

Figure 9-37 illustrates the operation of a bladder-type accumulator. The
hydraulic pump delivers oil into the accumulator and deforms the bladder. As the
pressure increases, the volume of gas decreases, thus storing hydraulic energy. In
the reverse case, where additional oil is required in the circuit, it comes from the
accumulator as pressure drops in the system by a corresponding amount. A charging and testing device for a bladder-type accumulator is shown in Fig. 9-38.

Figure 9-39 provides a photograph of an actual bladder-type accumulator,
which contains the following features:

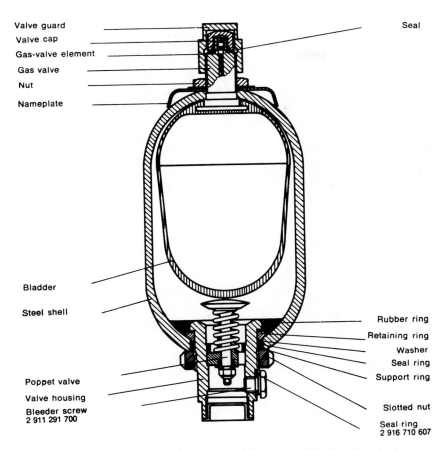

Valve guard
Valve cap
Gas-valve element
Gas valve
Nut
Nameplate

Seal

Bladder
Steel shell

Rubber ring
Retaining ring
Washer
Seal ring
Support ring

Poppet valve
Valve housing
Bleeder screw
2 911 291 700

Slotted nut

Seal ring
2 916 710 607

Figure 9-36.   Bladder-type accumulator. (*Courtesy of Robert Bosch Corp.,* *Broadview, Illinois.*)

1. The gas valve is integrally molded in the separator bag.
2. The spring-loaded poppet valve maintains the bag inside the shell. This increases volumetric efficiency.
3. There is a drain plug for bleeding air from the system.

One of the most common applications of accumulators is as an auxiliary power source. The purpose of the accumulator in this application is to store oil delivered by the pump during a portion of the work cycle. The accumulator then releases this stored oil upon demand to complete the cycle, thereby serving as a

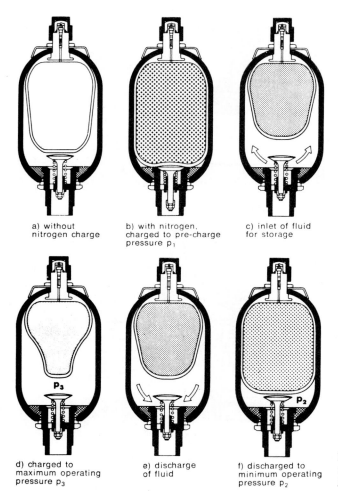

a) without
nitrogen charge

b) with nitrogen,
charged to pre-charge
pressure p_1

c) inlet of fluid
for storage

d) charged to
maximum operating
pressure p_3

e) discharge
of fluid

f) discharged to
minimum operating
pressure p_2

Figure 9-37. Operation of a bladder-type accumulator. (*Courtesy of Robert Bosch Corp., Broadview, Illinois.*)

secondary power source to assist the pump. In such a system where intermittent operations are performed, the use of an accumulator results in being able to use a smaller-sized pump.

This application is depicted in Fig. 9-40 in which a four-way valve is used in conjunction with an accumulator. When the four-way valve is manually actuated, oil flows from the accumulator to the blank end of the cylinder. This extends the piston until it reaches the end of its stroke. While the desired operation is occurring (the cylinder is in the fully extended position), the accumulator is being charged by the pump. The four-way valve is then deactivated for the retraction of the cylinder. Oil flows from the pump and accumulator to retract the cylinder

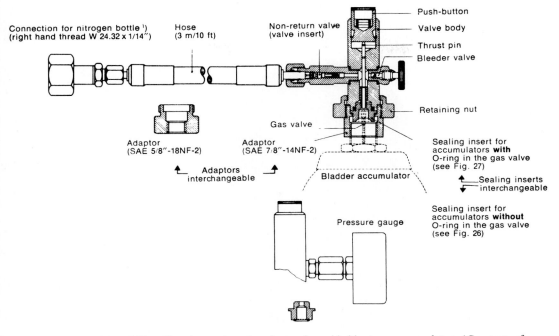

Figure 9-38.   Charging and testing device for a bladder-type accumulator. (*Courtesy of Robert Bosch Corp., Broadview, Illinois.*)

rapidly. The accumulator size is selected to supply adequate oil during the retraction stroke. The sizing of gas-loaded accumulators is presented in Chapter 10.

A second application for accumulators is as a compensator for internal or external leakage during an extended period of time during which the system is pressurized but not in operation. As shown in Fig. 9-41, for this application the pump charges the accumulator and system until the maximum pressure setting on the pressure switch is obtained. The contacts on the pressure switch then open to automatically stop the electric motor that drives the pump. The accumulator then supplies leakage oil to the system during a long period of time. Finally, when system pressure drops to the minimum pressure setting of the pressure switch, it closes the electrical circuit of the pump motor (not shown) until the system has been recharged. The use of an accumulator as a leakage compensator saves electrical power and reduces heat in the system. Electrical circuit diagrams such as those used in this application to control the pump motor are presented in Chapter 12.

In some hydraulic systems, safety dictates that a cylinder be retracted even though the normal supply of oil pressure is lost due to a pump or electrical power failure. Such an application requires the use of an accumulator as an emergency

Figure 9-39. Actual bladder-type accumulator. (*Courtesy of Greer Olaer Products Division/Greer Hydraulics Inc., Los Angeles, California.*)

power source, as depicted in Fig. 9-42. In this circuit, a solenoid actuated three-way valve is utilized in conjunction with the accumulator. When the three-way valve is energized, oil flows to the blank end of the cylinder and also through the check valve into the accumulator and rod end of the cylinder. The accumulator charges as the cylinder extends. If the pump fails due to an electrical failure, the solenoid will de-energize, shifting the valve to its spring offset mode. Then the oil stored under pressure is forced from the accumulator to the rod end of the cylinder. This retracts the cylinder to its starting position.

One of the most important industrial applications of accumulators is the elimination or reduction of high-pressure pulsations or hydraulic shock. Hydraulic shock (or water hammer, as it is frequently called) is caused by the sudden stoppage or deceleration of a hydraulic fluid flowing at relatively high velocity in a pipeline. This hydraulic shock creates a compression wave at the source, where the rapidly closing valve is located. This compression wave travels at the speed of sound upstream to the end of the pipe and back again, causing an increase in the

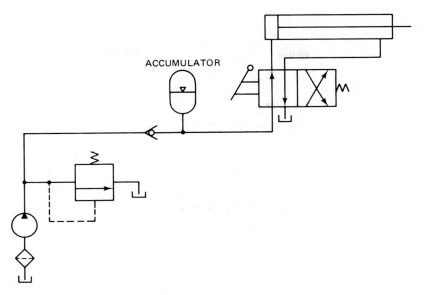

Figure 9-40.   Accumulator as an auxiliary power source.

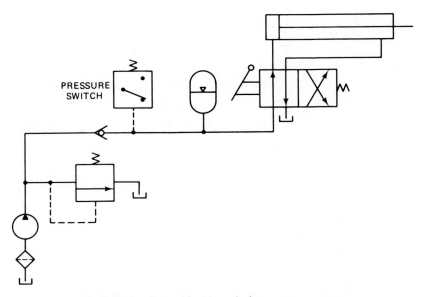

Figure 9-41.   Accumulator as a leakage compensator.

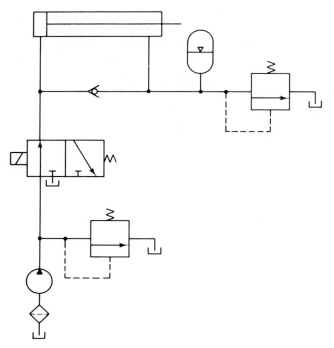

Figure 9-42. Accumulator as an emergency power source.

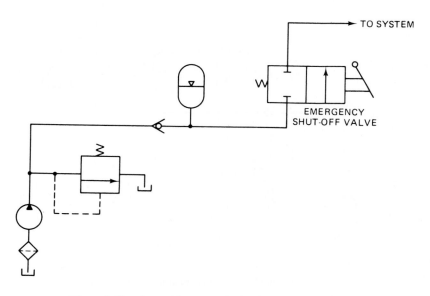

Figure 9-43. Accumulator as a hydraulic shock absorber.

line pressure. This wave travels back and forth along the entire pipe length until its energy is finally dissipated by friction. The resulting rapid pressure pulsations or high-pressure surges may cause damage to the hydraulic system components. If an accumulator is installed near the rapidly closing valve, as shown in Fig. 9-43, the pressure pulsations or high-pressure surges can be suppressed. In this application the accumulator serves as a shock-suppressing device.

9.20 MECHANICAL-HYDRAULIC SERVO SYSTEM

Figure 9-44 shows an automotive power-steering example of a mechanical-hydraulic servo system (closed-loop system). Operation is as follows:

1. The input or command signal is the turning of the steering wheel.
2. This moves the valve sleeve, which ports oil to the actuator (steering cylinder).
3. The piston rod moves the wheels via the steering linkage.
4. The valve spool is attached to the linkage and thus moves with it.
5. When the valve spool has moved far enough, it cuts off oil flow to the cylinder. This stops the motion of this actuator.
6. Thus, mechanical feedback recenters (nulls) the valve (actually a servo valve) to stop motion at the desired point as determined by the position of the steering wheel. Additional motion of the steering wheel is required to cause further motion of the output wheels.

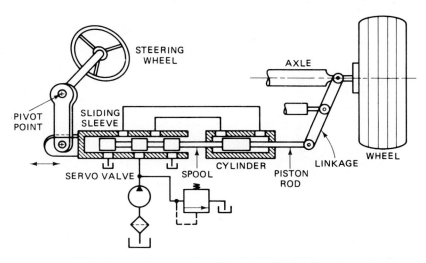

Figure 9-44.   Automotive example of mechanical-hydraulic servo system.

EXERCISES

9-1. When analyzing or designing a hydraulic circuit, what three important considerations must be taken into account?

9-2. What is the purpose of a regenerative circuit?

9-3. Why is the load-carrying capacity of a regenerative cylinder small if its piston rod area is small?

9-4. A double-acting cylinder is hooked up in the regenerative circuit of Fig. 9-4. The relief valve setting is 1500 psi. The piston area is 20 in.^{2}, and the rod area is 10 in.^{2} If the pump flow is 25 gpm, find the cylinder speed and load-carrying capacity for the

    (a) Extending stroke

    (b) Retracting stroke

9-5. For the system of Fig. 9-15, what pump pressure is required if the cylinder loads are 5000 lb each and cylinder 1 has a piston area of 10 in.^{2}?

9-6. What is a fail-safe circuit?

9-7. Under what condition is a hydraulic motor braking system desirable?

9-8. What is the difference between closed-circuit and open-circuit hydrostatic transmission?

9-9. What is meant by an air-over-oil system?

9-10. Name the three basic types of accumulators.

9-11. Name the three major classifications of the separator accumulator. Give one advantage of each classification.

9-12. Describe four applications of accumulators.

9-13. What is a mechanical-hydraulic servo system? Name one application.

9-14. What is the purpose of each relief valve of Fig. 9-42?

9-15. Properly complete the circuit diagram of Fig. 9-45. The clamp cylinder is to extend first, and then the work cylinder extends by the action of a directional control valve. By further action of the DCV, the work cylinder retracts, and then the clamp cylinder retracts.

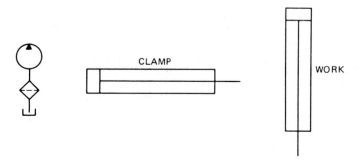

Figure 9-45. Partial circuit for Exercise 9-15.

9-16. What is wrong with the circuit of Fig. 9-46?

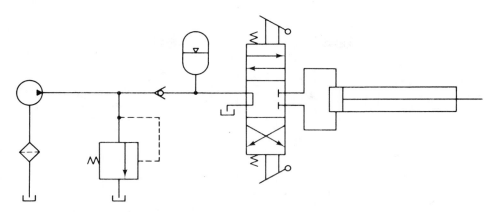

Figure 9-46.   Circuit for Exercise 9-16.

9-17. What is wrong with the circuit of Fig. 9-47?

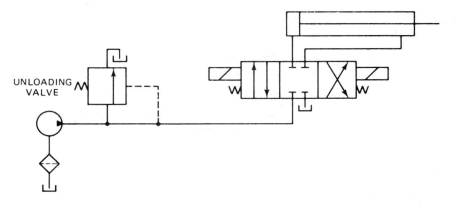

Figure 9-47.   Circuit for Exercise 9-17.

9-18. For the circuit of Fig. 9-48, give the sequence of operation of cylinders 1 and 2 when the pump is turned on. Assume both cylinders are initially fully retracted.

9-19. What safety feature does the circuit of Fig. 9-49 possess in addition to the pressure relief valve?

9-20. Assuming that the two double-rodded cylinders of Fig. 9-50 are identical, what unique feature does this circuit possess?

9-21. A double-acting cylinder is hooked up in the regenerative circuit of Fig. 9-4. The relief valve setting is 105 bars. The piston area is 130 cm², and the rod area is 65 cm².

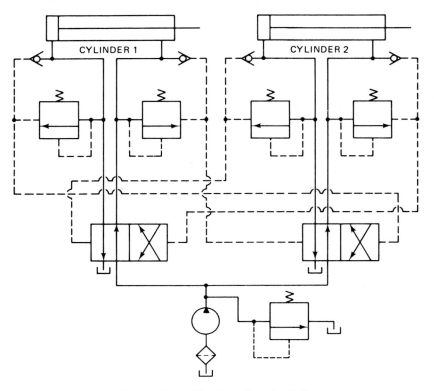

Figure 9-48. Circuit for Exercise 9-18.

If the pump flow is 0.0016 m³/s, find the cylinder speed and load-carrying capacity for the

a. Extending stroke

b. Retracting stroke

9-22. For the system of Fig. 9-15, what pump pressure is required if the cylinder loads are 22,000 N each and cylinder 1 has a piston area of 65 cm²?

9-23. Can a hydraulic cylinder be designed so that for the same pump flow, the extending and retracting speeds will be equal? Explain your answer.

9-24. Which type of accumulator operates at constant pressure? How can the pressure be changed?

9-25. For the fluid power system shown in Fig. 9-51, determine the external load (F_1 and F_2) that each hydraulic cylinder can sustain while moving in the extending direction. Take frictional pressure losses into account. The pump produces a pressure increase

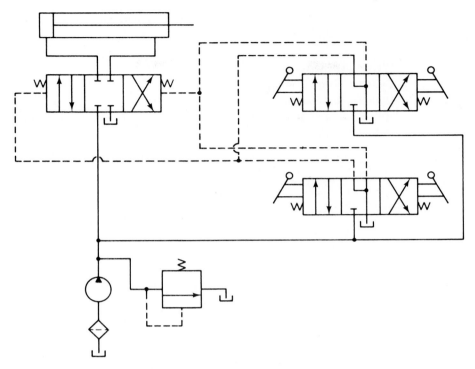

Figure 9-49.   Circuit for Exercise 9-19.

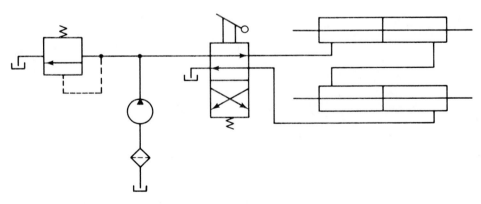

Figure 9-50.   Circuit for Exercise 9-20.

of 1000 psi from the inlet port to the discharge port and a flow rate of 40 gpm. The following data is applicable:

Kinematic viscosity of oil = 0.001 ft²/sec

Weight density of oil = 50 lb/ft³

Cylinder piston diameter = 8 in.

Cylinder rod diameter = 4 in.

All elbows are 90° with K factor = 0.75. Pipe lengths and inside diameters are given as follows:

Pipe No.	Length(ft)	Dia(in.)	Pipe No.	Length(ft)	Dia(in.)
1	6	1.5	6	10	0.75
2	30	1.0	7	10	0.75
3	20	1.25	8	40	1.25
4	10	1.0	9	40	1.25
5	10	1.0			

9-26. For the system of Exercise 9-25, as shown in Fig. 9-51, convert the data to metric units and solve for the external load that each cylinder can sustain while moving in the retraction direction.

9-27. For the system of Exercise 9-25, as shown in Fig. 9-51, determine the heat generation rate due to frictional pressure losses.

9-28. For the system of Exercise 9-25, as shown in Fig. 9-51, determine the retracting and extending speeds of both cylinders. Assume that the actual cylinder loads are equal and are less than the loads that can be sustained during motion.

9-29. For the system of Exercise 9-25, as shown in Fig. 9-51, if the load on cylinder 1 is greater than the load on cylinder 2, how will the cylinders move when the DCV is shifted into the extending or retracting mode? Explain your answer.

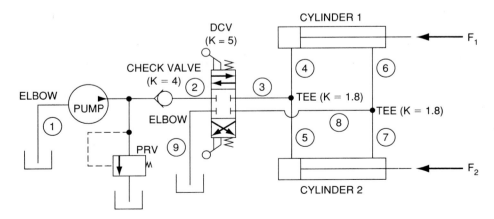

Figure 9-51. System for Exercise 9-25.

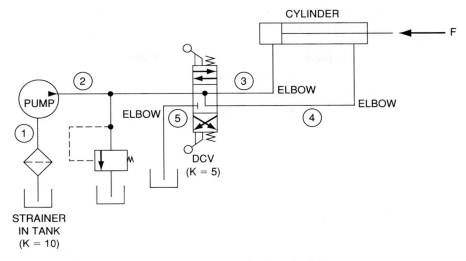

Figure 9-52.   System for Exercise 9-30.

9-30. Figure 9-52 shows a regenerative system in which a 25 HP electric motor drives a 90% efficient pump. Determine the external load F that the hydraulic cylinder can sustain in the regenerative mode (spring-centered position of the DCV). Pump discharge pressure is 1000 psi. Take frictional pressure losses into account. The following data is applicable:

Kinematic viscosity of oil = 0.001 ft²/sec

Weight density of oil = 50 lb/ft³

Cylinder piston diameter = 8 in.

Cylinder rod diameter = 4 in.

All elbows are 90° with K factor = 0.75. Pipe lengths and inside diameters are given as follows:

Pipe No.	Length(ft)	Dia(in.)
1	2	1.5
2	20	1.0
3	30	2.0
4	30	1.5
5	20	1.0

9-31. For the system of Exercise 9-30, as shown in Fig. 9-52, convert the data to metric units and determine the external load F that the hydraulic cylinder can sustain in the regenerative mode.

9-32. For the system of Exercise 9-30, as shown in Fig. 9-52, determine the heat generation rate (English units) due to frictional pressure losses in the regenerative mode.

9-33. For the system of Exercise 9-31, as shown in Fig. 9-52, determine the heat generation rate (metric units) due to frictional pressure losses in the regenerative mode.

9-34. For the system of Exercise 9-30, as shown in Fig. 9-52, determine the cylinder speed for each position of the DCV.

9-35. In a mechanical-hydraulic servo system, what part of the servo valve moves with the load? What part moves with the input?

10

Pneumatic Components and Circuits

10.1 INTRODUCTION

Pneumatic systems use pressurized gases to transmit and control power. As the name implies, pneumatic systems typically use air (rather than some other gas) as the fluid medium, because air is a safe, low-cost, and readily available fluid. It is particularly safe in environments where an electrical spark could ignite leaks from system components.

There are several reasons for considering the use of pneumatic systems instead of hydraulic systems. Liquids exhibit greater inertia than do gases. Therefore, in hydraulic systems the weight of oil is a potential problem when accelerating and decelerating actuators and when suddenly opening and closing valves. Due to Newton's law of motion (force equals mass multiplied by acceleration), the force required to accelerate oil is many times greater than that required to accelerate an equal volume of air. Liquids also exhibit greater viscosity than do gases. This results in larger frictional pressure and power losses. Also, since hydraulic systems use a fluid foreign to the atmosphere, they require special reservoirs and no-leak system designs. Pneumatic systems use air that is exhausted directly back into the surrounding environment. Generally speaking, pneumatic systems are less expensive than hydraulic systems.

However, because of the compressibility of air, it is impossible to obtain precise, controlled actuator velocities with pneumatic systems. Also, precise positioning control is not obtainable. In applications where actuator travel is to be smooth and steady against a variable load, the air exhaust from the actuator is normally metered. Whereas pneumatic pressures are quite low due to compressor

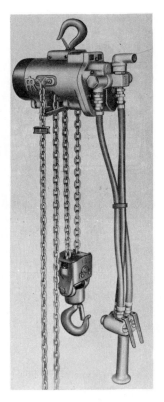

Figure 10-1. Pneumatic-powered impact tool. (*Courtesy of Ingersoll-Rand Co., Washington, New Jersey.*)

Figure 10-2. Pneumatic-powered hoist. (*Courtesy of Ingersoll-Rand Co., Washington, New Jersey.*)

design limitations (less than 250 psi), hydraulic pressures can be as high as 10,000 psi. Thus, hydraulics can be high-power systems, whereas pneumatics are confined to low-power applications. Principal applications for pneumatics include circuits where end conditions are of prime importance (piston rod fully extended or fully retracted). Pneumatic systems can be readily applied to drive rotary actuators as well as linear cylinders. Figure 10-1 shows an air motor-driven impact tool that can handle up to 6-in.-diameter bolts. Another industrial application is the air-powered hoist of Fig. 10-2, which has a 10-ton capacity. Industrial applications of pneumatic systems are growing at a rapid pace. Typical examples include stamping, drilling, hoisting, punching, clamping, assembling, riveting, materials handling, and logic controlling operations.

In pneumatic systems, compressors are used to compress and supply the necessary quantities of air. Compressors are typically of the piston, vane, or screw type. Basically a compressor increases the pressure of a gas by reducing its volume as described by the perfect gas laws. Pneumatic systems normally use a large centralized air compressor, which is considered to be an infinite air source similar to an electrical system where you merely plug into an electrical outlet for electricity. In this way, pressurized air can be piped from one source to various locations throughout an entire industrial plant. The compressed air is piped to each circuit through an air filter to remove contaminants, which might harm the closely fitting parts of pneumatic components such as valves and cylinders. The air then flows through a pressure regulator, which reduces the pressure to the desired level for the particular circuit application. Because air is not a good lubricant (contains about 20% oxygen), pneumatic systems require a lubricator to inject a very fine mist of oil into the air discharging from the pressure regulator. This prevents wear of the closely fitting moving parts of pneumatic components.

Free air from the atmosphere contains varying amounts of moisture. This moisture can be harmful in that it can wash away lubricants and thus cause excessive wear and corrosion. Hence, in some applications, air dryers are needed to remove this undesirable moisture. Since pneumatic systems exhaust directly into the atmosphere, they are capable of generating excessive noise. Therefore, mufflers are mounted on exhaust ports of air valves and actuators to reduce noise and prevent operating personnel from possible injury resulting not only from exposure to noise but also from high-speed airborne particles.

10.2 PROPERTIES OF AIR

Air is actually a mixture of gases containing about 21% oxygen, 78% nitrogen, and 1% other gases such as argon and carbon dioxide. The preceding percentage values are based on volume. Air also contains up to 4% water vapor depending on the humidity. The percent of water vapor in atmospheric air can vary constantly from hour to hour even at the same location.

The earth is surrounded by a blanket of air called the atmosphere. Because air has weight, the atmosphere exerts a pressure at any point due to the column of air above that point. The reference point is sea level, where atmospheric pressure exerts a pressure of 14.7 psi. Figure 10-3 shows how the atmospheric pressure decreases with altitude. For the region up to an altitude of 20,000 ft, the relationship is nearly linear, with a drop in pressure of about 0.5 psi per 1000-ft change in altitude.

When making pneumatic circuit calculations, atmospheric pressure of 14.7 psia is used as a standard. The corresponding standard weight density value for air is 0.0807 lb/ft^3 at 14.7 psia and 32°F. Appendix C gives weight density values of air for temperatures varying from 30°F to 300°F and pressure variations from 0 to 140 psig. Thus, as shown in Sec. 10.3 in a discussion of perfect gas laws, the density of a gas depends not only on its pressure but also on its temperature.

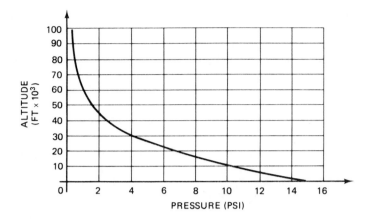

Figure 10-3.   Pressure variation in the atmosphere.

Air is not only readily compressible, but its volume will vary to fill the vessel containing it because the air molecules have substantial internal energy and are at a considerable distance from each other. This accounts for the sensitivity of density changes with respect to changes in pressure and temperature.

Free air is considered to be air at normal atmospheric conditions. Since atmospheric pressure and temperature vary from day to day, the characteristics of free air vary accordingly. Thus, when making pneumatic circuit calculations, the term *standard air* is used. Standard air is sea-level air having a temperature of 68°F, a pressure of 14.7 psia, and a relative humidity of 36%.

Circuit calculations dealing with volume and pressure changes of air must be performed using absolute pressure and absolute temperature values. Equations (10-1) and (10-2) permit the calculation of absolute pressures and temperatures, respectively:

$$\text{absolute pressure (psia)} = \text{gage pressure (psig)} + 14.7 \qquad (10\text{-}1)$$

$$\text{absolute temperature (°R)} = \text{temperature (°F)} + 460 \qquad (10\text{-}2)$$

The units of absolute temperature are degrees Rankine, abbreviated °R. A temperature of 0°R (—460°F) is the temperature at which all molecular motion ceases to exist and the volume and pressure of a gas theoretically become zero.

10.3   THE PERFECT GAS LAWS

During the sixteenth century, scientists discovered the laws that determine the interactions of pressure, volume, and temperature of a gas. These laws are called the perfect gas laws because they were derived on the basis of a perfect gas. Even though perfect gases do not exist, air behaves very closely to that predicted by Boyle's law, Charles' law, Gay-Lussac's law, and the general gas law. Each of these laws is defined and applied to a particular problem.

Boyle's law states that if the temperature of a given amount of gas is held constant, the volume of the gas will change inversely with the absolute pressure of the gas:

$$\frac{V_1}{V_2} = \frac{P_2}{P_1} \qquad (10\text{-}3)$$

EXAMPLE 10-1

The 2-in.-diameter piston of the pneumatic cylinder of Fig. 10-4 retracts 4 in. from its present position ($P_1 = 20$ psig, $V_1 = 20$ in.^{3}) due to the external load on the rod. If the port at the blank end of the cylinder is blocked, find the new pressure, assuming the temperature does not change.

Solution

$$V_1 = 20 \text{ in.}^3$$

$$V_2 = 20 \text{ in.}^3 - \frac{\pi}{4}(2)^2(4) = 7.43 \text{ in.}^3$$

$$P_1 = 20 + 14.7 = 34.7 \text{ psia}$$

Substituting into Eq. (10-3), which defines Boyle's law, we have

$$\frac{20}{7.43} = \frac{P_2}{34.7}$$

$$P_2 = 93.4 \text{ psia} = 78.7 \text{ psig}$$

Charles' law states that if the pressure on a given amount of gas is held constant, the volume of the gas will change in direct proportion to the absolute temperature:

$$\frac{V_1}{V_2} = \frac{T_1}{T_2} \qquad (10\text{-}4)$$

EXAMPLE 10-2

The cylinder of Fig. 10-4 has an initial position where $P_1 = 20$ psig and $V_1 = 20$ in.^{3} as controlled by the load on the rod. The air temperature is 60°F. The load on the rod is

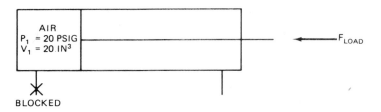

Figure 10-4.   System for Example 10-1.

held constant to maintain constant air pressure, but the air temperature is increased to 120°F. Find the new volume of air at the blank end of the cylinder.

Solution

$$T_1 = 60 + 460 = 520°R$$

$$T_2 = 120 + 460 = 580°R$$

Substituting into Eq. (10-4), which defines Charles' law, yields the answer:

$$\frac{20}{V_2} = \frac{520}{580}$$

$$V_2 = 22.3 \text{ in.}^3$$

Gay-Lussac's law states that if the volume of a given gas is held constant, the pressure exerted by the gas is directly proportional to its absolute temperature:

$$\frac{P_1}{P_2} = \frac{T_1}{T_2} \tag{10-5}$$

EXAMPLE 10-3

The cylinder of Fig. 10-4 has a locked position (V_1 = constant). P_1 = 20 psig, and T_1 = 60°F. If the temperature increases to 160°F, what is the new pressure in the blank end?

Solution

$$P_1 = 20 + 14.7 = 34.7 \text{ psia}$$

$$T_1 = 60 + 460 = 520°R \text{ and } T_2 = 160 + 460 = 620°R$$

Substituting into Eq. (10-5), which defines Gay-Lussac's law, we obtain

$$\frac{34.7}{P_2} = \frac{520}{620}$$

or

$$P_2 = 41.4 \text{ psia} = 26.7 \text{ psig}$$

Boyle's, Charles', and Gay-Lussac's laws can be combined into a single general gas law, as defined by

$$\frac{P_1 V_1}{T_1} = \frac{P_2 V_2}{T_2} \tag{10-6}$$

Although this equation is used later in this chapter to size gas-loaded accumulators, the following example illustrates its use.

EXAMPLE 10-4

Gas at 1000 psig and 100°F is contained in the 2000-in.^{3} cylinder of Fig. 10-5. A piston compresses the volume to 1500 in.^{3} while the gas is heated to 200°F. What is the final pressure in the cylinder?

Solution   Solve Eq. (10-6) for P_2 and substitute known values:

$$P_2 = \frac{P_1 V_1 T_2}{V_2 T_1}$$

$$P_2 = \frac{(1000 + 14.7)(2000)(200 + 460)}{(1500)(100 + 460)} = \frac{(1014.7)(2000)(660)}{1500(560)}$$

$$P_2 = 1594.5 \text{ psia} = 1579.8 \text{ psig}$$

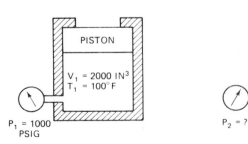

Figure 10.5   System for Example 10-4.

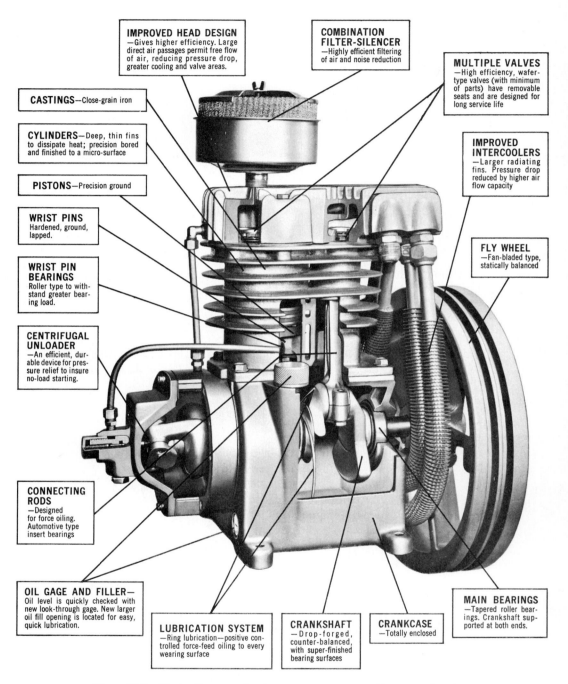

IMPROVED HEAD DESIGN
—Gives higher efficiency. Large direct air passages permit free flow of air, reducing pressure drop, greater cooling and valve areas.

COMBINATION FILTER-SILENCER
—Highly efficient filtering of air and noise reduction

MULTIPLE VALVES
—High efficiency, wafer-type valves (with minimum of parts) have removable seats and are designed for long service life

CASTINGS—Close-grain iron

CYLINDERS—Deep, thin fins to dissipate heat; precision bored and finished to a micro-surface

PISTONS—Precision ground

WRIST PINS
Hardened, ground, lapped.

WRIST PIN BEARINGS
Roller type to withstand greater bearing load.

CENTRIFUGAL UNLOADER
—An efficient, durable device for pressure relief to insure no-load starting.

CONNECTING RODS
—Designed for force oiling. Automotive type insert bearings

IMPROVED INTERCOOLERS
—Larger radiating fins. Pressure drop reduced by higher air flow capacity

FLY WHEEL
—Fan-bladed type, statically balanced

OIL GAGE AND FILLER—
Oil level is quickly checked with new look-through gage. New larger oil fill opening is located for easy, quick lubrication.

LUBRICATION SYSTEM
—Ring lubrication—positive controlled force-feed oiling to every wearing surface

CRANKSHAFT
—Drop-forged, counter-balanced, with super-finished bearing surfaces

CRANKCASE
—Totally enclosed

MAIN BEARINGS
—Tapered roller bearings. Crankshaft supported at both ends.

Figure 10-6. Design features of a piston-type compressor. (*Courtesy of Kellogg-American Inc., Oakmont, Pennsylvania.*)

10.4 *COMPRESSORS*

A compressor is a machine that compresses air or another type of gas from a low inlet pressure (usually atmospheric) to a higher desired pressure level. This is accomplished by reducing the volume of the gas. Air compressors are generally positive displacement units and are either of the reciprocating piston type or the rotary screw or rotary vane types. In Fig. 10-6 we see a compressor with its ANSI symbol. This figure illustrates many of the design features of a piston-type compressor. Such a design contains pistons sealed with piston rings operating in precision-bored close-fitting cylinders. Notice that the cylinders have air fins to help dissipate heat. Cooling is necessary with compressors to dissipate the heat generated during compression. When air is compressed, it picks up heat as the molecules of air come closer together and bounce off each other at faster and faster rates. Excessive temperature can damage the metal components as well as increase input power requirements. Portable and small industrial compressors are normally air-cooled, whereas larger units must be water-cooled. Figure 10-7 shows a typical small-sized, two-stage compressor unit. Observe that it is a complete system containing not only a compressor but also the compressed air tank (receiver), electric motor and pulley drive, pressure controls, and instrumentation for quick hookup and use. This particular compressor unit also contains an air dryer, which provides a constant supply of high-quality dry air for applications where moisture would be a problem. It is driven by a 10-hp motor, has a 120-gal receiver, and is designed to operate in the 145–175-psi range with a capacity of 46.3 cfm (cubic ft per min).

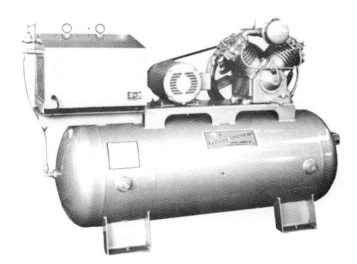

Figure 10-7.   Complete piston-type, two-stage compressor unit. (*Courtesy of Kellogg-American Inc., Oakmont, Pennsylvania.*)

Figure 10-8.   Direct-drive, fan-cooled, piston-type compressor. (*Courtesy of Gast Manufacturing Corp., Benton Harbor, Michigan.*)

Figure 10-8 gives a cutaway view of a direct-drive, two-cylinder piston-type compressor. The fan in the forefront accelerates the air cooling of the compressor by providing forced air flow.

A single-piston compressor can provide pressure up to about 150 psi. Above 150 psi, the compression chamber size and heat of compression prevent efficient pumping action. For compressors having more than one cylinder, staging can be used to improve pumping efficiency. Staging means dividing the total pressure among two or more cylinders by feeding the exhaust from one cylinder into the inlet of the next. Because effective cooling can be implemented between stages, multistage compressors can dramatically increase the efficiency and reduce input power requirements. In multistage piston compressors, successive cylinder sizes decrease, and the intercooling removes a significant portion of the heat of compression. This increases air density and the volumetric efficiency of the compressor. This is shown in Fig. 10-9 by the given pressure capacities for the various number of stages of a piston-type compressor.

An air compressor must start, run, deliver air to the system as needed, stop, and be ready to start again without the attention of an operator. Since these functions usually take place after a compressed air system has been brought up to pressure, automatic controls are required to work against the air pressure already established by the compressor.

NUMBER OF STAGES	PRESSURE CAPACITY (PSI)
1	150
2	500
3	2500
4	5000

Figure 10-9.   Effect of number of stages on pressure capacity.

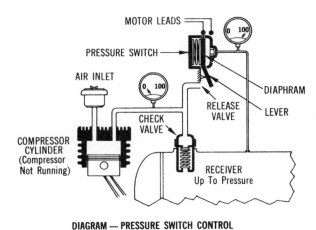

MOTOR LEADS

PRESSURE SWITCH

AIR INLET

DIAPHRAM

RELEASE
VALVE

LEVER

CHECK
VALVE

COMPRESSOR
CYLINDER
(Compressor
Not Running)

RECEIVER
Up To Pressure

DIAGRAM — PRESSURE SWITCH CONTROL

Figure 10-10. Pressure-switch type unloader control. (*Courtesy of Kellogg-American Inc., Oakmont, Pennsylvania.*)

If an air compressor is started for the very first time, there is no need for a starting unloader control since there is not yet an established pressure against which the compressor must start. However, once a pressure has been established in the compressed air piping, a starting unloader is needed to prevent the established air pressure from pushing back against the compressor, preventing it from coming up to speed. Figure 10-10 shows a pressure-switch-type unloader control. When the pressure switch shuts the electric motor off, pressure between the compressor head and the check valve is bled off to the atmosphere through the release valve. The compressor is then free to start again whenever needed.

Figure 10-11 illustrates the operation of the centrifugal-type unloader control. To provide a greater degree of protection for motors and drives, an unloader valve operates by the air compressor itself rather than by the switch. This type is

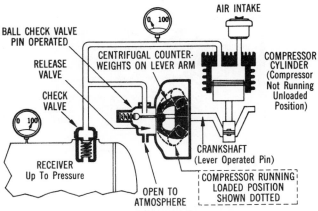

AIR INTAKE

BALL CHECK VALVE
PIN OPERATED

CENTRIFUGAL COUNTER-
WEIGHTS ON LEVER ARM

COMPRESSOR
CYLINDER
(Compressor
Not Running
Unloaded
Position)

RELEASE
VALVE

CHECK
VALVE

RECEIVER
Up To Pressure

CRANKSHAFT
(Lever Operated Pin)

OPEN TO
ATMOSPHERE

COMPRESSOR RUNNING
LOADED POSITION
SHOWN DOTTED

DIAGRAM — CENTRIFUGAL UNLOADER

Figure 10-11. Centrifugal-type unloader control. (*Courtesy of Kellogg-American Inc., Oakmont, Pennsylvania.*)

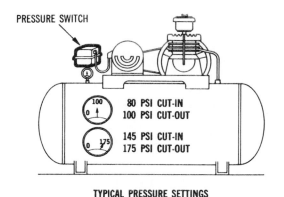

PRESSURE SWITCH

| 80 PSI CUT-IN |
| 100 PSI CUT-OUT |

| 145 PSI CUT-IN |
| 175 PSI CUT-OUT |

TYPICAL PRESSURE SETTINGS

Figure 10-12. Typical pressure settings for pressure switch. (*Courtesy of Kellogg-American Inc., Oakmont, Pennsylvania.*)

preferred on larger compressors. A totally enclosed centrifugal unloader operated by and installed on the compressor crankshaft is best for this purpose.

Once an air compressor is equipped with a starting unloader, it may be operated automatically by the pressure switch, as depicted in Fig. 10-12. This is the normal method, using an adjustable start-stop control switch. Normal air compressor operation calls for 50% to 80% running time when using pressure switch controls. An air compressor that cycles too often (more than once each 6 min) or one that runs more than 80% of the time delivering air to the tank should be regulated by a constant-speed control.

There is a present trend toward increased use of the rotary-type compressor due to technological advances, which have produced stronger materials and better manufacturing processes. Figure 10-13 shows a cutaway view of a single-stage screw-type compressor, which is very similar to a screw pump that was previously discussed. Compression is accomplished by rolling the trapped air into a

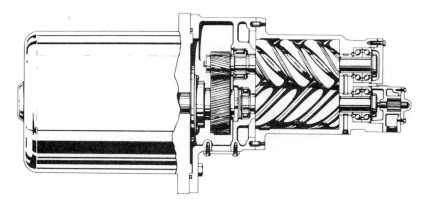

Figure 10-13. Single-stage screw compressor. (*Courtesy of Ingersoll-Rand Co., Washington, New Jersey.*)

Figure 10-14. Unsymmetrical profile of screw rotors. (Courtesy of Ingersoll-Rand Co., Washington, New Jersey.)

progressively smaller volume as the screws rotate. Figure 10-14 illustrates the unsymmetrical profile of the two rotors. The rotors turn freely, with a carefully controlled clearance between both rotors and the housing, protected by a film of oil. Rotor wear will not occur since metal-to-metal contact is eliminated. A precisely measured amount of filtered and cooled air is injected into the compression chamber, mixing with the air as it is compressed. The oil lubricates the rotors, seals the rotor clearances for high-compression efficiency, and absorbs heat of compression, resulting in low discharge air temperatures. Single-stage screw compressors are available with capacities up to 1450 cfm and pressures of 120 psi.

Figure 10-15 shows a cutaway view of the sliding-vane-type rotary compressor. In this design, a cylindrical slotted rotor turns inside of a stationary outer casing. Each rotor slot contains a rectangular vane, which slides in and out of the slot due to centrifugal force. As the rotor turns, air is trapped and compressed between the vanes and then discharged through a port to the receiver. Rotary sliding vane compressors can operate up to approximately 50 psi in a single stage and up to 150 psi in a two-stage design. This low-pressure, low-volume type of compressor is normally used for instrument and other laboratory-type air needs.

Figure 10-15. Sliding-vane-type rotary compressor. (*Courtesy of Gast Manufacturing Corp., Benton Harbor, Michigan.*)

Air compressors are generally rated in terms of cfm of free air. Free air is defined as air at standard atmospheric conditions. Therefore, a calculation is necessary to determine the compressor capacity in terms of cfm for a given application.

EXAMPLE 10-5

Air is used at a rate of 30 cfm from a receiver at 90°F and 125 psi. If the atmospheric pressure is 14.7 psia and the atmospheric temperature is 70°F, how many cfm of free air must the compressor provide?

Solution   Solve Eq. (10-6) for V_1 and let subscript 1 represent atmospheric conditions:

$$V_1 = \frac{V_2 P_2 T_1}{P_1 T_2} = \text{cfm of free air}$$

$$V_1 = \frac{(30)(125 + 14.7)(70 + 460)}{(14.7)(90 + 460)} = \frac{(30)(139.7)(530)}{(14.7)(550)}$$

$$V_1 = 275 \text{ cfm of free air}$$

In other words, the compressor must receive atmospheric air (14.7 psi and 70°F) at a rate of 275 cfm in order to deliver air (125 psi and 90°F) at 30 cfm.

The sizing of air receivers requires taking into account parameters such as system pressure and flow-rate requirements, compressor output capability, and the type of duty of operation. Basically a receiver is an air reservoir. Its function is to supply air at essentially constant pressure. It also serves to dampen pressure pulses either coming from the compressor or the pneumatic system during valve shifting and component operation. Frequently a pneumatic system demands air at a flow rate that exceeds the compressor capability. The receiver must be capable of handling this transient demand.

$$V_r = \frac{14.7t(Q_r - Q_c)}{P_{\max} - P_{\min}} \tag{10-7}$$

where t = time (min) that receiver can supply required amount of air
Q_r = consumption rate of pneumatic system (cfm)
Q_c = output flow rate of compressor (cfm)
$P_{\max}$ = maximum pressure level in receiver (psi)
$P_{\min}$ = minimum pressure level in receiver (psi)
V_r = receiver size (ft³)

EXAMPLE **10-6**

   a. Calculate the required size of a receiver that must supply air to a pneumatic system consuming 20 cfm for 6 min between 100 and 80 psi before the compressor resumes operation.

   b. What size is required if the compressor is running and delivering air at 5 cfm?

Solution

a.
$$V_r = \frac{14.7 \times 6 \times (20 - 0)}{100 - 80} = 88.2 \text{ ft}^3 = 660 \text{ gal}$$

b.
$$V_r = \frac{14.7 \times 6 \times (20 - 5)}{100 - 80} = 66.2 \text{ ft}^3 = 495 \text{ gal}$$

It is common practice to increase the calculated size of the receiver by 25% for unexpected overloads and by another 25% for possible future expansion needs.

10.5 FLUID CONDITIONERS

The purpose of fluid conditioners is to make air a more acceptable fluid medium for the pneumatic system as well as operating personnel. Fluid conditioners include filters, regulators, lubricators, mufflers, and air dryers.

   The function of a filter is to remove contaminants from the air before it reaches pneumatic components such as valves and actuators. Generally speaking, in-line filters contain filter elements that remove contaminants in the 5- to 50-micron range. Figure 10-16 shows a cutaway view of a filter that uses 5-micron cellulose felt, reusable, surface-type elements. These elements have gaskets molded permanently to each end to prevent air bypass and make element servicing foolproof. These elements have a large ratio of air to filter media and thus can hold an astonishing amount of contamination on the surface without suffering significant pressure loss. The baffling system used in these filters mechanically separates most of the contaminants before they reach the filter element. In addition, a quiet zone prevents contaminants collected in the bowl from reentering the airstream. Also shown in Fig. 10-16 is the ANSI symbol for an air filter.

   The compressor control system maintains system air pressure within a given range. For example, the compressor may automatically start when the pressure drops to 100 psi and automatically stop when the pressure in the receiver reaches 125 psi. This is generally accomplished with pressure switches, and a relief valve is used to protect the system if the compressor fails to shut down when required.

   So that a constant pressure is available for a given pneumatic system, a pressure regulator is used. Figure 10-17 illustrates a pressure regulator that uses a spring-loaded diaphragm and features balanced valves for superior regulation

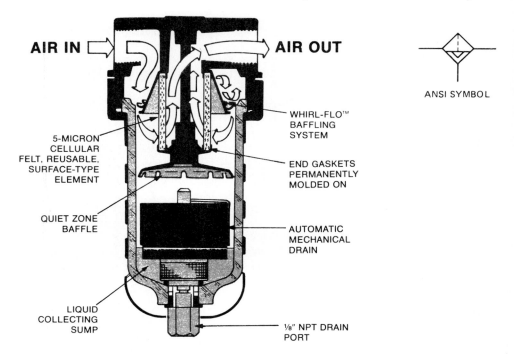

AIR IN   **AIR OUT**

ANSI SYMBOL

5-MICRON CELLULAR FELT, REUSABLE, SURFACE-TYPE ELEMENT

WHIRL-FLO™ BAFFLING SYSTEM

END GASKETS PERMANENTLY MOLDED ON

QUIET ZONE BAFFLE

AUTOMATIC MECHANICAL DRAIN

LIQUID COLLECTING SUMP

⅛″ NPT DRAIN PORT

Figure 10-16.   Operation of air filter. (*Courtesy of Wilkerson Corp., Englewood, Colorado.*)

characteristics. Large main valve seats and precisely positioned aspirator tubes provide for excellent flow characteristics and minimal pressure drop. The relationship between the diaphragm, valve size, and valve travel assure the controlling and maintaining of secondary pressures whether for circuit requirements, safety, or energy-saving needs. Units are provided with gage ports on both front and back, and either port can be used as an additional regulated outlet. Maintenance is easily performed when required without removing regulators from the air line. The regulator contains an adjustable upper spring, which allows the valve to hold a given pressure on the downstream side. The force of the spring is set for the required downstream pressure. This force holds the valve open until downstream pressure, acting on the diaphragm, starts to exceed the spring force. As a result, the push rod is allowed to move up and the spring-loaded valve at the bottom begins to close to throttle the air supply to the controlled pressure side. The ANSI symbol of an air pressure regulator is also shown in Fig. 10-17.

A lubricator ensures proper lubrication of internal moving parts of pneumatic components. Figure 10-18 illustrates the operation of a lubricator, which inserts every drop of oil leaving the drip tube, as seen through the sight dome, directly into the airstream. These drops of oil are transformed into an oil mist prior

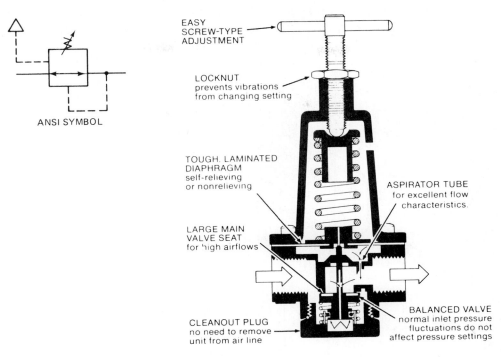

EASY
SCREW-TYPE
ADJUSTMENT

LOCKNUT
prevents vibrations
from changing setting

ANSI SYMBOL

TOUGH. LAMINATED
DIAPHRAGM
self-relieving
or nonrelieving

ASPIRATOR TUBE
for excellent flow
characteristics.

LARGE MAIN
VALVE SEAT
for high airflows

BALANCED VALVE
normal inlet pressure
fluctuations do not
affect pressure settings

CLEANOUT PLUG
no need to remove
unit from air line

Figure 10-17.   Air pressure regulator. (*Courtesy of Wilkerson Corp., Englewood, Colorado.*)

to their being transported downstream. This oil mist consists of both coarse and fine particles. The coarse particles may travel distances of 20 ft or more, while the fine particles often reach distances as great as 300 ft from the lubricator source. These oil mist particles are created when a portion of the incoming air passes through the center of the variable orifice and enters the mist generator, mixing with the oil delivered by the drip tube. This air-oil mixture then rejoins any air that has bypassed the center of the variable orifice and continues with that air toward its final destination. Oil reaching the mist generator was first pushed up the siphon tube, past the adjustment screw to the drip tube located within the sight dome. This is accomplished by diverting a small amount of air from the mainstream through the bowl pressure control valve, into the bowl or reservoir. This valve is so located that it will close, shutting off the air supply to the bowl when the fill plug is loosened or removed, permitting refilling of the bowl or reservoir without shutting off the air supply line. Upon replacement of the fill plug, the bowl pressure control valve will open automatically, causing the bowl to be pressurized once again and ready to supply lubrication where it is needed.

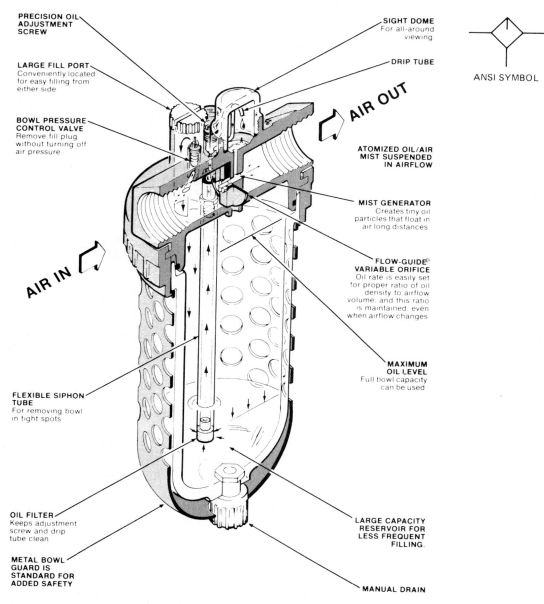

PRECISION OIL
ADJUSTMENT
SCREW

LARGE FILL PORT
Conveniently located
for easy filling from
either side

BOWL PRESSURE
CONTROL VALVE
Remove fill plug
without turning off
air pressure

FLEXIBLE SIPHON
TUBE
For removing bowl
in tight spots

OIL FILTER
Keeps adjustment
screw and drip
tube clean

METAL BOWL
GUARD IS
STANDARD FOR
ADDED SAFETY

SIGHT DOME
For all-around
viewing

DRIP TUBE

AIR OUT

ATOMIZED OIL/AIR
MIST SUSPENDED
IN AIRFLOW

MIST GENERATOR
Creates tiny oil
particles that float in
air long distances

FLOW-GUIDE©
VARIABLE ORIFICE
Oil rate is easily set
for proper ratio of oil
density to airflow
volume, and this ratio
is maintained, even
when airflow changes

MAXIMUM
OIL LEVEL
Full bowl capacity
can be used

AIR IN

LARGE CAPACITY
RESERVOIR FOR
LESS FREQUENT
FILLING.

MANUAL DRAIN

ANSI SYMBOL

Figure 10-18.   Air lubricator. (*Courtesy of Wilkerson Corp., Englewood, Colorado.*)

Also shown in Fig. 10-18 is the ANSI symbol for an air lubricator.

Figure 10-19 shows an individual filter, two individual pressure regulators, and an individual lubricator. In contrast, in Fig. 10-20 we see a combination filter-regulator-lubricator unit (FRL). Also shown is its ANSI symbol. In both Figs. 10-19 and 10-20, the units with the pressure gages are the pressure regulators.

Figure 10-21(a) shows a pneumatic indicator that provides a two-color, two-position visual indication of air pressure. The rounded lens configuration provides 180° view of the indicator status, which is a fluorescent signal visible from the front and side. This indicator is easily panel-mounted using the same holes as standard electrical pilot lights. However, they are completely pneumatic, requiring no electrical power.

These pneumatic indicators are field adjustable for either one input with spring return or two inputs with memory. This memory does not require continuous pressure to maintain its last signal input. Field conversion may be made to select either single-input, spring return, or two-input maintained modes of opera-

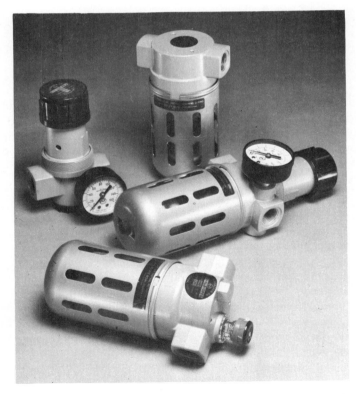

Figure 10-19. · Individual filter, regulator, lubricator units. (*Courtesy of C. A. Norgren Co., Littleton, Colorado.*)

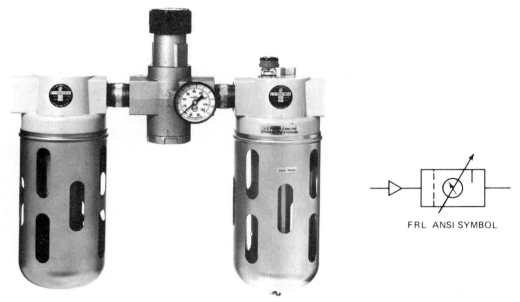

FRL ANSI SYMBOL

Figure 10-20. Combination filter, regulator, lubricator unit. (*Courtesy of C. A. Norgren Co., Littleton, Colorado.*)

tion. Figure 10-21(b) shows the adjustment on the rear of the indicator housing. By using the same adjustment, either of the two display colors and its individual input may be selected for single-input operation. In the center position, this adjustment allows the indicator to accept two inputs for a maintained (memory) mode of

(a)   (b)

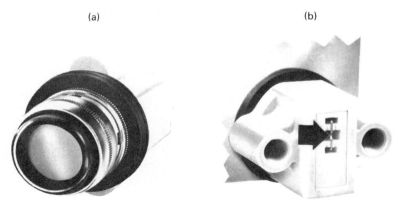

Figure 10-21. Pneumatic indicator. (a) Front view, (b) rear view. (*Courtesy of Numatics Incorp., Highland, Michigan.*)

operation. If both inputs are on simultaneously, the indicator will assume an intermediate position and show parts of both colors.

These indicators come in a variety of color combinations and are completely compatible with pneumatic systems. They are available with pressure ranges of 0.5 to 30 psi, 25 to 150 psi, and 45 to 150 psi. The smallest pressure value of each pressure range (0.5, 25, and 45 psi) is the pressure at which the indicator has fully transferred to the second color. The actuation time, or time elapsed until the indicator has fully transferred to the second color, is less than 1 sec.

A pneumatic exhaust silencer (muffler) is used to control the noise caused by a rapidly exhausting airstream flowing into the atmosphere. The increased use of compressed air in industry has created a noise problem. Compressed air exhausts generate high-intensity sound energy, much of it in the same frequency ranges as normal conversation. Excessive exposure to these noises can cause loss of hearing without noticeable pain or discomfort. Noise exposure also causes fatigue and lowers production. It blocks out warning signals, thus causing accidents. This noise problem can be solved by installing a pneumatic silencer at each pneumatic exhaust port. Figure 10-22 depicts several types of exhaust silencers, which are designed not to build up back pressure with continued use.

Air from the atmosphere contains varying amounts of moisture in the form of water vapor. Compressors do not remove this moisture. Cooling of compressed air in piping causes condensation of moisture, much of which is eventually carried along into air-operated tools and machines. Water washes away lubrication, causing excessive wear and decreased efficiency. In addition, the temperature of the compressed air discharge from virtually all air compressors should be reduced to approximately 100°F before entering the piping system. If an aftercooler is placed

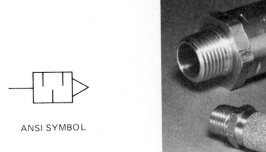

ANSI SYMBOL

Figure 10-22.   Pneumatic silencers. (*Courtesy of C. A. Norgren Co., Littleton, Colorado.*)

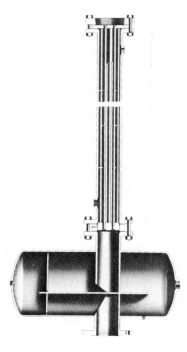

Figure 10-23. Aftercooler. (*Courtesy of Ingersoll-Rand Co., Washington, New Jersey.*)

Figure 10-24. Chiller air dryer. (*Courtesy of Ingersoll-Rand Co., Washington, New Jersey.*)

immediately downstream of a compressor, it will remove most of the moisture. An aftercooler is essential to reduce the air temperature to convenient levels and to act as a first stage in the removal of moisture prior to entering an air dryer. Figure 10-23 shows an aftercooler that is installed in the air line between the compressor and the air receiver. Water flow is opposite to air flow with internal baffles to provide proper water velocity and turbulence for high heat transfer rates. After passing through the tubes, the cooled air enters the moisture separating chamber, which effectively traps out condensed moisture. The tube is easily removed for inspection and cleaning.

Aftercoolers remove only about 80% of the moisture from the air leaving the compressor. Figure 10-24 shows a chiller air dryer, which removes virtually all moisture by lowering the temperature of the pressurized air to a dew point of 50°F. It uses a mechanical refrigeration system which is shipped completely assembled, piped, and wired. All that is needed are the connections to the air line, the electric power system, the cooling water circuit, and the condensate discharge line.

10.6 AIR PRESSURE LOSSES IN PIPES

As is the case for liquids, when air flows through a pipe, it loses energy due to friction. The energy loss shows up as a pressure loss, which can be calculated using the Harris formula:

$$P_f = \frac{cLQ^2}{CRd^5} \tag{10-8}$$

where P_f = pressure drop (psi)
   c = experimentally determined coefficient
   L = length of pipe (ft)
   Q = flow rate (ft^3/sec of free air)
   CR = compression ratio = pressure in pipe/atmospheric pressure
   d = inside diameter of pipe (in.)

For schedule 40 commercial pipe, the experimentally determined coefficient can be represented as a function of the pipe inside diameter:

$$c = \frac{0.1025}{d^{0.31}} \tag{10-9}$$

Substituting Eq. (10-9) into the Harris formula yields a single usable equation for calculating pressure drops in air pipelines: SCHEDULE 40

$$P_f = \frac{0.1025LQ^2}{CRd^{5.31}} \tag{10-10}$$

Tabulated values of $d^{5.31}$ are given in Fig. 10-25 to expedite calculations.

Nominal Pipe Size	Inside Diameter (d)	$d^{5.31}$	Nominal Pipe Size	Inside Diameter (d)	$d^{5.31}$
$\frac{3}{8}$	0.493	0.0234	$1\text{-}\frac{1}{2}$	1.610	12.538
$\frac{1}{2}$	0.622	0.0804	2	2.067	47.256
$\frac{3}{4}$	0.824	0.3577	$2\text{-}\frac{1}{2}$	2.469	121.419
1	1.049	1.2892	3	3.068	384.771
$1\text{-}\frac{1}{4}$	1.380	5.5304	$3\text{-}\frac{1}{2}$	3.548	832.550

Figure 10-25.   Tabulated values of $d^{5.31}$ for several pipe sizes.

EXAMPLE 10-7

A compressor delivers 100 cfm of free air through a 1-in. pipe at a receiver pressure of 150 psi. Find the pressure drop for a 250-ft length of pipe.

Solution   First solve for the compression ratio:

$$CR = \frac{150 + 14.7}{14.7} = 11.2$$

Next find the value of $d^{5.31}$ from Fig. 10-25:

$$d^{5.31} = 1.2892$$

Finally, using the Harris formula, the pressure drop is found:

$$P_f = \frac{0.1025 \times 250 \times (100/60)^2}{11.2 \times 1.2892} = 4.93 \text{ psi}$$

The frictional losses in pneumatic fittings can be computed using the Harris formula if the equivalent lengths of the fittings are known. The term L in the Harris formula would then represent the total equivalent length of the pipeline including its fittings. Figure 10-26 gives equivalent length values in feet for various types of fittings.

FITTING	NOMINAL PIPE SIZE						
	3/8	1/2	3/4	1	1–1/4	1–1/2	2
GATE VALVE (FULLY OPEN)	0.30	0.35	0.44	0.56	0.74	0.86	1.10
GLOBE VALVE (FULLY OPEN)	14.0	18.6	23.1	29.4	38.6	45.2	58.0
TEE (THROUGH RUN)	0.50	0.70	1.10	1.50	1.80	2.20	3.30
TEE (THROUGH BRANCH)	2.50	3.30	4.20	5.30	7.00	8.10	10.44
90° ELBOW	1.40	1.70	2.10	2.60	3.50	4.10	5.20
45° ELBOW	0.50	0.78	0.97	1.23	1.60	1.90	2.40

Figure 10-26.   Equivalent length of various fittings (ft).

EXAMPLE 10-8

If the pipe of Example 10-7 has two gate valves, three globe valves, five tees (through run), four 90° elbows, and six 45° elbows, find the pressure drop.

Solution   The total equivalent length of the pipe is

$$L = 250 + 2(0.56) + 3(29.4) + 5(1.50) + 4(2.60) + 6(1.23)$$

$$L = 250 + 1.12 + 88.2 + 7.5 + 10.4 + 7.38 = 364.6 \text{ ft}$$

Substituting into the Harris formula yields the answer:

$$P_f = \frac{0.1025 \times 364.6 \times (100/60)^2}{11.2 \times 1.2892} = 7.19 \text{ psi}$$

10.7 CONTROL WITH ORIFICES

Since a valve is a variable orifice, it is important to evaluate the flow rate of air through an orifice. Such a relationship is discussed for liquid flow in Chapter 5. However, because of the compressibility of air, the relationship describing the flow rate of air is more complex.

Equation (10-11) provides for the calculation of air flow rates through orifices:

$$Q = 22.67 C_v \sqrt{\frac{(P_1 - P_2)(P_2)}{T}} \tag{10-11}$$

where Q = flow rate (cfm) of free air
   C_v = capacity constant
   P_1 = upstream pressure (psia)
   P_2 = downstream pressure (psia)
   T = absolute temperature (°R)

The preceding equation is valid when P_2 is more than $0.53P_1$ or when P_2 is more than 53% of P_1. Beyond this region, the flow through the orifice is said to be choked. Thus, the flow through the orifice increases as the pressure drop $P_1 - P_2$ increases until P_2 becomes equal to $0.53P_1$. Any lowering of P_2 to values below $0.53P_1$ does not produce any increase in flow rate, as would be predicted by Eq. (10-11), because the speed of sound has been reached.

From a practical point of view, this means that a downstream pressure of 53% of the upstream pressure is the limiting factor for passing air through a valve to an actuator. Thus, for example, with 100-psia line pressure, if the pressure at the inlet of an actuator drops to 53 psi, the flow rate is at its maximum. No higher flow rate can be attained even if the pressure at the inlet of the actuator drops below 53 psia. Assuming an upstream pressure of 100 psi, the flow rate must be calculated for a downstream pressure of 53 psia using Eq. (10-11) even though the downstream pressure may be less than 53 psia. By the same token, if P_2 is less than $0.53P_1$, increasing the value of P_1 will result in a greater pressure drop across the valve but will not produce an increase in flow, because the orifice is already choked and increasing the ratio of P_1/P_2 beyond 0.53 does not produce any increase in flow rate. Values of C_v are determined experimentally by valve manufacturers and are usually given in table form for various sizes of valves.

EXAMPLE 10-9

Air at 80°F passes through a $\frac{1}{2}$-in.-diameter orifice having a capacity constant of 7.4. If the upstream pressure is 80 psi, what is the maximum flow rate in units of cfm of free air?

Solution

$$T = 80 + 460 = 540°R$$

$$P_1 = 80 + 14.7 = 94.7 \text{ psia}$$

$$P_2 = 0.53 \times 94.7 = 50.2 \text{ psia}$$

Substituting directly into Eq. (10-11) yields

$$Q = 22.67 \times 7.4 \sqrt{\frac{(94.7 - 50.2)(50.2)}{540}} = 22.67 \times 7.4 \times 2.03$$

$$Q = 340 \text{ cfm of free air}$$

10.8 AIR CONTROL VALVES

Air control valves are used to control the pressure, flow rate, and direction of air in pneumatic circuits. Pneumatic pressure control valves are air line regulators that are installed at the inlet of each separate pneumatic circuit. As such, they establish the working pressure of the particular circuit. Sometimes air line regulators are installed within a circuit to provide two or more different pressure levels for separate portions of the circuit. A cutaway view of an actual pressure regulator (whose operation is discussed in Sec. 10.5) is given in Fig. 10-27. The desired pressure level is established by the T-handle, which exerts a compressive force on the spring. The spring transmits a force to the diaphragm, which regulates the opening and closing of the control valve. This regulates the air flow rate to establish the desired downstream pressure.

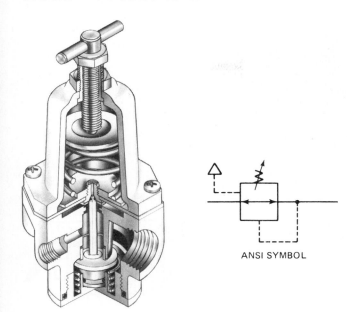

ANSI SYMBOL

Figure 10-27.   Cutaway of pneumatic pressure regulator. (*Courtesy of Aro Corp., Bryan, Ohio.*)

In Fig. 10-28 we see a check valve that shuts off instantaneously against reverse flow and opens at low cracking pressures in the forward direction. As shown in the schematic views, the disk seals before reverse flow is established, thus avoiding fluid shock on reversal of pressure differential. Although the design shown has a metal body, lightweight plastic body designs with fittings suitable for plastic or metal tubing are also available.

Figure 10-29(a) is a photograph of a pneumatic shuttle valve that automatically selects the higher of two input pressures and connects that pressure to the output port while blocking the lower pressure. This valve has two input ports and one output port and employs a free-floating spool with an open-center action. At

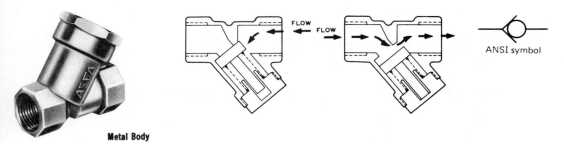

Metal Body

Figure 10-28.   Pneumatic check valve. (*Courtesy of Automatic Switch Co., Florham Park, New Jersey.*)

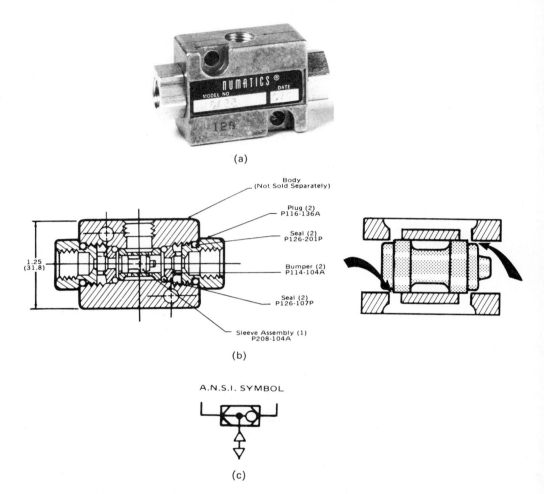

Figure 10-29.   Shuttle valve. (a) External view, (b) internal view and spool-port configuration, (c) ANSI symbol. (*Courtesy of Numatics Incorp., Highland, Michigan.*)

one end of the spool's travel, it connects one input with the output port. At the other end of its travel, it connects the second input with the output port.

When a pressure is applied to an input port, the air shifts the spool and then moves through the sleeve ports and out the output port. When the pressure is removed from the input port, the air in the output port exhausts back through the shuttle valve and out one of the input ports. It normally exhausts out the input port through which it entered, but there is no guarantee and it may exhaust out the other. If a signal is applied to the second input port, a similar action takes place.

If while one input is pressurized, the second input port receives a pressure that is 1.5 psig greater than the first, the higher pressure will appear at the output.

If the second input is the same as the first, no change will take place until the first signal is exhausted. Then, as it drops in pressure, the second input will predominate.

The open-center action of the shuttle valve is shown in Fig. 10-29(b), where the arrows indicate the clearance in the center position. As the spool shifts through the center position, it lacks a few thousandths of an inch of blocking the ports. Thus, it is not possible for the spool to find a center position that will block all exhaust action. It will always move to one position or the other.

Figure 10-29(c) gives the ANSI symbol for a pneumatic shuttle valve.

In Fig. 10-30 we see an air-operated (air-piloted), two-way, pneumatic valve. As shown, this valve is available to operate either normally open or normally

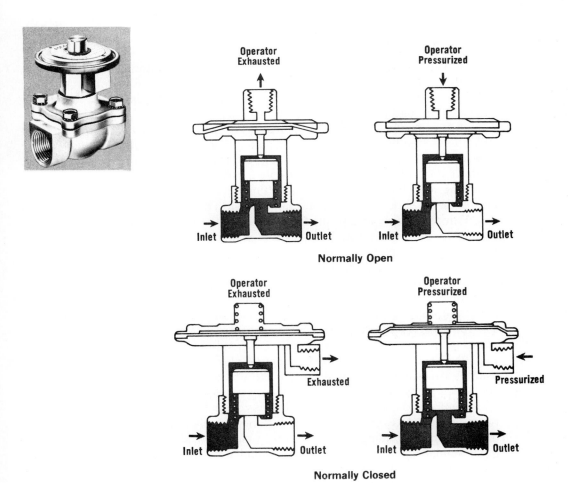

Figure 10-30.   Two-way, air-piloted valve. (*Courtesy of Automatic Switch Co., Florham Park, New Jersey.*)

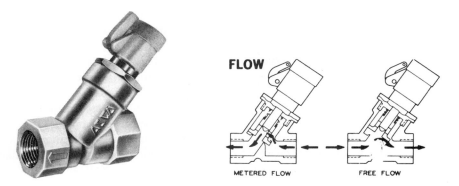

Figure 10-31. Flow control valve. (*Courtesy of Automatic Switch Co., Florham Park, New Jersey.*)

closed. The poppet-type construction provides a tight shutoff, and variations in the pilot air pressure or main line pressure do not affect the operation of these valves. The pilot pressure need not be constant. These valves will handle dry or lubricated air and provide long life.

A flow control valve is illustrated in Fig. 10-31. As shown, a spring-loaded disk allows free flow in one direction and an adjustable or controlled flow in the opposite direction. Flow adjustment is performed by a tapered brass stem that controls the flow through the cross hole in the disk.

The adjustable knob contains a unique locking device that consists of a plastic metering knob and thumb latch pawl. The valve bonnet is scribed with graduations to serve as a position indicator for the stem. When the pawl is in the up position, it creates a friction lock on the knurled bonnet, and the knob cannot rotate. When the pawl is at 90° to the knob, the knob is free to rotate. Mounting in any position will not affect operation.

Figure 10-32 is a photograph of a multipurpose three-way or open-exhaust four-way, push-button directional control valve. The three-way valves are three-port, multipurpose valves, and any port can be pressurized [see Fig. 10-33(a)]. The four-way valves may also be used as normally open or closed three-ways by plugging the appropriate cylinder port. Exhaust is through two screened ports [see Fig. 10-33(b)]. Since these ports cannot be plugged, four-way valves may not be used as two-ways. The force required to operate these valves is 2.5 lb. Figure 10-33 also gives the ANSI symbols for these valves.

Figure 10-34 shows a palm-button directional control valve. The large mushroom heads are extra heavy duty operators specifically designed to survive the day-after-day pounding of heavy, gloved hands in stamping press, foundry, and other similar applications. The large, rounded button is padded with a soft synthetic rubber cover, which favors the operator's hand.

In Fig. 10-35, we see a limit valve that uses a roller-level actuator. These directional control valves are available as multipurpose three-ways or open-

Figure 10-32. Push-button directional control valve. (*Courtesy of Numatics Incorp., Highland, Michigan.*)

exhaust four-ways. This type of valve is normally actuated by a cylinder piston rod at the ends or limits of its extension or retraction strokes.

In Fig. 10-36, we see a hand-lever-operated four-way directional control valve. The hand lever is used with two- or three-position valves. Hand movement of the lever causes the spool to move. The lever is directly connected to the spool. Detents, which provide a definite "feel" when the spool is in a specific position, are available.

Figure 10-37 illustrates the internal construction features of a four-way, two-position, solenoid-actuated directional control valve. The single-solenoid operator shown will move the spool when energized, and a spring will return the spool when the solenoid is deenergized.

Using two solenoids, a two-position valve can be shifted by energizing one solenoid momentarily. The valve will remain in the shifted position until the opposite solenoid is energized momentarily.

Three-position valves will remain in the spring-centered position until one of the solenoids is energized. Energizing the solenoid causes the spool to shift and stay shifted until the solenoid is deenergized. When the solenoid is deenergized, the spool will return to the center position.

10.9 PNEUMATIC ACTUATORS

Pneumatic systems make use of actuators in a fashion similar to that of hydraulic systems. However, because air is the fluid medium rather than hydraulic oil, pressures are lower, and hence pneumatic actuators are of lighter construction.

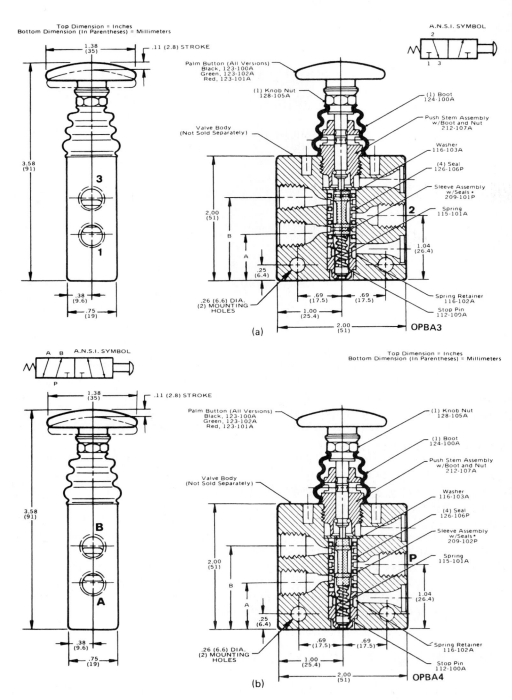

Figure 10-33. Operation of three-way and four-way push-button valves. (a) Three-way operation, (b) four-way operation. (*Courtesy of Numatics Incorp., Highland, Michigan.*)

Figure 10-34.   Palm-button valve. (*Courtesy of Numatics Incorp., Highland, Michigan.*)

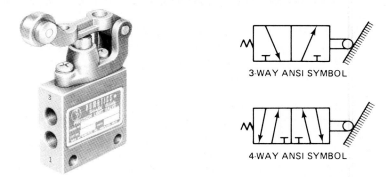

Figure 10-35.   Limit valve. (*Courtesy of Numatics Incorp., Highland, Michigan.*)

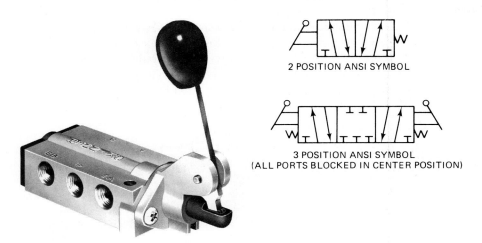

Figure 10-36.   Hand-lever-operated four-way valve. (*Courtesy of Skinner Precision Industries Inc., New Britain, Connecticut.*)

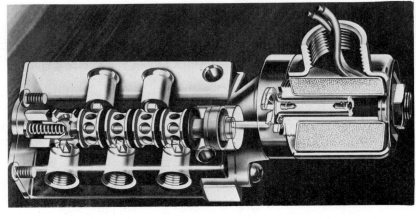

 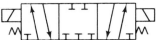

ANSI SYMBOL
2 POSITION-SINGLE SOLENOID

ANSI SYMBOL
3 POSITION-DOUBLE SOLENOID

Figure 10-37.   Solenoid-actuated directional control valve. (*Courtesy of Skinner Precision Industries Inc., New Britain, Connecticut.*)

For example, air cylinders make extensive use of aluminum and other nonferrous alloys to reduce weight, improve heat transfer characteristics, and minimize corrosive action of air.

   Figure 10-38 illustrates the internal construction features of a typical double-acting pneumatic cylinder. The piston uses wear-compensating, pressure-energized U-cup seals to provide low-friction sealing and smooth chatter-free movement of this 200-psi pressure-rated cylinder. The end plates use ribbed aluminum alloy to provide strength while minimizing weight. Self-aligning Buna-N seals provide a positive leakproof cushion with check valve action, which reverts

ANSI SYMBOL

Figure 10-38.   Construction of pneumatic cylinder. (*Courtesy of Aro Corp., Bryan, Ohio.*)

Figure 10-39.   Air cylinder-drive rotary index table. (*Courtesy of Allenair Corp., Mineola, New York.*)

to free flow upon cylinder reversal. The cushion adjustment, which uses a tapered self-locking needle at each end, provides positive control over the stroke, which can be as large as 20 in.

Figure 10-39 depicts a rotary index table driven by a double-acting pneumatic cylinder. The inlet pressure can be adjusted to provide exact force for moving the load and to prevent damage in case of accidental obstructions. A rack and gear drive transmits the straight-line motion of the air cylinder to the rotary motion with full power throughout its cycle. Through the use of different cams, the table can be indexed in 90°, 60°, 45°, 30°, or 15° increments.

In Fig. 10-40 we see a pneumatic rotary actuator, which is available in five basic models to provide a range of torque outputs from 100 to 10,000 in. · lb using 100-psi air. Standard rotations are 94°, 184°, and 364°. The cylinder heads at each end serve as positive internal stops for the enclosed floating pistons. The linear motion of the piston is modified into rotary motion by a rack and pinion made of hardened steel for durability.

Figure 10-40.   Pneumatic rotary actuator. (*Courtesy of Flo-Tork Inc., Orrville, Ohio.*)

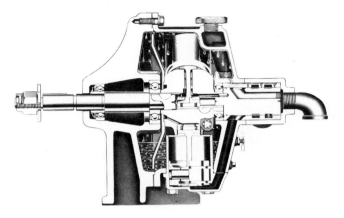

Figure 10-41. Radial piston air motor. (*Courtesy of Gardner-Denver Co., Quincy, Illinois.*)

Rotary air motors can be utilized to provide a smooth source of power. They are not susceptible to overload damage and can be stalled for long periods of time without any heat problems. They can be started and stopped very quickly and with pressure regulation and metering of flow can provide infinitely variable torque and speed.

Figure 10-41 shows a radial piston air motor. The five-cylinder piston design provides even torque at all speeds due to overlap of the five power impulses occurring during each revolution of the motor. At least two pistons are on the power stroke at all times. The smooth overlapping power flow and accurate balancing make these motors vibrationless at all speeds. This smooth operation is especially noticeable at low speeds when the flywheel action is negligible. This air motor has relatively little exhaust noise, and this can be further reduced by use of an exhaust muffler. It is suitable for continuous operation using 100-psi air pressure and can deliver up to 15 hp.

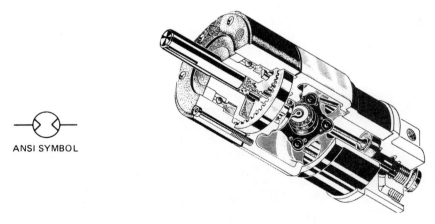

ANSI SYMBOL

Figure 10-42. Axial piston air motor. (*Courtesy of Gardner-Denver Co., Quincy, Illinois.*)

In Fig. 10-42 we see an axial piston air motor, which can deliver up to 3 hp using 100-psi air. The power pulses for these five-piston axial design motors is the same as those for the radial piston design. At least two pistons are on the power stroke at all times, providing even torque at all speeds.

10.10 BASIC PNEUMATIC CIRCUITS

In this section we present a number of basic pneumatic circuits utilizing pneumatic components that have been previously discussed. Pneumatic circuits are similar to their hydraulic counterparts. One difference is that no return lines are used in pneumatic circuits because the exhausted air is released directly into the atmosphere. This is depicted by a short dashed line leading from the exhaust port of each valve. Also, no input device (such as a pump in a hydraulic circuit) is shown, because most pneumatic circuits use a centralized compressor as their source of energy. The input to the circuit is located at some conveniently located manifold, which leads directly into the filter-regulator-lubricator (FRL) unit.

Figure 10-43 shows a simple pneumatic circuit, which consists of a three-way valve controlling a single-acting cylinder. The return stroke is accomplished by a compression spring located at the rod end of the cylinder. When the push-button valve is actuated, the cylinder extends. It retracts when the valve is deactivated. Needle valves $V1$ and $V2$ permit speed control of the cylinder extension and retraction strokes, respectively.

In Fig. 10-44 we see the directional control of a double-acting cylinder using a four-way valve. Notice that control of a double-acting cylinder requires a DCV with four different functioning ports (each of the two exhaust ports perform the same function). Thus a four-way valve has four different functioning ports. In contrast, the control of a single-acting, spring-return cylinder requires a DCV with only three ports. Hence a three-way valve has only three ports, as shown in Fig. 10-43.

Actuation of the push-button valve extends the cylinder. The spring offset mode causes the cylinder to retract under air power.

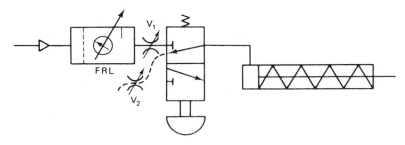

Figure 10-43. Operation of a single-acting cylinder.

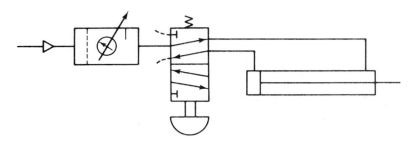

Figure 10-44.   Operation of a double-acting cylinder.

In Fig. 10-45 we see a circuit in which a double-acting cylinder can be remotely operated through the use of an air-pilot-actuated DCV. Push-button valves $V1$ and $V2$ are used to direct air flow (at low pressure such as 10 psi) to actuate the air-piloted DCV, which directs air at high pressure such as 100 psi to the cylinder. Thus, operating personnel can use low-pressure push-button valves to remotely control the operation of a cylinder that requires high-pressure air for performing its intended function. When $V1$ is actuated and $V2$ is in its spring offset mode, the cylinder extends. Deactivating $V1$ and then actuating $V2$ retracts the cylinder.

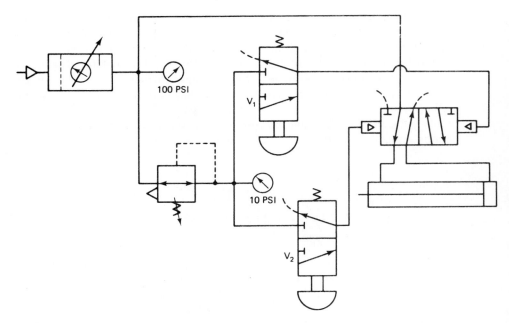

Figure 10-45.   Air pilot control of a double-acting cylinder.

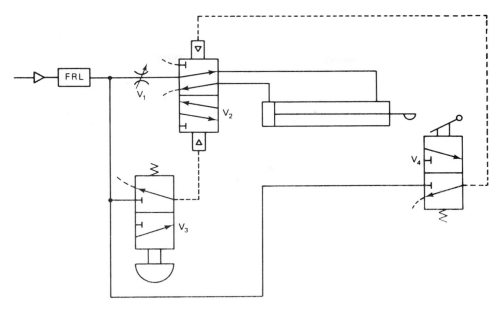

Figure 10-46. Cylinder cycle timing system.

Figure 10-46 shows a circuit that employs a limit valve to provide a timed cylinder extend and retract cycle. When push-button valve V3 is momentarily actuated, valve V2 shifts to extend the cylinder. When the piston rod cam actuates limit valve V4, it shifts V2 into its opposite mode to retract the cylinder. Flow control valve V1 controls the flow rate and thus the cylinder speed.

A two-step speed control system is shown in Fig. 10-47. The operation is as follows, assuming that flow control valve V3 is adjusted to allow a greater flow rate than valve V4. Initially the cylinder is fully retracted. When push-button valve V1 is actuated, air flow goes through valves V2, V3, and the shuttle valve (V5) to extend the cylinder at high speed. When the piston rod cam actuates valve V6, valve V2 shifts. The flow is therefore diverted to valve V4 and through the shuttle valve. However, due to the low flow setting of valve V4, the extension speed of the cylinder is reduced. After the cylinder has fully extended, valve V1 is released by the operator to cause retraction of the cylinder.

Figure 10-48 shows a two-handed safety control circuit. Both palm-button valves (V1 and V2) must be actuated to cause the cylinder to extend. Retraction of the cylinder will not occur unless both palm buttons are released.

If both palm-button valves are not operated together, the pilot air to the three-position valve is vented. Hence, this three-way valve goes into its spring-centered mode, and the cylinder is locked.

In Fig. 10-49 we see a circuit used to control an air motor. The operation is as follows. When the START push-button valve is actuated momentarily, the air

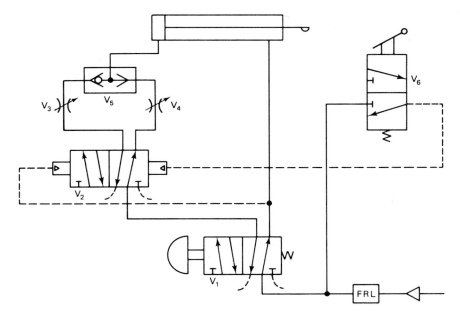

Figure 10-47.   Two-step speed control.

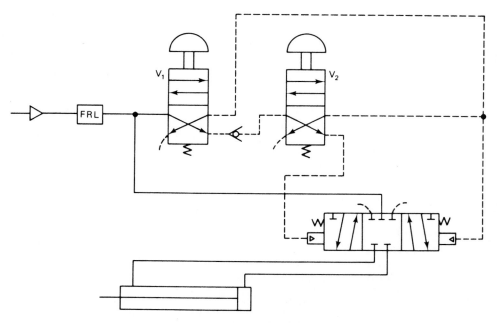

Figure 10-48.   Two-handed safety control circuit.

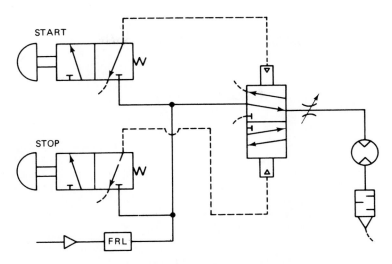

Figure 10-49.   Control of an air motor.

pilot valve shifts to supply air to the motor. When the STOP push-button valve is actuated momentarily, the air pilot valve shifts into its opposite mode to shut off the supply of air to the motor. The flow control valve is used to adjust the speed of the motor.

Figure 10-50 shows a riveting assembly machine, which performs continuous, high-speed, repetitive production of riveted components. The control system contains many pneumatic components such as regulators, filters, lubricators, solenoid valves, and cylinders. These machines were designed to operate under tough production-line conditions with a minimum of downtime for maintenance and adjustment.

10.11 ACCUMULATOR SYSTEMS ANALYSIS

Example 10-10 illustrates the operation and analysis of accumulators in hydraulic circuits. Boyle's law is used, assuming that the gas temperature change inside the accumulator is negligibly small.

EXAMPLE 10-10

The circuit of Fig. 10-51 has been designed to crush a car body into bale size using a 6-in.-diameter hydraulic cylinder. The hydraulic cylinder is to extend 100 in. during a period of 10 sec. The time in between crushing strokes is 5 min. The following accumulator gas pressures are given:

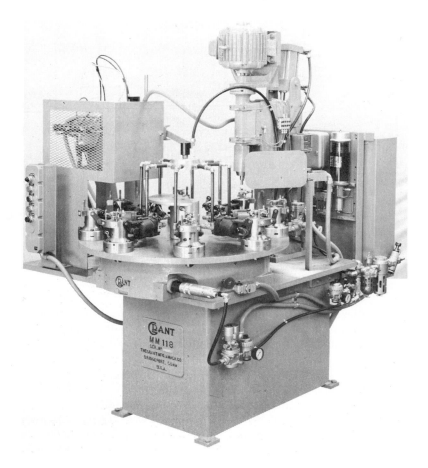

Figure 10-50. Pneumatically controlled riveting assembly machine. (*Courtesy of C. A. Norgren Co., Littleton, Colorado.*)

P_1 = gas precharge pressure = 1200 psi

P_2 = gas charge pressure when pump is turned on

  = 3000 psi = pressure relief valve setting

P_3 = minimum pressure required to actuate load = 1800 psi

a. Calculate the required size of the accumulator.

b. What are the pump horsepower and flow requirements with and without an accumulator?

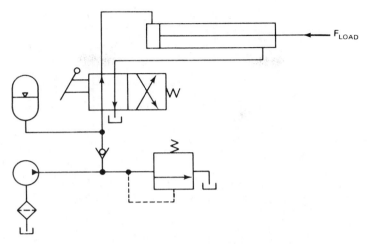

Figure 10-51.   Hydraulic system for crushing car bodies.

Solution   Figure 10-52 shows the three significant accummulator operating conditions:

1. Preload [Fig. 10-52(a)]. This is the condition just after the gas has been introduced into the top of the accumulator. Notice that the piston (assuming a piston design) is all the way down to the bottom of the accumulator.

2. Charge [Fig. 10-52(b)]. The pump has been turned on, and hydraulic oil is pumped into the accumulator since P_2 is greater than P_1. During this phase, the four-way valve is in its spring offset position. Thus, system pressure builds up to the 3000-psi level of the pressure relief valve setting.

3. Final position of accumulator piston at end of cylinder stroke [Fig. 10-52(c)]. The four-way valve is actuated to extend the cylinder against its load. When the system pressure drops below 3000 psi, the accumulator 3000-psi gas pressure forces oil out of the accumulator into the system to assist the pump during the rapid extension of the cylinder. The accumulator gas pressure reduces to a minimum value of P_3, which must not be less than the minimum value of 1800 psi required to drive the load.

    a. Use Eq. (10-3):

$$P_1V_1 = P_2V_2 = P_3V_3$$

where V_1 = required accumulator size, and also use $V_{\text{hydraulic cylinder}} = V_3 - V_2$ (assuming negligible assistance from the pump).

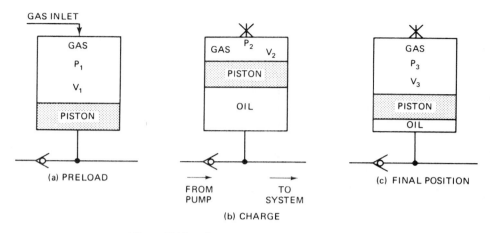

Figure 10-52. Operation of piston accumulator.

Thus, we have

$$V_3 = \frac{P_2 V_2}{P_3} = \frac{3000 V_2}{1800} = 1.67 V_2$$

$$V_{\text{hydraulic cylinder}} = \frac{\pi}{4}(6)^2 \times 100 = 2830 \text{ in.}^3 = V_3 - V_2$$

Solving the preceding equations yields

$$V_2 = 4230 \text{ in.}^3, \qquad V_3 = 7060 \text{ in.}^3$$

Therefore, we have a solution:

$$V_1 = \frac{P_2 V_2}{P_1} = \frac{(3000)(4230)}{1200} = 10{,}550 \text{ in.}^3 = 45.8\text{-gal accumulator}$$

Note that this is a very large accumulator because it is required to do a big job. For example, if the cylinder stroke were only 10 in., a 4.58-gal accumulator would suffice.

b. *With accumulator* (pump charges accumulator in 5 min):

$$Q_{\text{pump}} = \frac{45.8 \text{ gal}}{5 \text{ min}} = 9.16 \text{ gpm} \quad \text{(a small-sized pump)}$$

$$HP_{\text{pump}} = \frac{(3000)(9.16)}{1714} = 16.0 \quad \text{(a small horsepower requirement)}$$

Without accumulator (pump extends cylinder in 10 sec):

$$Q_{pump} = \frac{\left(\dfrac{2830}{231}\right) \text{ gal}}{\left(\dfrac{1}{6}\right) \text{ min}} = 73.4 \text{ gpm} \qquad \text{(a very large pump)}$$

$$HP_{pump} = \frac{(1800)(73.4)}{1714} = 76.6 \qquad \text{(a very large horsepower requirement)}$$

The results show that an accumulator, by handling large transient demands, can dramatically reduce the size and power requirements of the pump.

10.12 PNEUMATIC CIRCUIT ANALYSIS USING METRIC UNITS

As stated in Sec. 10.2, circuit calculations with volume and pressure changes of air must be performed using absolute pressure and absolute temperature values. Equations 10-12 and 10-13 permit the calculation of absolute pressures and temperatures in metric units as follows:

$$\text{absolute pressure (Pa)} = \text{gage pressure (Pa)} + 101,000 \qquad (10\text{-}12)$$

$$\text{absolute temperature (°K)} = \text{temperature (°C)} + 273 \qquad (10\text{-}13)$$

The units of absolute temperature in the metric system are degrees Kelvin, abbreviated °K. From Eq. (10-13), we note that a temperature of 0°K (absolute zero) equals — 273 degrees Celsius (°C).

Boyle's, Charles', Gay-Lussac's, and the general gas law are applicable for use with English or metric units. The following example illustrates the application of the general gas law using metric units (see Example 10-4, which solves the same problem using English units).

EXAMPLE 10-11

Gas at 70 bars gage pressure and 37.8°C is contained in the 12,900 cm³ cylinder of Fig. 10.5. A piston compresses the volume to 9,680 cm³ while the gas is heated to 93.3°C. What is the final pressure in the cylinder?

Solution   Solve Eq. (10-6) for P_2 and substitute known values.

$$P_2 = \frac{P_1V_1T_2}{V_2T_1} = \frac{(70 \times 10^5 + 1 \times 10^5)(12,900)(93.3 + 273)}{(9680)(37.8 + 273)}$$

$$P_2 = 111.5 \times 10^5 \text{ Pa absolute} = 111.5 \text{ bars absolute}$$

Example 10-12 illustrates the metric units analysis of accumulators in hydraulic circuits. Boyle's law is used, assuming that the gas temperature change inside the accumulator is negligibly small (see Example 10-10, which solves the same problem using English units).

EXAMPLE 10-12

The circuit of Fig. 10-51 has been designed to crush a car body into bale size using a 152 mm diameter hydraulic cylinder. The hydraulic cylinder is to extend 2.54 m during a period of 10 s. The time in between crushing strokes is 5 min. The following accumulator gas pressures are given:

P_1 = gas precharge pressure = 84 bars

P_2 = gas charge pressure when pump is turned on

    = 210 bars = pressure relief valve setting

P_3 = minimum pressure required to actuate load = 126 bars

a. Calculate the required size of the accumulator.

b. What are the pump kW power and the flow requirements with and without an accumulator?

Solution   Figure 10-52 shows the three significant accumulator operating conditions (preload, charge, and final position of accumulator piston at end of cylinder stroke).

a. Use Eq. (10-3).

$$P_1V_1 = P_2V_2 = P_3V_3$$

where V_1 = required accumulator size, and also use

$$V_{\text{hydraulic cylinder}} = V_3 - V_2$$

Thus, we have:

$$V_3 = \frac{P_2V_2}{P_3} = \frac{210}{126} V_2 = 1.67 V_2$$

$$V_{\text{hydraulic cylinder}} = \frac{\pi}{4} (0.152)^2 \times 2.54 = 0.0461 \text{ m}^3 = V_3 - V_2$$

Solving the preceding equations yields:

$$V_2 = 0.0688 \text{ m}^3, \qquad V_3 = 0.115 \text{ m}^3$$

Therefore, we have a solution

$$V_1 = \frac{P_2 V_2}{P_1} = \frac{(210)(0.0688)}{84} = 0.172 \text{ m}^3 = 172 \text{ liters}$$

   b. With accumulator (pump charges accumulator in 5 min):

$$Q_{\text{pump}} \quad = \frac{172 \, l}{300 \text{ s}} = 0.573 \, l/s \qquad \text{(a small-sized pump)}$$

$$kW_{\text{pump}} \quad = \frac{(210 \times 10^5)(57.3 \times 10^{-5})}{1000} = 12.0 \, kW \quad \text{(a small power requirement)}$$

Without accumulator (pump extends cylinder in 10 s):

$$Q_{\text{pump}} = \frac{46.1 \, l}{10 \, s} = 4.61 \, l/s \qquad \text{(a very large pump)}$$

$$kW_{\text{pump}} = \frac{(126 \times 10^5)(461 \times 10^{-5})}{1000} = 58.1 \, kW \quad \text{(a very large power requirement)}$$

EXERCISES

10-1. Name three reasons for considering the use of pneumatics instead of hydraulics.

10-2. State Boyle's, Charles', and Gay-Lussac's laws.

10-3. The 2-in.-diameter piston of the pneumatic cylinder of Fig. 10-4 retracts 5 in. from its present position ($P_1 = 30$ psig, $V_1 = 20$ in.^{3}) due to the external load on the rod. If the port at the blank end of the cylinder is blocked, find the new pressure, assuming the temperature does not change.

10-4. The cylinder of Fig. 10-4 has an initial position where $P_1 = 30$ psig and $V_1 = 20$ in^3. as controlled by the load on the rod. The air temperature is 80°F. The load on the rod is held constant to maintain constant air pressure, but the air temperature is increased to 150°F. Find the new volume of air at the blank end of the cylinder.

10-5. The cylinder of Fig. 10-4 has a locked position ($V_1 = $ constant). $P_1 = 30$ psig, and $T_1 = 80$°F. If the temperature increases to 160°F, what is the new pressure in the blank end?

10-6. Gas at 1200 psig and 120°F is contained in the 2000-in.^{3} cylinder of Fig. 10-5. A piston compresses the volume to 150 in.^{3} while the gas is heated to 250°F. What is the final pressure in the cylinder?

10-7. Name three types of air compressors.

10-8. What is a multistage compressor?

10-9. Air is used at a rate of 30 cfm from a receiver at 100°F and 150 psi. If the atmospheric pressure is 14.7 psia and the atmospheric temperature is 80°F, how many cfm of free air must the compressor provide?

10-10. (a) Calculate the required size of a receiver which must supply air to a pneumatic system consuming 30 cfm for 10 min between 120 psi and 100 psi before the compressor resumes operation.

   (b) What size is required if the compressor is running and delivering air at 6 cfm?

10-11. Describe the function of an air filter.

10-12. Describe the function of an air pressure regulator.

10-13. Why would a lubricator be used in a pneumatic system?

10-14. What is a pneumatic indicator?

10-15. Why are exhaust silencers used in pneumatic systems?

10-16. What is the difference between an aftercooler and a chiller air dryer?

10-17. A compressor delivers 150 cfm of free air through a 1-in. pipe at a receiver pressure of 125 psi. Find the pressure drop for a 150-ft length of pipe.

10-18. If the pipe of Exercise 10-17 has three gate valves, two globe valves, four tees (through run), and five 90° elbows, find the pressure drop.

10-19. Air at 100°F passes through a $\frac{1}{2}$-in.-diameter orifice having a capacity constant of 7. If the upstream pressure is 125 psi, what is the maximum flow rate in units of cfm of free air?

10-20. How do pneumatic actuators differ from hydraulic actuators?

10-21. What does the circuit of Fig. 10-53 accomplish when the manual shutoff valve $V1$ is opened?

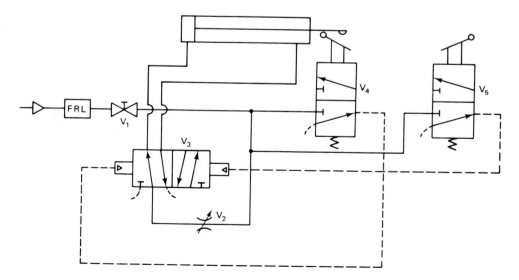

Figure 10-53. Circuit for Exercise 10-21.

10.22 For the circuit of Fig. 10-54,
   (a) What happens to the cylinder when valve $V4$ is depressed?
   (b) What happens to the cylinder when valve $V5$ is depressed?

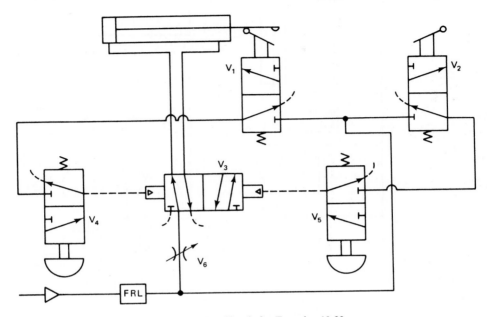

Figure 10-54.   Circuit for Exercise 10-22.

10-23. For the circuit of Fig. 10-55, what happens to the cylinders when
   (a) Valve $V1$ is actuated and held?
   (b) Valve $V1$ is released and valve $V2$ is actuated and held? Valves $V3$ and $V4$ are sequence valves?

10-24. For the circuit of Fig. 10-55, cylinder 1 will not hold against a load while cylinder 2 is retracting. Modify this circuit by adding a pilot check valve and appropriate piping so that cylinder 1 will hold while cylinder 2 is retracting.

10-25. The accumulator of Fig. 10-56 is to supply 450 in.^{3} of oil with a maximum pressure of 3000 psi and a minimum pressure of 1800 psi. If the nitrogen precharge pressure is 1200 psi, find the size of the accumulator.

10-26. For the accumulator of Exercise 10-25, find the load force F_{load} that the cylinder can carry over its entire stroke. What would be the total stroke of the cylinder if the entire output of the accumulator is used?

10-27. The 50-mm diameter piston of the pneumatic cylinder of Fig. 10-4 retracts 130 mm from its present position (P_1 = 2 bars gage, V_1 = 130 cm^3) due to the external load on the rod. If the port at the blank end of the cylinder is blocked, find the new pressure, assuming the temperature does not change.

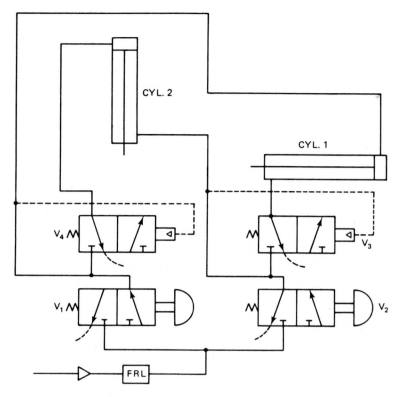

Figure 10-55.   Circuit for Exercise 10-23.

10-28. The cylinder of Fig. 10-4 has an initial position where $P_1 = 2$ bars gage and $V_1 = 130$ cm^3 as controlled by the load on the rod. The air temperature is 30°C. The load on the rod is held constant to maintain constant air pressure, but the air temperature is increased to 65°C. Find the new volume of air at the blank end of the cylinder.

10-29. The cylinder of Fig. 10-4 has a locked position (V_1 = constant). $P_1 = 2$ bars gage and $T_1 = 25$°C. If the temperature increases to 70°C, what is the new pressure in the blank end?

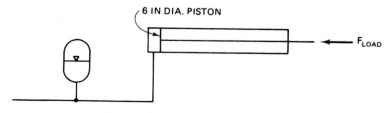

Figure 10-56.   Circuit for Exercise 10-25.

10-30. Gas at 80 bars gage and 50°C is contained in the 12,900 cm³ cylinder of Fig. 10-5. A piston compresses the volume to 1000 cm³ while the gas is heated to 120°C. What is the final pressure in the cylinder?

10-31. The accumulator of Fig. 10-56 is to supply 7370 cm³ of oil with a maximum pressure of 210 bars gage and a minimum pressure of 126 bars gage. If the nitrogen precharge is 84 bars gage, find the size of the accumulator. The hydraulic cylinder piston diameter is 152 mm.

10-32. For the accumulator to Exercise 10-31, find the load force F_{load} that the cylinder can carry over its entire stroke. What would be the total stroke of the cylinder if the entire output of the accumulator is used?

10-33. A single-acting air cylinder with a 2½-in. diameter piston and 12-in. stroke operates at 100 psi and cycles at 30 cycles per min. Compute the air consumption in ft³/min of free air.

10-34. A single-acting air cylinder with a 6 cm diameter piston and 30 cm stroke operates at 700 kPa and cycles at 30 cycles per min. Compute the air consumption in m³/min of free air.

10-35. Convert a temperature of 160°F to °C, °R, and °K.

10-36. An accumulator under a pressure of 10 MPa is reduced in volume from 0.04 m³ to 0.03 m³ while the temperature increases from 40°C to 180°C. Determine the final pressure.

10-37. A double-acting air cylinder has a 50-mm diameter piston, a 2.5-cm stroke, operates at 600 kPa, and cycles at 80 cycles/min. Determine the time it takes to consume 100 m³ of free air.

10-38. When an oscillating load comes to a stop at mid-stroke, the charge gas in an accumulator expands from 180 in.³ to 275 in.³ and cools from 200°F to 100°F. If the resulting pressure in the accumulator is 1000 psi, calculate the charge gas pressure.

10-39. Name the steps required to size an air compressor.

10-40. Name four functions of an air receiver.

10-41. A double-acting air cylinder with a 2-in. diameter piston and 12-in. stroke operates at 200 cycles/min using 100 psi air. What is the flow rate of free air to the cylinder? Ignore the piston rod cross-sectional area and assume the temperature remains constant.

10-42. A rotary vane air motor has a displacement volume of 4 in.³/rev and operates at 1750 rpm using 100 psi air. Calculate the free air rate of consumption and HP output of the motor. Assume the temperature remains constant.

10-43. The pneumatic system of Fig. 10-57 contains a double-rod cylinder controlled by two three-way, two-position directional control valves. Describe the four operating conditions for this system.

10-44. For the system of Exercise 10-43, as shown in Fig. 10-57, the cylinder is free (both ends vented to the atmosphere) in the unactuated (spring offset) position of the directional control valves. Redesign the system using the same components to accomplish the following operations:
a. Cylinder rod moves left when only *V*1 is actuated.
b. Cylinder rod moves right when only *V*2 is actuated.

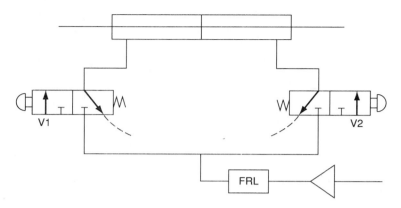

Figure 10-57.   System for Exercise 10-43.

    c. Cylinder rod stops moving when a single actuated valve is unactuated (both valves are unactuated).

    d. When both valves are actuated, the cylinder is free (both ends are vented to the atmosphere).

10-45. The pneumatic system in Figure 10-58 is designed to lift and lower two loads using single-acting cylinders connected in parallel. If the load on cylinder 1 is larger than the load on cylinder 2 ($F_1 > F_2$), what will happen when the directional control valve is shifted into the lift mode?

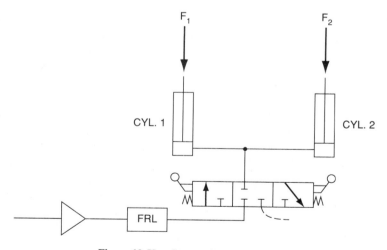

Figure 10-58.   System for Exercise 10-45.

10-46. Redesign the system of Exercise 10-45, as shown in Fig. 10-58, so that the cylinders will extend and retract together (synchronized at same speed).

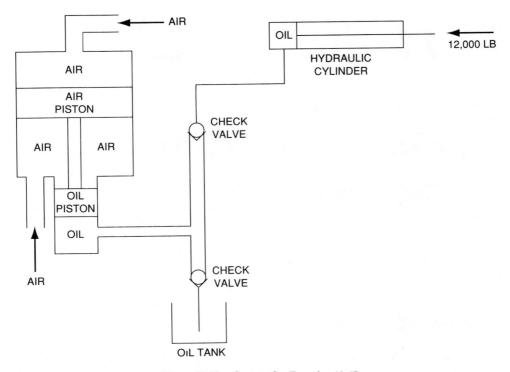

Figure 10-59.   System for Exercise 10-47.

10-47. An air-hydraulic intensifier is connected to a hydraulic cylinder driving a 12,000-lb load as shown in Fig. 10-59. The diameter of the hydraulic cylinder piston is 1.5 in. The following data applies to the intensifier:

Air piston diameter = 8 in.
Oil piston diameter = 1 in.
Intensifier stroke = 2 in.
Intensifier frequency = 1 stroke/sec

Determine the:
a. Volumetric displacement of the intensifier oil piston.
b. Volumetric displacement of the hydraulic cylinder piston per intensifier stroke.
c. Movement of the hydraulic cylinder piston per intensifier stroke.
d. Volumetric displacement of the blank end of the intensifier air cylinder piston.
e. Flow rate of oil from the intensifier.
f. Air consumption rate of intensifier in cfm.

11

Fluid Logic Control Systems

11.1 INTRODUCTION

Fluid logic control systems use logic devices that switch a fluid, usually air, from one outlet of the device to another outlet. Hence an output of a fluid logic device is either ON or OFF as it is rapidly switched from one state to the other by the application of a control signal. Devices that use a fluid as the power supply medium are broadly classified as either moving-part logic (MPL) devices or fluidic devices.

Moving-part logic devices are miniature valve-type devices, which by the action of internal moving parts, perform switching operations in fluid logic systems. MPL devices are typically available as spool, poppet, and diaphragm valves, which can be actuated by means of mechanical displacement, electric voltage, or fluid pressure.

Figure 11-1 shows the design features of a MPL valve containing a manifold body that encircles the central valve cavity with air passages. These passages can be used to allow air flow at any point along the axis of the valve. The passages terminate at the base of the body in a circular Octoport pattern (see Fig. 11-2). The body mates with a manifold subplate, which mounts on the complete module and provides tapped holes for hose fittings. Because of the easy availability of an air connection wherever it is required, the manifold body permits valve elements to be designed for maximum performance without the restrictive limitations of rigid port configurations. It also allows multiple porting using two or more ports as an inlet, outlet, supply, etc. This reduces the amount of external piping needed to complete a circuit. Furthermore, the manifold body enables the internal intercon-

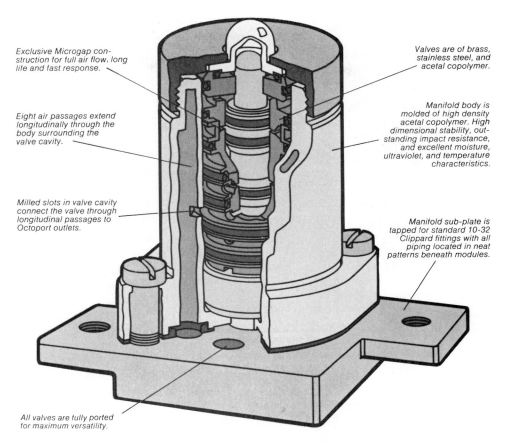

Exclusive Microgap con-
struction for full air flow, long
life and fast response.

Valves are of brass,
stainless steel, and
acetal copolymer.

Eight air passages extend
longitudinally through the
body surrounding the
valve cavity.

Manifold body is
molded of high density
acetal copolymer. High
dimensional stability, outstanding impact resistance,
and excellent moisture,
ultraviolet, and temperature
characteristics.

Milled slots in valve cavity
connect the valve through
longitudinal passages to
Octoport outlets.

Manifold sub-plate is
tapped for standard 10-32
Clippard fittings with all
piping located in neat
patterns beneath modules.

All valves are fully ported
for maximum versatility.

Figure 11-1.   MPL valve with modular manifold body. (*Courtesy of Clippard
Instrument Laboratory Inc., Cincinnati, Ohio.*)

nection of ports. This is especially valuable in a number of modules that contain
more than one valve (see Fig. 11-3). The separate elements are interconnected in
the same module to provide complete subcircuits such as OR/NOR and AND/
NAND gates. These functions further reduce external piping. Directional control
valve configurations include two-way, three-way, and four-way designs with pressure pilot, electronic, and mechanical actuation schemes. Other types of valves
incorporating this modular, manifold design include interface valves, flow control
valves, shuttle valves, pressure regulators, pulse valves, and sequence valves.

Figure 11-4 shows a torque wrench machine that uses a four module MPL
circuit. It operates as follows: a motor-driven socket wrench drives through a
pneumatic clutch, with the amount of pressure applied to the clutch determining
the torque. The entire wrench assembly is raised and lowered with air cylinders.

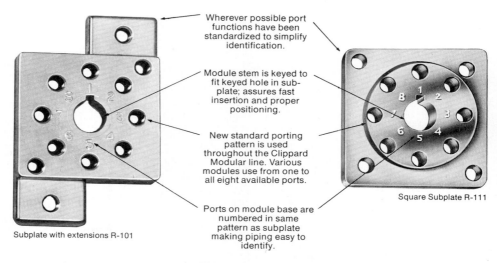

Wherever possible port functions have been standardized to simplify identification.

Module stem is keyed to fit keyed hole in sub-plate; assures fast insertion and proper positioning.

New standard porting pattern is used throughout the Clippard Modular line. Various modules use from one to all eight available ports.

Ports on module base are numbered in same pattern as subplate making piping easy to identify.

Square Subplate R-111

Subplate with extensions R-101

Figure 11-2.   Octoport subplate for MPL valve with modular manifold body. (*Courtesy of Clippard Instrument Laboratory Inc., Cincinnati, Ohio.*)

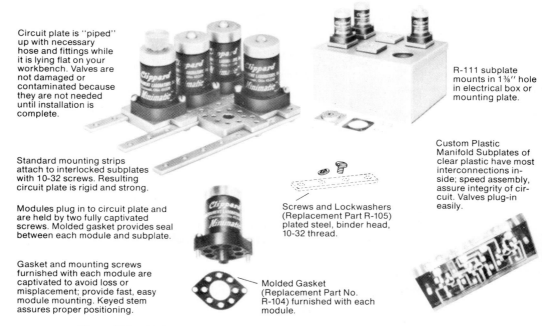

Circuit plate is "piped" up with necessary hose and fittings while it is lying flat on your workbench. Valves are not damaged or contaminated because they are not needed until installation is complete.

Standard mounting strips attach to interlocked subplates with 10-32 screws. Resulting circuit plate is rigid and strong.

Modules plug in to circuit plate and are held by two fully captivated screws. Molded gasket provides seal between each module and subplate.

Gasket and mounting screws furnished with each module are captivated to avoid loss or misplacement; provide fast, easy module mounting. Keyed stem assures proper positioning.

R-111 subplate mounts in 1⅜" hole in electrical box or mounting plate.

Custom Plastic Manifold Subplates of clear plastic have most interconnections inside; speed assembly, assure integrity of circuit. Valves plug-in easily.

Screws and Lockwashers (Replacement Part R-105) plated steel, binder head, 10-32 thread.

Molded Gasket (Replacement Part No. R-104) furnished with each module.

Figure 11-3.   Multiple assembly of MPL valves with modular manifold body. (*Courtesy of Clippard Instrument Laboratory Inc., Cincinnati, Ohio.*)

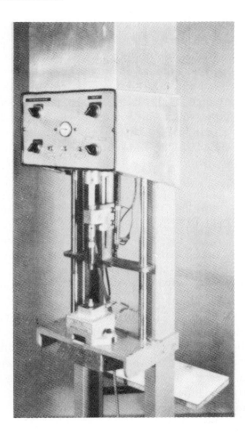

Figure 11-4.   MPL controlled torque wrench machine. (*Courtesy of Clippard Instrument Laboratory Inc., Cincinnati, Ohio.*)

The parts to be tightened are placed by the operator in a pneumatic collet fixture. The foot pedal is pushed and the collet closes, initiating the cycle. The wrench assembly lowers onto the part and tightens to the predetermined torque. When the wrench stops turning, the operator releases the pedal. The wrench raises, the collet vice opens, and the part is removed. By using this machine, rather than hand tightening, more accurate torques and a safer operation with three times the manual production rate are realized.

Fluidic devices use a completely different technique for providing control logic capability as compared to MPL devices. Fluidics is the technology that utilizes fluid flow phenomena in components and circuits to perform a wide variety of control functions. These include sensing, logic, memory, timing, and interfacing to other control media.

The history of fluidics goes back to 1959 and the research efforts of three engineers at the U.S. Army's Harry Diamond Laboratories. R. E. Bowles, R. W. Warren, and B. M. Horton devised a method of using flowing fluid to accomplish control functions. They were trying to find a way to overcome the deficiencies of existing electronic systems, which were adversely affected by temperature, shock, vibration, and radiation. A logic system was desired that would eliminate or reduce the number of moving parts and interface between electrical and mechanical components in both hydraulic and pneumatic systems.

The concepts of fluidics are basically simple. They involve the effect of one stream meeting another to change its direction of flow and the effect of a fluid sticking to a wall. This wall attachment phenomenon was first discovered by the Hungarian scientist Henri Coanda in 1932 but was never usefully employed prior to the Diamond Labs' experiments. The wall attachment phenomenon, which is frequently called the *Coanda effect*, is the basis of operation of many fluidic components.

When using fluidics to control fluid power systems, the fluidic control circuit performs the function of the "brains," and the fluid power system provides the "brawn," or muscle. Thus, fluidic components operate at low power and pressure levels (normally below 15 psi). Yet these low-pressure fluidic signals can reliably control hydraulic systems operating with up to 10,000-psi oil to provide the muscle to do useful work at rates up to several hundred horsepower.

Among the many advantages of fluidic control systems are unique sensing capabilities and environmental immunity.

Sensing needs frequently occur in locations such as a die cavity on a punch press, where it is impossible to place a mechanically actuated switch. However, a small bleed hole can be connected to a fluidic device to solve the problem. Fluidic proximity sensors can be used to provide a safe solution in applications where fragile items could be damaged by physical contact. The sensing of transparent films or light materials is very difficult for nonfluidic devices but can be easily handled using an interruptible jet sensor.

Since fluidic components have no moving parts, they virtually do not wear out. However, component malfunction can occur due to clogging of critical flow passageways if contaminants in the air supply are not eliminated by proper filtration. Thus strict maintenance procedures for providing a clean supply of air and unobstructed passageways must be established and rigidly followed.

Environmental immunity is a characteristic of fluidic components, which enables them to operate in environments unsuitable for other types of control equipment. Since fluidic systems cannot cause spark hazards, they are ideally suited for applications in an explosive-type atmosphere. Fluidic components do not generate electrical noise and therefore will not interfere with electronic equipment. By the same token, electrical noise has no effect on fluidic components. Fluidic systems can also operate satisfactorily in areas subject to nuclear radiation, magnetic flux, temperature extremes, vibration, and mechanical shock.

One of the principal reasons for selecting fluidic control systems for a given application is the reliability of properly maintained fluidic devices. A prime reason for this reliability is success in maintaining close tolerances in the internal air channels of fluidic components. One manufacturer uses a photographic process in producing layers of glass-ceramic material, which are fused into a monolithic

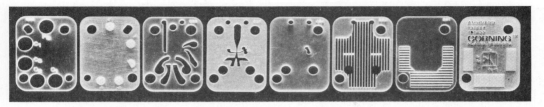

Figure 11-5.   Precision-etched, flat glass fluidic panels. (*Courtesy of Corning Glass Works, Corning, New York.*)

structure. Figure 11-5 shows a number of these precision-etched, flat glass parts, which can be stacked up to 20 layers thick and fusion-bonded into monolithic assemblies without using foreign adhesive substances. By designing each layer to contribute to the final, three-dimensional whole, a limitless variety of internal and external openings, channels, cavities, and contours can be created. The unique photosensitive glass allows up to 10,000 holes/in.^{2} in flat panels that can be as thin as 0.010 in. Intricate outlines, holes, slots, or cavities of almost any shape or size can be chemically machined into each panel. Tolerances of ± 0.001 in. and surface finishes of 1 μin. are attainable. The parts are made by selectively exposing the glass to ultraviolet light through a photographic mask. This is followed by a controlled heat treatment to form an etched crystalline image, which is removed by chemical etching to leave the desired finished part, which is a reproduction of the photographic mask. Additional heat treatments convert the glass into crystalline glass-ceramics without the loss of dimensional integrity. With the inertness, stability, and strength of the glass-ceramic and with no moving parts to cause wear or fatigue or to generate heat, the fluidic components are not subject to the common causes of failure.

Interfacing capability, which is a requirement in all control systems applications, can be easily accomplished with fluidics. Fluidic components may be interfaced to control pneumatic, electrical, or other systems. Pneumatic system compatibility is readily obtained since a fluidic control system can be powered from a pneumatic line by pressure reduction and additional filtration. The fluidic system in turn can operate pneumatic valves to control machines or processes.

11.2 MOVING PART LOGIC (MPL) CONTROL SYSTEMS

Moving-part logic (MPL) control systems use miniature valve-type devices, each small enough to fit in a person's hand. Thus an entire MPL control system can be placed in a relatively small space due to miniaturization of the logic components. Figure 11-6 shows a miniature three-way limit valve along with its outline dimensions of $1\frac{3}{16}$ in. long by $\frac{3}{4}$ in. by $\frac{1}{2}$ in. This valve, which is designed to give dependable performance in a small, rugged package, has a stainless steel stem ex-

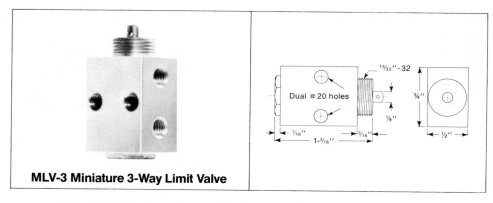

MLV-3 Miniature 3-Way Limit Valve

Figure 11-6.   Miniature three-way limit valve. (*Courtesy of Clippard Instrument Laboratory Inc., Cincinnati, Ohio.*)

tending $\frac{1}{8}$ in. from the top. The valve design is a poppet type with fast opening and high flow 7.0 CFM at 100 psi air (working range is 0 to 150 psi). Mounted on a machine or fixture, the valve is actuated by any moving part that contacts and depresses the stem.

Figure 11-7 shows a MPL pneumatic control package with a push button for ON/OFF operation. The subplate and the four valves mounted on it form a single push button input providing a binary four-way valve output that is pressure and speed regulated by restrictions on the exhaust ports. It is an ideal control for air collet vises, airclamps, assembly devices, indexing positioners, and other air-powered tools and devices.

Figure 11-7.   MPL pneumatic control package. (*Courtesy of Clippard Instrument Laboratory Inc., Cincinnati, Ohio.*)

Figure 11-8(a) shows a MPL circuit manifold, which is a self-contained modular subplate with all interconnections needed to provide a "two hand no tie down" pneumatic circuit. The manifold is designed to be used with three modular plug-in control valves and to eliminate the piping time and materials normally

(a)

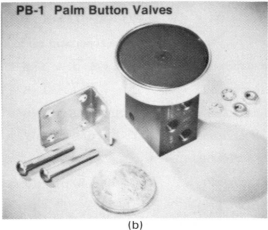

(b)

Figure 11-8.   MPL circuit manifold for two-hand, no tie down control. (*Courtesy of Clippard Instrument Laboratory Inc., Cincinnati, Ohio.*)

associated with circuitry. The main function of this control system is to require a machine operator to use both hands to actuate the machinery, thus ensuring that the operator's hands are not in a position to be injured by the machine as it is actuated. When used with two guarded palm button valves (see Fig. 11-8b), which have been properly positioned and mounted, the control system provides an output to actuate machinery when inputs indicate the operator's hands are safe.

Moving-part logic circuits utilize four major logic control functions: AND, OR, NOT, and MEMORY.

Figure 11-9(a) shows a circuit that provides the AND function, which requires that two or more control signals must exist in order to obtain an output. The circuit consists of three two-way, two-position, pilot-actuated, spring offset valves connected in series. If control signals exist at all three valves (A, B, and C), then output D will exist. If any one of the pilot signals is removed, output D will disappear.

A second method of implementing an AND function, shown in Figure 11-9(b), uses a single directional control valve and two shuttle valves. Pilot lines A, B, and C must be vented to shut off the output from S to P.

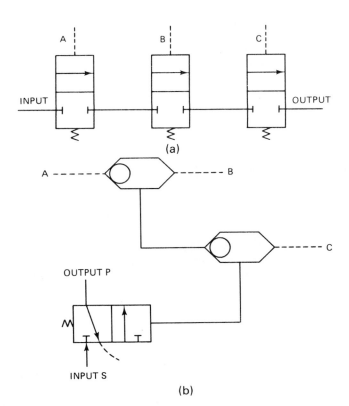

Figure 11-9. AND function. (a) Multiple directional control valves, (b) single directional control valve.

An OR circuit is one in which a control signal at any one valve will produce an output. Thus, all control signals must be off in order for the output not to exist. This is accomplished in Fig. 11-10(a) in which the three valves are now hooked in parallel. If any one of the valves picks up an air pilot signal, it will produce an output at *D*. Figure 11-10(b) shows how an OR function can be implemented using

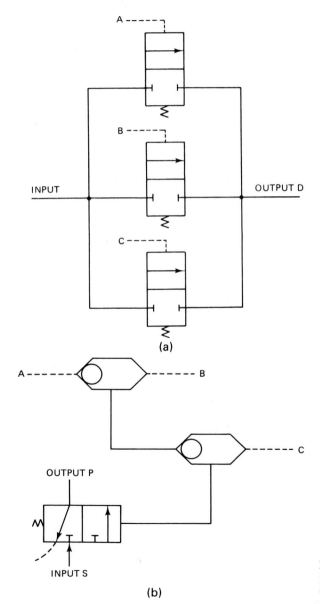

(a)

(b)

Figure 11-10.   OR function. (a) Multiple directional control valves, (b) single directional control valve.

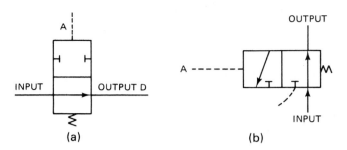

Figure 11-11.   NOT function. (a) Two-way valve, (b) three-way valve.

one directional valve and two shuttle valves. In this case, a signal applied at A, B, or C will produce an output from S to P.

In a NOT function, the output is ON only when the single input control signal A is OFF and vice versa. This is illustrated in Figure 11-11(a), which shows that the output will not exist if the control signal A is received. A second way to implement a NOT function is to use a three-way valve as shown in Figure 11-11(b).

MEMORY is the ability of a control system to retain information as to where a signal it has received originated. Figure 11-12(a) shows a MEMORY circuit, which operates as follows: If control signal A is momentarily applied, output C will come on. Conversely, if control signal B is momentarily applied, the output will exist at D. Thus, an output at C means the signal was applied at A, and an output at D means the signal was applied at B. The MEMORY circuit does not function if control signals A and B are applied simultaneously because both ends of the output pilot valve would be piloted at the same time.

A second way to implement a MEMORY function is to use two three-way, double-piloted valves as shown in Figure 11-12(b).

11.3 MPL CONTROL OF FLUID POWER CIRCUITS

In this section we show the use of MPL control in fluid power circuits. In Fig. 11-13 we have a MPL circuit, which controls the extension and retraction strokes of two double-acting cylinders. The operation is as follows, assuming that both cylinders are initially fully retracted: When the START valve $V1$ is momentarily depressed, pilot valve $V2$ shifts to extend cylinder 1. At full extension, limit valve $V4$ is actuated to shift valve $V5$ and extend cylinder 2. Upon full extension, limit valve $V6$ is actuated. This shifts valve $V2$ to retract cylinder 1. Upon full retraction, limit valve $V3$ is actuated. This shifts valve $V5$ to fully retract cylinder 2. Thus, the cylinder sequence is as follows: Cylinder 1 extends, cylinder 2 extends, cylinder 1 retracts, and finally cylinder 2 retracts. The cycle can be repeated by

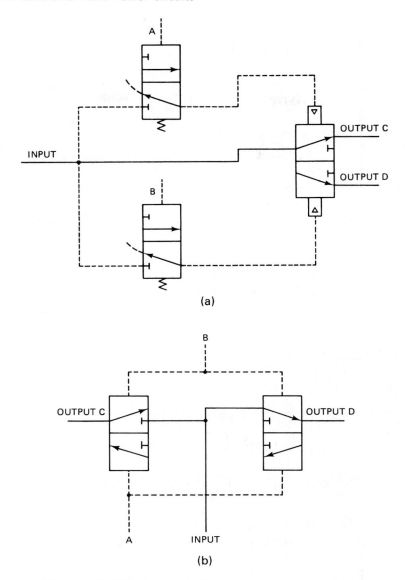

Figure 11-12. MEMORY function. (a) Three directional control valves, (b) two directional control valves.

subsequent momentary actuation of the START push-button valve. The sequence can be made continuous by removing the START valve and adding a limit switch to be actuated at the retraction end of cylinder 2. Upon actuation, this limit switch would pilot-actuate valve $V2$ to initiate the next cycle.

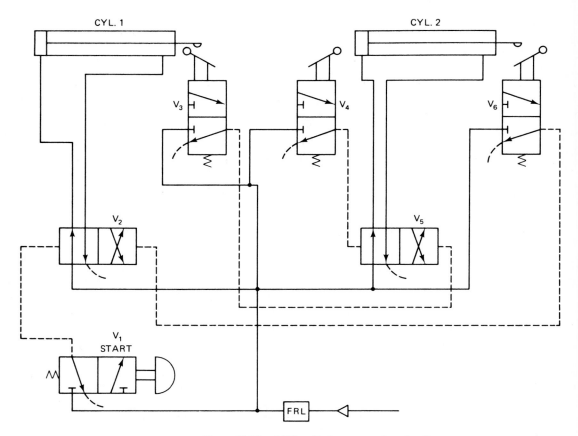

Figure 11-13.   MPL cylinder sequencing circuit.

In Figure 11-14 we have a MPL circuit that controls the extension of a double-acting cylinder by having the following features:

1. The system provides interlocks and alternative control position.
2. In order to extend the cylinder, either one of the two manual valves (A or B) must be actuated and valve C (controlled by a protective device such as a guard on a press) must also be actuated.
3. The output signal is memorized while the cylinder is extending.
4. At the end of the stroke, the signal in the MEMORY is cancelled.

The circuit operation is described as follows:

1. The input signals A and B are fed into an OR gate so that either A or B can be used to extend the cylinder. The OR gate consists of one shuttle and two three-way, button-actuated direction control valves.

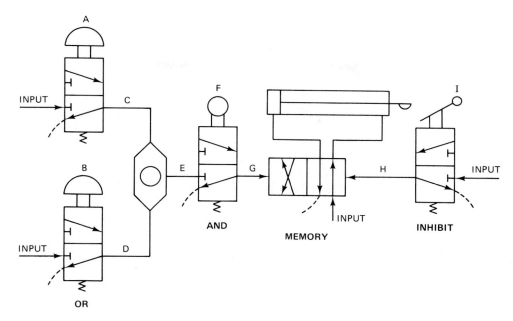

Figure 11-14.   MPL control of a single cylinder.

2. The output from the OR gate (C or D) is fed into an AND gate along with the mechanical control signal F (guard of press actuates valve). A single three-way directional control valve represents the AND gate in this system.

3. The output from the AND gate is fed into the MEMORY device, which remembers to keep pressure on the blank end of the cylinder during extension.

4. At the end of the stroke, the inhibit (cancel) limit valve is actuated to cancel the signal in the memory. This stops the extension motion and retracts the cylinder.

It is interesting to note that a single directional control valve (four-way, double-piloted) can function as a MEMORY device. Also note that for the limit valve to provide the inhibit (cancel) function, the operator must release the manual input A or B.

11.4 PRINCIPLES OF FLUIDIC LOGIC CONTROL

The wall attachment device was one of the earliest fluidic elements to be developed. It has since become the most important digital fluidic device. Its operation is based on the wall attachment or Coanda effect.

Drawings A and B of Fig. 11-15 illustrate the Coanda effect phenomenon. As shown in drawing A, a jet of air (from the supply port) is emitted into a confined region at a velocity high enough to produce turbulent flow. This turbulent jet functions very much like a jet pump and entrains air from its surroundings. As a result, a flow is established along the walls of the confined region. This air flow rushes in to replace the air being pumped out by the power jet.

Since a turbulent jet is dynamically unstable, it will veer rapidly back and forth. When the jet veers close to one wall, it interrupts the flow path along the wall on that side, as shown in drawing B. The result is that no more air is flowing on that side to replace the air being pumped out by the power jet. This constricted flow causes a lowering of pressure on that side of the power nozzle. This generates a low-pressure bubble next to the jet. The low-pressure bubble (separation bubble) causes the stream to become stable and remain attached to that wall.

To have a practical control device out of this Coanda effect, a method must be established to reliably control this wall attachment phenomenon. This is accomplished by cutting control ports into the confined region, as shown in drawing C of Fig. 11-15. These control ports permit the selection of which wall the stream will attach. When a puff of air is admitted to the separation bubble through the control port, the low-pressure bubble is destroyed. This causes the power jet to flip over and attach itself to the opposite wall. This jet will stay there until an air signal is admitted to the new low-pressure bubble to destroy it. When this is done, the device switches back. Such a component is called a digital device because a particular output is either fully on or fully off.

There are no intermediate output levels. Thus, digital systems have devices that have only two output states ON and OFF and that are switched rapidly from one state to the other by control signals applied to their inputs. The control signal is considered ON when it is at a level that is high enough to guarantee switching a digital device to which it is applied. Conversely, the signal is considered OFF when it is at a second level that is low enough to allow the device to switch back to its original state. Signals between these two levels are indeterminate and are avoided. Transitions from one level to the other are made as quickly as possible.

In contrast, analog systems are those in which the signals may assume a continuous range of meaningful values. Hence, the output of an analog device may take on any value within its output capability range. Most fluidic logic control devices are of the digital type and operate on the wall attachment or Coanda effect principle. The Coanda effect can be visually demonstrated by turning on a water faucet and holding a finger near the stream of water, as shown in Fig. 11-16(a). Next move your finger toward the stream of water. When your finger (slightly curved as shown) makes contact with the stream, the water attaches itself to the wall, which in this case is your finger [see Fig. 11-16(b)]. A low-pressure bubble (between the stream and your finger) keeps the stream attached until you move your finger away from the stream to overcome the low-pressure bubble. This low-pressure region is created due to particles of fluid (around the jet) becoming

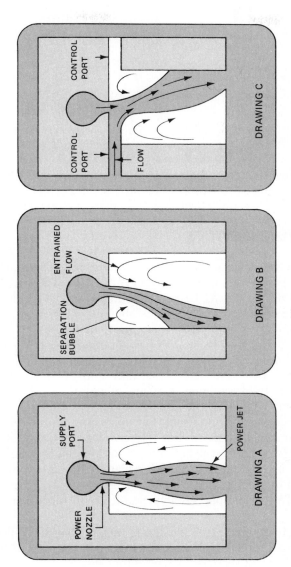

Figure 11-15. Coanda effect device. (*Courtesy of C. A. Norgren Co., Littleton, Colorado.*)

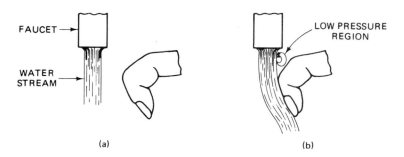

Figure 11-16. Simple demonstration of the Coanda effect.

entrained in the jet. The location of the finger prevents additional fluid from entering the low-pressure bubble once the stream becomes attached.

11.5 BASIC FLUIDIC DEVICES

In this section we discuss the operation of the following basic fluidic devices: BASIC FLIP-FLOP, FLIP-FLOP WITH START UP PREFERENCE, SRT FLIP-FLOP, OR/NOR, EXCLUSIVE OR, and AND/NAND.

A bistable digital control device is obtained if a splitter is added to our wall attachment device, as depicted in drawing D of Fig. 11-17. This provides controlled assurance as to which of the two output ports will deliver the power stream. Such a device is a basic bistable or flip-flop wall attachment device. The separation bubble provides the pressure differential that maintains the stability of this device. A certain minimum velocity of flow is required in the interaction region to properly maintain stability. However, to use either output, it must be restricted enough to develop a usable back pressure. It is therefore possible for the flow restriction to cause the stream velocity to drop below the critical minimum value required for a stable wall attachment stream. Such a characteristic produces what is called a load-sensitive device because the output can switch from the desired port to the second port whenever the load restricts output flow from the desired port. This is an unacceptable characteristic since the device is basically unstable.

Drawing E of Fig. 11-17 illustrates design changes within the flip-flop to eliminate the load sensitivity problem. Notice the addition of the two vent ports on the output legs. In the case where output flow is not significantly restricted, flow velocity past the vent can reduce the pressure sufficiently to allow flow in through the vent. When the output becomes restricted, air flows out through the vents. When the output becomes completely blocked (which happens, for example, when a load hits a stop), there is enough flow out through the vent to maintain stability. As a result, the output flow does not switch to the other output port, and

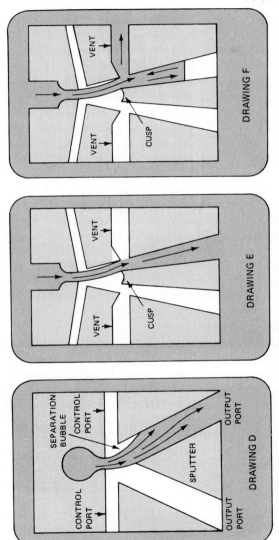

Figure 11-17. Operation of the basic bistable flip-flop. (*Courtesy of C. A. Norgren Co., Littleton, Colorado.*)

we have a predictable device, as shown in drawing F of Fig. 11-17. Also, observe that the flip-flop also contains a cusp on the splitter, which serves as an additional stability augmentation feature. The splitter cusp tends to increase the internal pressure on the high-pressure side of the power stream.

A flip-flop provides a memory function in fluidic control systems. This can be seen by examination of the symbol and truth table of an actual flip-flop, as shown in Fig. 11-18. A truth table tells how a particular device behaves. Number 0 means OFF and number 1 means ON for all devices. Therefore, when control signal $C1$ is ON and control signal $C2$ is OFF, the output is at 01. If $C1$ is then turned OFF, the device stays in its first stable state with the output of 01.

If $C2$ is then turned ON, the flip-flop switches to the 02 output. Removing signal $C2$ leaves the device in its second stable state with the output of 02. Thus, the flip-flop has two stable states when all control signals are OFF. However, each of the two stable states are predictable, and thus a flip-flop is a reliable bistable device. It should be pointed out, however, that when the basic flip-flop first has its power supply pressure P_s turned ON (and neither control signal has been turned ON), the output is indeterminant. Also, both control signals $C1$ and $C2$ should not be ON simultaneously. Under these conditions, the output flow would split because no low-pressure bubble can be sustained on either wall. Thus, with $C1$ and $C2$ both ON, the flip-flop does not produce useful digital ON/OFF output signals.

A flip-flop can have more than two control ports, as shown by the symbol of a flip-flop in Fig. 11-18. In such a design $C1$ and $C2$ perform the same function as do $C3$ and $C4$. For positive switching the control signal pressure must be at least 10% of the power supply pressure P_s but must not exceed 30% of P_s. The supply pressure should be between 3 and 10 psi.

In some applications it is necessary to have a specific output when the power supply is first turned ON and all control signals are OFF. For these applications a flip-flop with a start-up preference is used. Figure 11-19 shows such a flip-flop with its symbol (note the + sign) and truth table. As shown, when all control signals are OFF, the output is ON at 01 and OFF at 02. Otherwise it behaves exactly the same as a basic flip-flop.

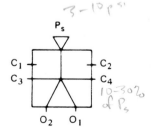

C1	C2	01	02
1	0	1	0
0	0	1	0
0	1	0	1
0	0	0	1

Figure 11-18. Basic flip-flop with its symbol and truth table. (*Courtesy of Corning Glass Works, Corning, New York.*)

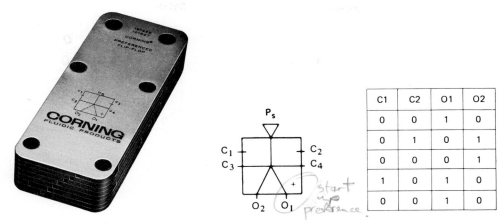

C1	C2	O1	O2
0	0	1	0
0	1	0	1
0	0	0	1
1	0	1	0
0	0	1	0

Figure 11-19.   Preferenced flip-flop with its symbol and truth table. (*Courtesy of Corning Glass Works, Corning, New York.*)

   A flip-flop with a preference start-up may have more than two control ports, as shown by the symbol in Fig. 11-19. In such a design $C1$ and $C2$ perform the basic function, as do $C3$ and $C4$. The $+$ sign on the symbol indicates the output 01 is preferred over output 02. One way to accomplish this is to build the splitter slightly off center. Thus, when the device first receives its power supply (control signals are OFF), the wall attachment is preferenced to the 01 output. An application circuit where a flip-flop with a start-up preference is desired is discussed in Sec. 11.7.

   An SRT flip-flop has the same capabilities as a basic flip-flop except it can switch by applying a signal to the trigger port. S and R stand for SET and RESET, respectively, and perform as regular control signals. T stands for trigger, and whenever it is applied, it complements (switches) the output. Figure 11-20 shows the symbol and truth table for an SRT flip-flop.

   The bistable function (memory) of a flip-flop does not provide all the control features required in a complete system. A monostable (which is analogous to spring return) function is also necessary. The two basic monostable devices are the OR/NOR and AND/NAND gates. These two devices are essentially the same basic monostable fluidic component but with different input passageways. Drawing G of Fig. 11-21 illustrates the operation of the monostable device. Observe the similarity of the monostable device in drawing G of Fig. 11-21 and the flip-flop in drawing E of Fig. 11-17. Differences are as follows: The flip-flop is symmetrical to provide the bistable characteristic (memory). However, the monostable device is slightly asymmetrical. Notice that the power nozzle is at a slight angle to the legs and that the vents are also different. As a result, the monostable device favors one leg. A control signal will switch the power stream, but the control signal must be

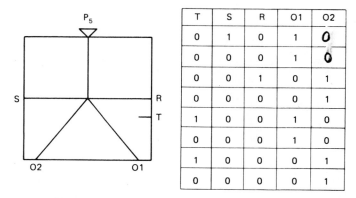

T	S	R	O1	O2
0	1	0	1	0
0	0	0	1	0
0	0	1	0	1
0	0	0	0	1
1	0	0	1	0
0	0	0	1	0
1	0	0	0	1
0	0	0	0	1

Figure 11-20. Symbol and truth table for SRT flip-flop.

maintained to keep it switched. Removal of the control signal causes the device to switch back to the favored leg.

A special input passageway is added to the basic monostable to produce the OR/NOR device, as illustrated in drawing H of Fig. 11-21. An input signal may be applied at any control port in any combination to switch the device. For example, a signal applied at port 2 will cross the narrow gap and appear at the control inlet of the monostable. Signals at ports 3 and 4 will do the same thing. Thus, any one of these signals will switch the device, as will any combination of these signals. The tiny gap in the input path is vented to the atmosphere. In this way, a signal on 2 cannot sneak out 3 or 4, and the reverse is also prevented.

Figure 11-22 shows an actual OR/NOR gate with its symbol and truth table. Although the truth table only shows control signals $C1$ and $C3$, the device actually has two additional control ports $C5$ and $C7$. A control signal at any one or any combination of these ports causes the device to switch to the 01 output. With all control signals OFF, the output is automatically 02.

The operation of the AND/NAND gate is illustrated in drawing I of Fig. 11-21. Observe that the input passageway is different from that of the OR/NOR gate. In the case of the AND/NAND gate, a control signal applied to port 2 will go straight through the passageway to the atmosphere. However, if control signals are present at both ports 2 and 3, they will intersect at a point. The interaction causes each to deflect the other into the control inlet of the monostable element. Thus, both control signals must be ON to switch the AND/NAND gate. Figure 11-23 shows an actual AND/NAND gate with its symbol and truth table.

It should be noted that NOR means NOT OR. Thus, if 01 is the output and 02 the vent (see Fig. 11-22), the device is OR, and conversely it is NOR. Specifically, an OR gate is one that will have an output if $C1$ *or* $C3$ *or* $C5$ *or* $C7$ *or* any combination of control signals is ON.

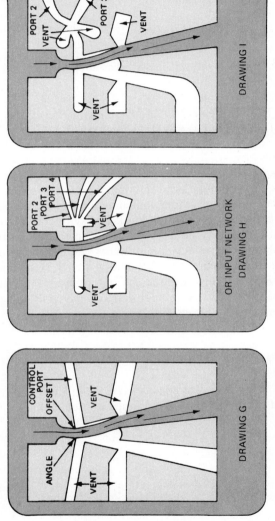

Figure 11-21. Operation of OR/NOR and AND/NAND monostable devices. (*Courtesy of C. A. Norgren Co., Littleton, Colorado.*)

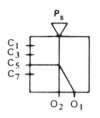

C1	C3	01	02
0	0	0	1
1	0	1	0
0	1	1	0
1	1	1	0

Figure 11-22. OR/NOR gate with its symbol and truth table. (*Courtesy of Corning Glass Works, Corning, New York.*)

Similarly NAND means NOT AND. Thus, if 01 is the output and 02 the vent (see Fig. 11-23), the device is AND, and conversely it is NAND. Specifically, an AND gate is one that will have an output only if control signal $C1$ *and* control signal $C3$ are ON.

An exclusive OR gate is one that will have an output only if $C1$ *or* $C3$ *or* $C5$ *or* $C7$ (but not any combination of control signals) is ON. As shown in Fig. 11-24(a), an exclusive OR gate is obtained by feeding the output of a jet interaction cavity into the control port of an OR gate. The symbol and truth table are given in Fig. 11-24(b). If $C1$ and $C3$ are both ON, their interaction causes them to vent to the atmosphere and not produce an output to the control port of the OR gate. If $C1$ or $C3$ is ON, an output from the jet interaction cavity exists and becomes a control signal of the OR gate. This produces an output from the OR gate to satisfy the truth table. Observe that the symbol for an exclusive OR gate contains a short line crossing the control jet passageway within the symbol.

11.6 FLUIDIC SENSORS

A fluidic sensor is a device that senses a change in some parameter and as a result causes a related change in another parameter (such as the switching of a flip-flop)

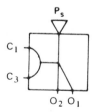

C1	C3	01	02
0	0	0	1
1	0	0	1
0	1	0	1
1	1	1	0

Figure 11-23. AND/NAND gate with its symbol and truth table. (*Courtesy of Corning Glass Works, Corning, New York.*)

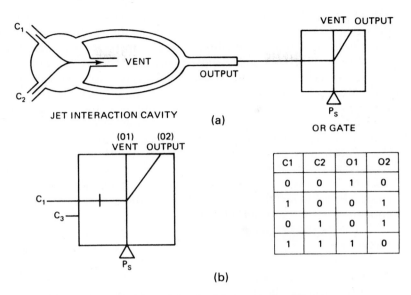

C1	C2	O1	O2
0	0	1	0
1	0	0	1
0	1	0	1
1	1	1	0

Figure 11-24.   Operation of an exclusive OR gate.

which can be recognized and interpreted. In this section we discuss the following types of fluidic sensors: back-pressure sensor, proximity sensor, interruptible jet sensor, and contact sensor.

Probably the most widely used method of switching a fluidic gate is the sensing of the buildup of back pressure. It works on the principle that pressure drop across a nozzle or orifice increases with flow through the nozzle or orifice.

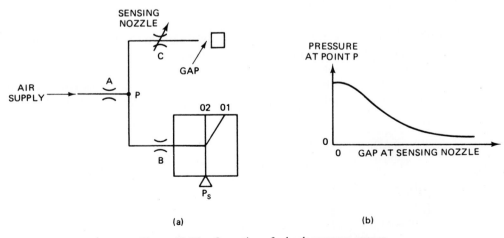

Figure 11-25.   Operation of a back-pressure sensor.

For operation, refer to Fig. 11-25(a), which gives a circuit consisting of three such flow restrictions—one connected in series with the other two in parallel. Restrictor A is a calibrated orifice with a constant upstream pressure ahead of it. Restrictor B is an OR/NOR gate, and restrictor C is a variable restriction called the sensing nozzle. When the sensing nozzle is unobstructed (no external object present), restrictor C has a low resistance compared to that of A or B, and the flow through A to C is relatively high. Therefore, as shown in Fig. 11-25(b) (for a large value of gap between external object and sensing nozzle), the pressure at point P is low due to the high pressure drop across A. Thus, the pressure signal leading to the control port of the OR/NOR gate is not large enough to switch the gate. However, when flow is impeded by some external object close to the sensing nozzle, the pressure at point P increases. If restrictor A has a low resistance compared to that of B (the OR/NOR gate), the pressure increases at P will be great enough to switch the OR/NOR gate.

The OR/NOR key is a back-pressure switch, which incorporates nozzle B as part of the device, as illustrated in Fig. 11-26. If the pressure at control port S is low, the output is 02. When the pressure at A becomes large enough, it will cause the OR/NOR key to switch the output to 01. The device switches when the control pressure reaches about 10% of the supply pressure. When S is blocked, the control pressure is approximately equal to 50% of P_s.

A proximity sensor permits the detection of objects at greater distances than is possible with back-pressure devices. The cone-jet proximity sensor uses an annular nozzle, surrounding a sensing hole, connected to the output port [see Fig. 11-27(a)]. The high-velocity jet of supply air converges after leaving the nozzle, enclosing a low-pressure bubble within the flow cone. Hence, the pressure at the sensing opening into the base of the cone is normally slightly below atmospheric. When an object comes near the flow cone, a portion of the supply jet is reflected from the object back into the low-pressure bubble region. This increases the pressure at the output port to permit the switching of an appropriate fluidic component, as shown in Fig. 11-27(b). A cone-jet sensor can sense gaps 10 times larger than the back-pressure sensor for an equivalent amount of flow consumption. Figure 11-27(c) shows an actual cone-jet sensor.

The interruptible jet sensor uses a nozzle to transmit an air stream across a gap to a receiver. When the flow is unobstructed, adequate pressure is recovered by the receiver to hold the fluidic element in a particular mode [see Fig. 11-28(a)].

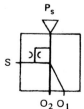

Figure 11-26. OR/NOR key and its symbol. (*Courtesy of Corning Glass Works, Corning, New York.*)

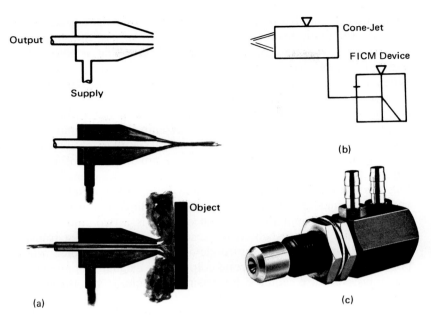

Figure 11-27.   Operation of a cone-jet sensor. (*Courtesy of Corning Glass Works, Corning, New York.*)

However, when an object enters the jet area, the output pressure in the receiver is reduced to cause the fluidic element to switch to the opposite mode [see Fig. 11-28(b)]. A separate air supply, reduced in pressure by a restrictor, allows purging to prevent contamination by particles that enter the receiver along with ambient air entrained by the jet. An actual interruptible jet sensor is shown in Fig. 11-28(c). It operates with supply pressures in the range of 3 to 10 psi and consumes

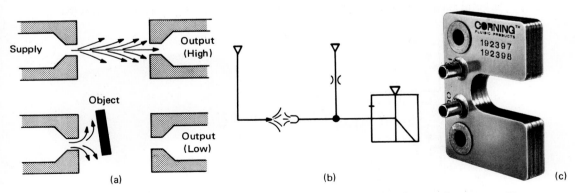

Figure 11-28.   Interruptible jet sensor. (*Courtesy of Corning Glass Works, Corning, New York.*)

0.60 scfm at a supply pressure of 5 psi. The dead-headed (output of fluidic component blocked) output pressure is greater than 15% of the supply pressure.

Contact sensing is accomplished by the sensing of objects by physical contact. It is usually done by a moving-part device called a limit valve or limit switch. When an object contacts the actuator [see Fig. 11-29(a)], the output switches from the normally ON output to the normally OFF output. When the contacting object is removed, the output switches back. This type of limit switch operation is similar to that of an OR/NOR gate, except a mechanical input signal is used rather than a fluidic one. Figure 11-29(b) shows an actual limit switch with its fluidic symbol.

In addition to sensing devices, there are a number of interface components used in fluidic control systems. Included among these are push-button switches, toggle switches, and two-position and three-position selector switches. These switches, which are manually actuated, are illustrated in Fig. 11-30. Figure 11-31 shows a panel-mounted fluidic indicator. It is a two-color, two-position device that displays ambient-lighted fluorescent colors behind a transparent window. This single-input, spring-return device displays one color when no signal is present and switches to a second color when a signal is applied. Twelve inches of water pressure are required for positive switching, but the maximum operating pressure should not exceed 8.0 psi.

11.7 FLUIDIC CONTROL OF FLUID POWER SYSTEMS

Fluidic circuits operate at pressures normally less than 15 psi to provide the necessary logic functions to pilot-operate pneumatic or hydraulic valves for the control of fluid power. In all such systems, interface devices are used to interact between the fluidic air medium and the pneumatic air medium or hydraulic oil medium. The clean, dry fluidic air and air lines are kept separate from the 100—150-psi air supply utilized by pneumatic systems. This is due to the more stringent clean and dry air requirements of fluidic devices. In this section we present several basic circuits where fluidics is used to control fluid power systems.

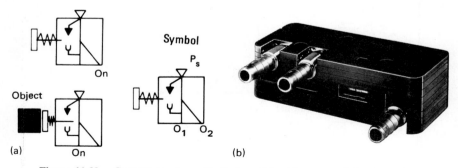

Figure 11-29. Contact sensing with limit switch. (*Courtesy of Corning Glass Works, Corning, New York.*)

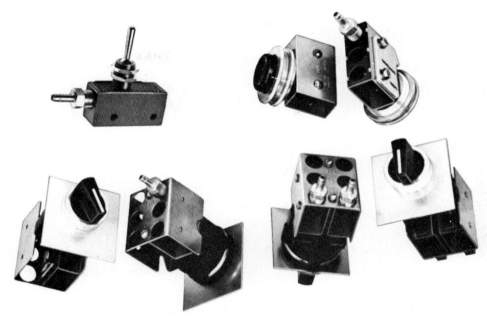

Figure 11-30.   Manual input switches. (*Courtesy of Corning Glass Works, Corning, New York.*)

Figure 11-32 shows an application that provides for the push-button start and air limit switch reversal of an air cylinder. When the start button is pressed momentarily, the flip-flop (using 8-psi air) pilot-actuates interface valve *V*1 to extend the cylinder with 100-psi air. When the normally closed (N.C.) limit switch

Figure 11-31.   Fluidic indicator. (*Courtesy of Corning Glass Works, Corning, New York.*)

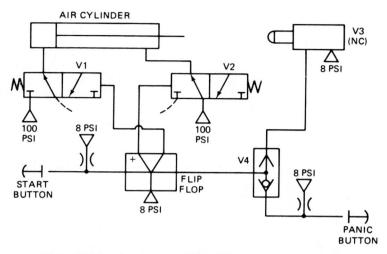

Figure 11-32.   Control of air cylinder using preferenced flip-flop.

(*V3*) is actuated, it supplies an 8-psi signal through the shuttle valve (*V4*) to switch the flip-flop. This causes valve *V2* to become pilot-actuated, and the cylinder is retracted. The panic button is used to override the air limit switch. If the panic button is momentarily pressed (while the cylinder is extending), it will cause the cylinder to retract instantly. The flip-flop has a start-up preference to ensure that the cylinder will not extend initially when the pneumatic and fluidic air supplies are first turned on. Notice that the start and panic buttons represent back-pressure sensors to allow the 8-psi nozzle-restricted air supplies to reach the control ports of the flip-flop with adequate strength to switch it.

The sequence control of two air cylinders is accomplished by the circuit given in Fig. 11-33. The system consists of two interface valves (*V1* and *V2*), a preferenced flip-flop, an OR/NOR gate, two normally closed (N.C.) air limit switches (*V3* and *V4*), and two pneumatic cylinders. When the push-button valve is momentarily released, the following sequence cycle occurs:

1. Cylinder 1 extends
2. Cylinder 2 extends
3. Both cylinders retract together

The cycle is repeated each time the push button is momentarily pressed. A flip-flop with a start-up preference is used to prevent the cycle from starting automatically as soon as the fluidic and pneumatic air supplies are turned on.

Figure 11-34 shows a circuit that controls the continuous reciprocation of a double-rod-ended hydraulic cylinder. The circuit consists of two NOR gates, two interruptible jets, a three-position, four-way pilot-actuated interface valve, a pref-

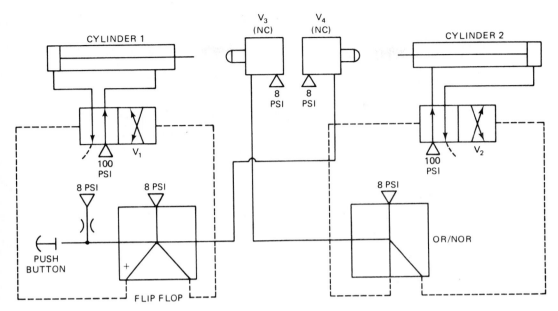

Figure 11-33.   Fluidic sequencing control of two pneumatic cylinders.

erenced flip-flop, a selector switch, and the hydraulic cylinder. With the selector switch in the OFF position, the interface valve assumes its spring-centered position because there is no output from the flip-flop to pilot the interface valve. The cylinder is therefore hydraulically locked. When the selector switch is placed into the ON position, the right side of the interface valve is piloted. Thus, the cylinder moves to the right until the rod interrupts the jet. This causes the right side NOR gate to switch to output 01 to switch the flip-flop. This pilots the left side of the interface valve to cause the cylinder to begin moving to the left. When the rod on the right end retracts out of the interruptible jet, the right side NOR gate switches back to output 02. But this does not switch the flip-flop because it possesses memory.

When the rod on the left end of the cylinder interrupts its jet, the left side NOR gate switches to output 01. This switches the flip-flop output to once again pilot the right side of the interface valve to drive the cylinder to the right. Therefore, the cylinder reciprocates continuously until the selector switch is placed in the OFF position. This stops the hydraulic cylinder motion instantly as the interface valve shifts into its spring-centered, blocked-port flow path configuration. It should be noted that each interruptible jet could be replaced by a normally open air limit switch whose output is connected to the control port of its corresponding NOR gate.

In Fig. 11-35 we see a production line application in which boxes of two sizes (high and low) are moving on a conveyor. At a given location the high boxes are to

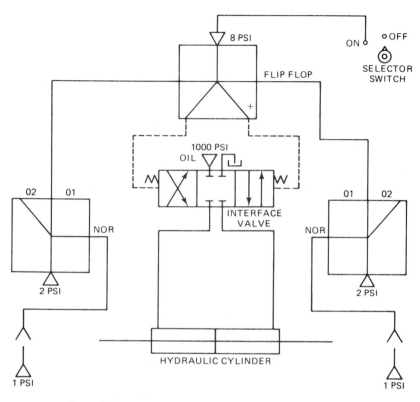

Figure 11-34.   Continuous reciprocation of hydraulic cylinder.

be pushed onto a second conveyor while the low boxes continue moving on the same conveyor. This fluidic box-sorting system uses a double-rod-ended pneumatic sorting cylinder, two normally closed (N.C.) air switches, an interface valve, and a flip-flop with a preference start-up. When a high box actuates limit switch $V1$, the flip-flop is switched to output 01. This pilot shifts the interface valve $V3$ to move the cylinder to the right to push the high box onto the second conveyor. The cylinder rod button then actuates limit switch $V2$, which switches the flip-flop output back to 01. This pilot shifts the interface valve to move the cylinder back to the left as it waits for another high box to repeat the cycle.

11.8 INTRODUCTION TO BOOLEAN ALGEBRA

The foundations of formal logic were developed by the Greek philosopher Aristotle during the third century B.C. The basic premise of Aristotle's logic is "a statement is either true or false; it cannot be both and it cannot be neither." Many

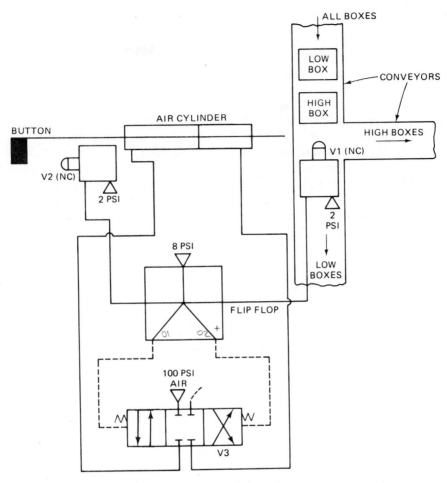

Figure 11-35.   Fluidic box-sorting system.

philosophers have tried without success to create a suitable mathematical model of the preceding sentence based on the logical reasoning process of Aristotle. This was finally accomplished in 1854 when George Boole, an English mathematician, developed a two-valued algebra that could be used in the representation of true-false propositions.

In developing this logical algebra (called Boolean algebra), Boole let a variable such as A represent whether a statement was true or false. An example of such a statement is "the valve is closed." The variable A would have a value of either zero (0) or one (1). If the statement is true, the value of A would equal one ($A = 1$). Conversely, if the statement is false, then the value of A would equal zero ($A = 0$).

Boolean algebra serves two useful functions relative to fluid power systems:

1. It provides a means by which a logic circuit can be reduced to its simplest form.

2. It allows for the quick synthesis of a circuit that is to perform desired logic operations.

In Boolean algebra, all variables have only two possible states (0 or 1). Multiplication and addition of variables are permitted. Division and subtraction are not defined and thus cannot be performed.

The following shows how Boolean algebra can be used to represent the basic logic functions (OR, AND, NOT, NOR, NAND, EXCLUSIVE-OR, and MEMORY). The components that perform these functions are called gates.

An OR function can be represented in fluid flow systems by the case where an outlet pipe receives flow from two lines containing MPL valves controlled by input signals A and B as shown in Figure 11-36(a).

Thus fluid will flow in the outlet pipe (output exists) if input signal A is ON, OR input signal B is ON, OR both input signals are ON. Representing the flow in the outlet pipe by Z, we have

$$Z = A + B \qquad (11\text{-}1)$$

where the plus sign ($+$) is used to represent the OR function. In this case we are dealing with only two possible output conditions (either fluid is flowing or it is not). We give the logical value one (1) to the state when output fluid flows and zero (0) when it does not. Thus, $Z = 1$ when output fluid flows and $Z = 0$ when it does not. Also $A = 1$ when signal A is ON and $A = 0$ when A is OFF. The same is true

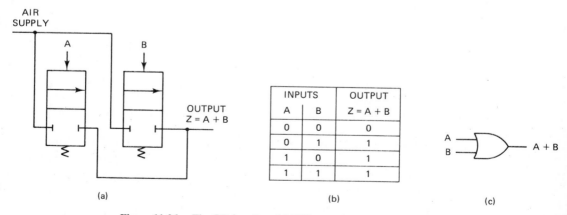

Figure 11-36.   The OR function. (a) MPL components, (b) truth table, (c) symbol.

for signal B. Applying all the possible states of values for input signals A and B to logical equation (11-1) we obtain:

$$A \text{ OFF}, B \text{ OFF}; Z = 0 + 0 = 0$$

$$A \text{ OFF}, B \text{ ON}; Z = 0 + 1 = 1$$

$$A \text{ ON}, B \text{ OFF}; Z = 1 + 0 = 1$$

$$A \text{ ON}, B \text{ ON}; Z = 1 + 1 = 1$$

For these two inputs, the logical equation is: $Z = A + B$. For X inputs it is:

$$Z = A + B + \cdots + W + X \tag{11-2}$$

Since each input variable has two possible states (ON and OFF), for n input variables there are 2^n possible combinations of MPL valve settings. For our two input variable case, there are $2^2 = 4$ combinations as shown in the truth table of Fig. 11-36(b). Figure 11-36(c) shows the graphic symbol used to represent an OR gate. This is a general symbol that is used regardless of the type of component involved (electrical relay, MPL valve, or fluidic element) in the logic system.

The AND function can be represented for fluid flow systems by the case where we have a number of MPL valves connected in series in a pipeline. The simplest case is for two valves with input signals A and B as shown in Fig. 11-37(a).

As can be seen, the output flow is zero ($Z = 0$) when either valve signal is OFF or both valve signals are OFF. Thus fluid flows in the outlet pipe (there is an output) only when both A and B are ON. Figure 11-37(b) and (c) show the truth table and general symbol for the AND gate, respectively. The logic AND function for two variables is represented by the equation

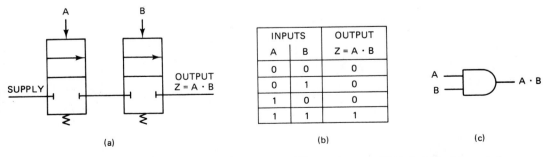

Figure 11-37. The AND function. (a) MPL components, (b) truth table, (c) symbol.

$$Z = A \cdot B \tag{11-3}$$

and for X variables by

$$Z = A \cdot B \cdots W \cdot X \tag{11-4}$$

The dot ($\cdot$) is used to indicate the logic AND connective and this form of equation is known as the logic product function. Inspection of each row of the truth table shows that the numerical value of the logic product function is also equal to the numerical value of the arithmetic product of the variables.

The NOT function is the process of logical inversion. This means that the output signal is NOT equal to the input signal. Since we have only two signal states (0 and 1), then an input of "1" gives an output of "0" and vice versa. Figure 11-38 gives the MPL valve, the truth table, and general symbol for a NOT gate.

A NOT operation is also known as logical complementing or logical negation in addition to logical inversion. It is represented in Boolean algebra by placing a bar over the variable as follows:

$$Z = \text{NOT } A = \overline{A} \tag{11-5}$$

The NOR function has its name derived from the following relationship:

$$\text{NOR} = \text{NOT OR} = \overline{\text{OR}} \tag{11-6}$$

Thus the NOR function is an inverted OR function whose MPL valve system, truth table, and general symbol are provided in Fig. 11-39. Also as shown in Figure 11-39(d), a NOR function can be created by placing a NOT gate in series with an OR gate. The Boolean relationship is:

$$Z = \text{NOT } (A + B) = \overline{A + B} \tag{11-7}$$

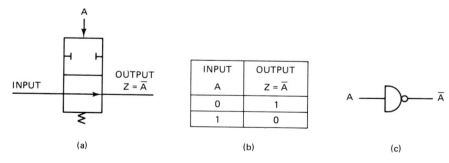

Figure 11-38.   The NOT function. (a) MPL components, (b) truth table, (c) symbol.

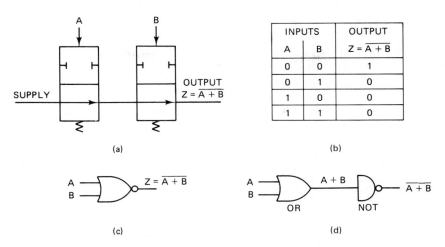

(a)

INPUTS		OUTPUT
A	B	$Z = \overline{A + B}$
0	0	1
0	1	0
1	0	0
1	1	0

(b)

(c)

(d)

Figure 11-39.   The NOR function. (a) MPL components, (b) truth table, (c) symbol, (d) OR/NOT combination = NOR.

As shown by the truth table, the output of a NOR gate is ON (1) only when all inputs are OFF (0). One significant feature of a NOR gate is that it is possible to generate any logic function (AND, OR, NOT and MEMORY) using only NOR gates.

The NAND function has its name derived from the relationship:

$$\text{NAND} = \text{NOT AND} = \overline{\text{AND}} \qquad (11\text{-}8)$$

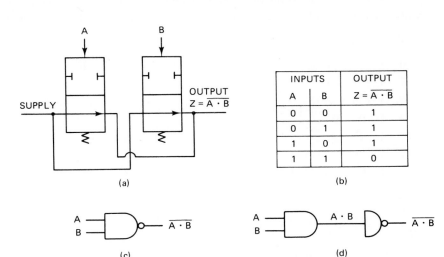

(a)

INPUTS		OUTPUT
A	B	$Z = \overline{A \cdot B}$
0	0	1
0	1	1
1	0	1
1	1	0

(b)

(c)

(d)

Figure 11-40.   The NAND function. (a) MPL components, (b) truth table, (c) symbol, (d) AND/NOR combination = NAND.

Thus the NAND function is an inverted AND function whose MPL valve system, truth table, and general symbol are provided in Fig. 11-40. As can be seen, both signals must be ON to cause a loss of output. Also as shown in Fig. 11-40(d), a NAND function can be created by placing a NOT gate in series with an AND gate. The Boolean relationship is:

$$Z = \text{NOT } (A \cdot B) = \overline{A \cdot B} \tag{11-9}$$

There are a number of laws of Boolean algebra that can be used in the analysis and design of fluid logic systems. These laws are presented as follows:

1. Commutative law:
$A + B = B + A$
$A \cdot B = B \cdot A$

2. Associative law:
$A + B + C = (A + B) + C = A + (B + C) = (A + C) + B$
$A \cdot B \cdot C = (A \cdot B) \cdot C = A \cdot (B \cdot C) = (A \cdot C) \cdot B$

3. Distributive law:
$A + (B \cdot C) = (A + B) \cdot (A + C)$
$A \cdot (B + C) = (A \cdot B) + (A \cdot C)$

4. DeMorgan's theorem:
$\overline{A + B + C} = \overline{A} \cdot \overline{B} \cdot \overline{C}$
$\overline{A \cdot B \cdot C} = \overline{A} + \overline{B} + \overline{C}$

Additional theorems that can be used to simplify complex equations and thus minimize the number of components required in a logic system are:

5. $A + A = A$
6. $A \cdot A = A$
7. $A + 1 = 1$
8. $A + 0 = A$
9. $A \cdot 0 = 0$
10. $A \cdot 1 = A$
11. $A + (A \cdot B) = A$
12. $A \cdot (A + B) = A$
13. $\overline{(\overline{A})} = A$
14. $A \cdot \overline{A} = 0$
15. $A + \overline{A} = 1$

It should be noted that all the preceding laws can be proven by the use of truth tables. Within the truth table, all combinations of the variables are listed and values of both sides of the equation to be proven are computed using the definitions of the operators. If both sides of the equation have exactly the same values for every combination of the inputs, the theorem is proven.

11.9 ILLUSTRATIVE EXAMPLES USING BOOLEAN ALGEBRA

In this section we show how to use Boolean algebra to provide logic control of fluid power.

EXAMPLE 11-1

Prove that $A + (A \cdot B) = A$ using a truth table.

Solution

A	B	$A \cdot B$	$A + (A \cdot B)$
0	0	0	0
1	0	0	1
0	1	0	0
1	1	1	1

The column $A \cdot B$ is obtained based on an AND function whereas the column $A + (A \cdot B)$ is obtained based on an OR function.

EXAMPLE 11-2

Prove the first DeMorgan theorem using two variables: $\overline{(A + B)} = \overline{A} \cdot \overline{B}$

Solution

A	B	$\overline{A}$	$\overline{B}$	$\overline{A} \cdot \overline{B}$	$A + B$	$\overline{A + B}$
0	0	1	1	1	0	1
1	0	0	1	0	1	0
0	1	1	0	0	1	0
1	1	0	0	0	1	0

EXAMPLE 11-3

Generate the truth table for the function $Z = A \cdot \overline{B} + \overline{A} \cdot B$. Draw the logic circuit diagram representing the function using OR, AND, and NOT gates.

Solution Since there are two variables A and B, there are $2^2 = 4$ combinations of input variable values. These are shown in the following truth table along with intermediate values for $A \cdot \overline{B}$ and $\overline{A} \cdot B$ and the values of the output Z.

Inputs				Output
A	B	$A \cdot \overline{B}$	$\overline{A} \cdot B$	$Z = A \cdot \overline{B} + \overline{A} \cdot B$
0	0	0	0	0
0	1	0	1	1
1	0	1	0	1
1	1	0	0	0

Let's examine the first row of the truth table. Since $A = 0$, $A \cdot \overline{B}$ is zero. Likewise since $B = 0$, $\overline{A} \cdot B$ is zero giving an output $Z = 0$. In the second row, $A \cdot \overline{B}$ remains at zero since $A = 0$. However the product $\overline{A} \cdot B$ equals one since $A = 1$ (from theorem 10). Finally $Z = 0 + 1 = 1$ from theorem 8. The rest of the truth table is completed in a similar fashion.

Further examination of the truth table of Example 11-3 reveals that the function is an EXCLUSIVE OR function. This is the case because an EXCLUSIVE OR function gives an output only if input A or input B is ON. It differs from an OR function (also called INCLUSIVE OR), which gives an output when A or B is ON or both A and B are ON. Thus the Boolean relationship for an EXCLUSIVE OR gate is:

$$Z = A \cdot \overline{B} + \overline{A} \cdot B$$

whereas for an OR gate (INCLUSIVE OR) it is:

$$Z = A + B$$

The logic circuit diagram representing the function Z is given in Fig. 11-41. Note that inputs A and $\overline{B}$ are applied to the upper AND gate, so that its output is the function $A \cdot \overline{B}$. These two signals are applied to the OR gate to produce a system output: $Z = A \cdot \overline{B} + \overline{A} \cdot B$.

EXAMPLE 11-4

Determine the logic function generated by the circuit in Fig. 11-42. Simplify the function expression developed as much as possible using the theorems and laws of logical algebra.

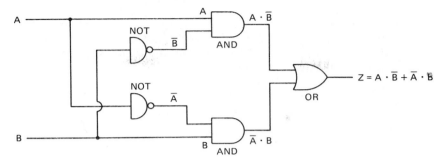

Figure 11-41.   EXCLUSIVE-OR logic circuit.

Solution   First establish the intermediate outputs.

$$01 = A + B$$

$$02 = \overline{C}$$

$$03 = (01) \cdot (02) = (A + B) \cdot \overline{C}$$

$$04 = A \cdot B$$

$$05 = \overline{04} = \overline{A \cdot B}$$

$$06 = A \cdot (02) = A \cdot \overline{C}$$

Then the output from the entire circuit is:

$$Z = 03 + 05 + 06$$

$$Z = (A + B) \cdot \overline{C} + \overline{A \cdot B} + A \cdot \overline{C}$$

The simplification of the function is accomplished next.

$$Z = A \cdot \overline{C} + B \cdot \overline{C} + \overline{A \cdot B} + A \cdot \overline{C} \qquad \text{(distributive law)}$$

$$= A \cdot \overline{C} + B \cdot \overline{C} + \overline{A \cdot B} \qquad \text{(theorem 5)}$$

$$= A \cdot \overline{C} + B \cdot \overline{C} + \overline{A} \cdot \overline{B} \qquad \text{(DeMorgan's theorem)}$$

$$= (\overline{A} + A \cdot \overline{C}) + (\overline{B} + B \cdot \overline{C}) \qquad \text{(associative law)}$$

$$= (\overline{A} + A) \cdot (\overline{A} + \overline{C}) + (\overline{B} + B) \cdot (\overline{B} + \overline{C}) \qquad \text{(distributive law)}$$

$$= 1 \cdot (\overline{A} + \overline{C}) + 1 \cdot (\overline{B} + \overline{C}) \qquad \text{(theorem 15)}$$

$$= \overline{A} + \overline{C} + \overline{B} + \overline{C} \qquad \text{(theorem 10)}$$

$$= \overline{A} + \overline{B} + \overline{C} \qquad \text{(theorem 5)}$$

This example shows how an entire circuit can possess many redundant logic components since $\overline{A} + \overline{B} + \overline{C}$ requires only four gates (three NOTS and one OR) as shown in Fig. 11-43.

A MEMORY function (SR FLIP-FLOP) can be generated by the use of two NOR gates as shown in Fig. 11-44(a). The inputs are S (SET) and R (RESET) and the outputs are P and $\overline{P}$.

The truth table for a SR FLIP-FLOP is shown in Fig. 11-44(b) and the general symbol in Fig. 11-44(c). The truth table shows that when S is ON (R is OFF), P is the output ($P = 1$, $\overline{P} = 0$). Turning S OFF and then ON repeatedly does not shift the output from P (system has memory). Turning S OFF and R ON causes the output to shift to $\overline{P}$ ($\overline{P} = 1$, $P = 0$), which represents the second of two stable states possessed by a SR FLIP-FLOP.

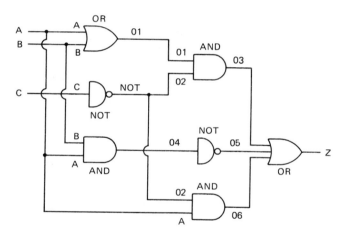

Figure 11-42. Logic circuit for Example 11-4.

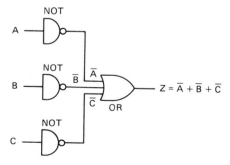

$$Z = \overline{A} + \overline{B} + \overline{C}$$

Figure 11-43. Simplified circuit for Example 11-4.

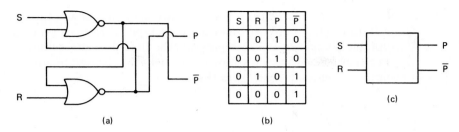

S	R	P	P̄
1	0	1	0
0	0	1	0
0	1	0	1
0	0	0	1

(a)   (b)

Figure 11-44. A MEMORY function (SR flip-flop). (a) Two NOR gate combination, (b) truth table, (c) symbol.

Figure 11-45 shows a two-handed press safety control application of fluid logic. From a safety consideration, it is necessary to ensure that both of an operator's hands are used to initiate the operation of a press. If either hand is removed during the operation of the press cycle, the cylinder retracts. Both hands

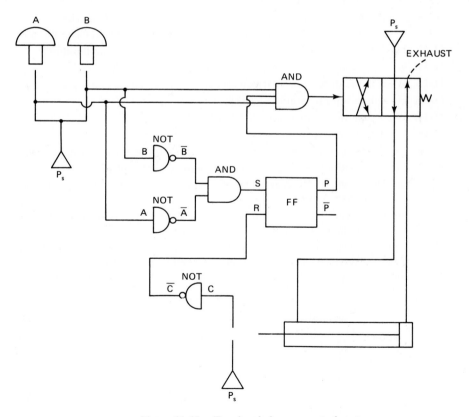

Figure 11-45. Two-handed press control system.

must be removed from the push button controls A and B before the next operation can be started. Push buttons A and B are back-pressure sensors, element C is an interruptable jet sensor and P_s represents the fluid pressure supply sources.

In Section 12.11 we show how Boolean algebra is used to implement the use of programmable logic controllers (PLCs).

EXERCISES

11-1. What is fluidics?

11-2. Name four advantages of fluidics.

11-3. Why is fluidics called the "brains" and fluid power the "brawn" when using fluidics to control fluid power systems?

11-4. Define an AND function.

11-5. Define an OR function.

11-6. Explain the Coanda effect.

11-7. What is a flip-flop? Explain how it works.

11-8. What is the difference between a bistable and monostable device?

11-9. How does a preferenced flip-flop differ in operation from a basic flip-flop?

11-10. How does an SRT flip-flop differ in operation from a basic flip-flop?

11-11. What is the difference between an OR gate and an EXCLUSIVE OR gate?

11-12. What is the difference between a back-pressure sensor and a proximity sensor?

11-13. How does an interruptible jet sensor work?

11-14. What type of sensor is an OR/NOR key?

11-15. What is meant by the expression *contact sensing?*

11-16. How important is clean dry air to the successful operation of fluidic systems?

11-17. For the continuous reciprocation hydraulic cylinder of Fig. 11-34.

    (a) How can the NOR gates be eliminated without adding any additional components to the circuit?

    (b) How would the operation be affected if the preferenced flip-flop were replaced by a regular flip-flop?

11-18. Figure 11-46 shows a potentially explosive area, which therefore cannot contain any electrical devices. It is desired to have a remote monitor with two fluid indicators to show whether or not sliding doors 1 and/or 2 are open or closed. If door 1 is open,

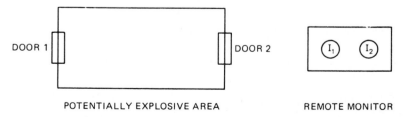

Figure 11-46.   System for Exercise 11-18.

indicator 1 lights up and vice versa. The same is true for door 2 and indicator 2. Design a fluidics circuit to accomplish this remote indication.

11-19. For the circuit of Fig. 11-47, what happens to the cylinder when the

    (a) Push-button switch is momentarily depressed?

    (b) The selector switch is closed?

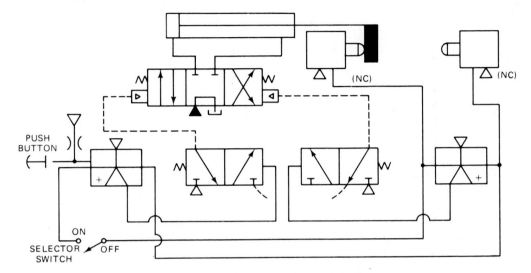

Figure 11-47. Circuit for Exercise 11-19.

11-20. For the circuit of Fig. 11-48, what happens to the cylinder if

    (a) 1-PB is pushed and held?

    (b) 1-PB and 2-PB are both pushed and held?

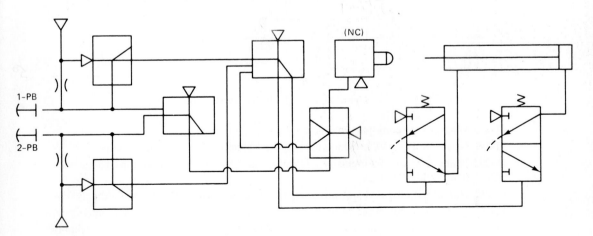

Figure 11-48. Circuit for Exercise 11-20.

11-21. For the circuit of Fig. 11-48, if 1-PB is tied in the actuated position, will the system function properly by the actuation of 2-PB whenever operation is required? Explain your answer.

11-22. For the circuit of Fig. 11-49, what happens to the two cylinders when the start button is momentarily pressed? Assume both cylinders are initially retracted.

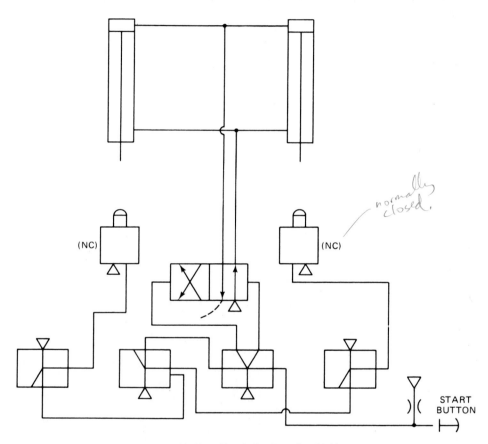

Figure 11-49. Circuit for Exercise 11-22.

11-23. What are moving-part logic devices?

11-24. Name three ways in which moving-part logic devices can be actuated.

11-25. Name a major distinguishing feature that differentiates moving-part logic devices from fluidic devices.

11-26. Name one advantage of moving-part logic systems compared to fluidic systems.

11-27. Name one advantage of fluidics systems compared to moving-part logic systems.

11-28. What must be done to fluidic systems to ensure that they will operate reliably?

11-29. Name two useful functions provided by Boolean algebra.

11-30. What algebraic operations are permitted in Boolean algebra.

11-31. Define the term *logic inversion*.

11-32. What is the difference between the commutative and associative laws?

11-33. State DeMorgan's theorem.

11-34. Prove that $A \cdot (A + B) = A$ using a truth table.

11-35. Prove that $A + (B + C) = (A + B) + C$ the first associative law, using a truth table.

11-36. Prove that $\overline{A \cdot B} = \overline{A} + \overline{B}$ DeMorgan's theorem No. 2 using a truth table.

11-37. Prove that $A \cdot (B + C) = (A \cdot B) + (A \cdot C)$ the second distributive law, using a truth table.

11-38. Prove that $A \cdot (\overline{A} + B) = A \cdot B$ using a truth table.

11-39. Using DeMorgan's theorem, show how the AND function can be developed using NOR gates.

11-40. Using DeMorgan's theorem, design a NOR circuit to generate the function: $Z = \overline{A} + \overline{B} + \overline{C}$.

11-41. Describe the operation of the fluid logic system of Fig. 11-50.

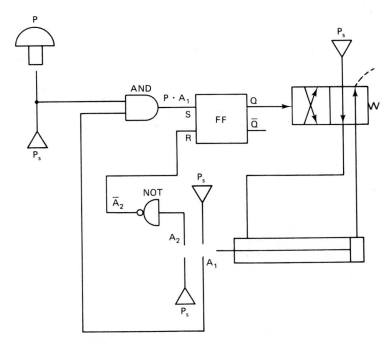

Figure 11-50.   Circuit for Exercise 11-41.

11-42. Describe the operation of the fluid logic circuit system of Fig. 11-51, when the machine guard blocks the back pressure sensor and then unblocks.

11-43. For the logic circuit of Fig. 11-52, use Boolean algebra to show that valve 3 is not needed.

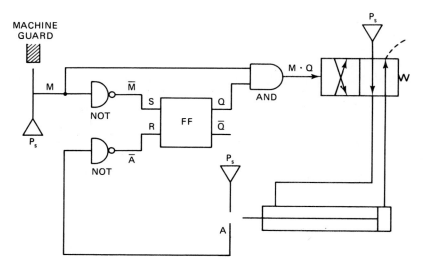

Figure 11-51.   Circuit for Exercise 11-42.

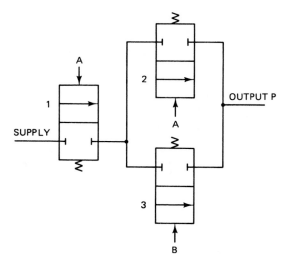

Figure 11-52.   Circuit for Exercise 11-43.

12

Electrical Controls for Fluid Power Circuits

12.1 INTRODUCTION

Electrical devices have proven to be an important means of improving the overall control flexibility of fluid power systems. In recent years, the trend has been toward electrical control of fluid power systems and away from manual control. One of the reasons for this trend is that more machines are being designed for automatic operation to be controlled with electrical signals from computers.

There are seven basic electrical devices commonly used in the control of fluid power systems: manually actuated switches, limit switches, pressure switches, solenoids, relays, timers, and temperature switches. By the use of a simple push-button switch, an operator can cause sophisticated equipment to begin performing complex operations. These push-button switches are used mainly for starting and stopping the operation of machinery as well as providing for manual override when an emergency arises.

Solenoids provide a push or pull force to remotely operate fluid power valves. Relays are switches whose contacts open or close when their corresponding coils are energized. These relays are commonly used for the energizing and deenergizing of solenoids because they operate at a high current level. In this way a manually actuated switch can be operated at low voltage levels to protect the operator. This low-voltage circuit can be used to energize relay coils that control high-voltage contacts used to open and close circuits containing the solenoids. The use of relays also provides interlock capability, which prevents the accidental energizing of two solenoids at the opposite ends of a valve spool. This safety feature can, therefore, prevent the burnout of one or both of these solenoids.

Pressure switches open or close their contacts based on system pressure. They generally have a high-pressure setting and a low-pressure setting. For example, it may be necessary to start or stop a pump to maintain a given pressure. The low-pressure setting would start the pump, and the high-pressure setting would stop it. Figure 12-1 is a photograph of a pressure switch that can be wired either normally open (N.O.) or normally closed (N.C.), as marked on the screw terminals. Observe the front scale that is used for visual check of the pressure setting, which is made by the self-locking, adjusting screw located behind the scale.

Limit switches open and close circuits when they are actuated either at the end of the retraction or extension strokes of hydraulic or pneumatic cylinders. Figure 12-2 shows a hydraulic cylinder that incorporates its own limit switches (one at each end of the cylinder). Either switch can be wired normally open or normally closed. The limit switch on the cap end of the cylinder is actuated by an internal cam when the rod is fully retracted. The cam contacts the limit switch about $\frac{3}{16}$ in. from the end of the stroke. At the end of the cylinder stroke, the cam has moved the plunger and stem of the limit switch about $\frac{1}{16}$ in. for complete actuation. The limit switch on the head end of the cylinder is similarly actuated. Since these limit switches are built into the cylinder end plates, they are not susceptible to accidental movement, which can cause them to malfunction.

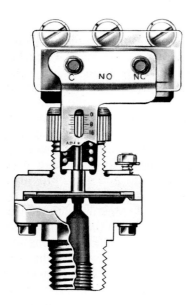

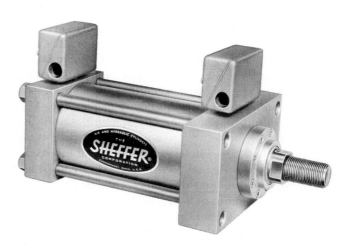

Figure 12-1. Pressure switch. (*Courtesy of DeLaval Turbine Inc., Barksdale Control Division, Los Angeles, California.*)

Figure 12-2. Cylinder with built-in limit switches. (*Courtesy of Sheffer Corp., Cincinnati, Ohio.*)

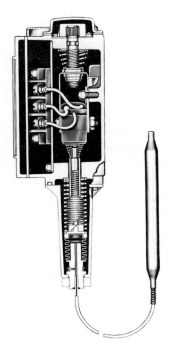

Figure 12-3.   Remotely located temperature switch. (*Courtesy of DeLaval Turbine Inc., Barksdale Control Division, Los Angeles, California.*)

Figure 12-3 shows a temperature switch, which is an instrument that automatically senses a change in temperature and opens or closes an electrical switch when a predetermined temperature is reached. This switch can be wired either normally open or normally closed. Notice that at its upper end there is an adjustment screw to change the actuation point.

The capillary tube (which comes in standard lengths of 6 or 12 ft) and bulb permit remote temperature sensing. Thus, the actual temperature switch can be located at a substantial distance from the oil whose temperature is to be sensed.

Temperature switches can be used to protect a fluid power system from serious damage when a component such as a pump or strainer or cooler begins to malfunction. The resulting excessive buildup in oil temperature is sensed by the temperature switch, which shuts off the entire system. This permits troubleshooting of the system to repair or replace the faulty component.

Time delay devices are used to control the time duration of a working cycle. In this way a dwell can be provided where needed. For example, a dwell can be applied to a drilling machine operation, which allows the drill to pause for a predetermined time at the end of the stroke to clean out the hole. Most timers can be adjusted to give a specified dwell to accommodate change in feed rates and other system variables.

When drawing electrohydraulic or electropneumatic circuits, a separate circuit is drawn for the fluid system and a separate circuit is drawn for the electrical system. Each component is labeled to show exactly how the two systems interface. Electrical circuits use ladder diagrams with the power connected to the left leg and the ground connected to the right leg. It is important to know the symbols used to represent the various electrical components. The operation of the total system can be ascertained by examination of the fluid power circuit and electrical diagram, as they show the interaction of all components.

In recent years, the electrohydraulic servo system has arrived on the industrial scene because these closed-loop systems can provide very precise control of the position, velocity, or acceleration of large loads. The overall operation of an electrohydraulic servo system is described in the following paragraph.

An electrical signal from a device called a transducer (which is mechanically connected to the load) is subtracted (negative feedback) from the input electrical command signal by a device called a summer (adds negative of feedback signal to input signal). This difference between the two signals (called the error signal) is electronically amplified to a higher power level to drive the torque motor of a servo valve. The torque motor shifts the spool of the servo valve, which produces hydraulic flow output to drive the load actuator. The velocity or position of the load is fed back in electrical form via the transducer, which is usually either a potentiometer or a tachometer. Thus, the feedback signal (which electrically measures load position or velocity) is continuously compared to the command input signal. If the load position or velocity is not that called for by the command signal, an error signal is generated by the summer to correct the discrepancy. When the desired output is achieved, the feedback and command signals are equal, producing a zero error signal. Thus no change in output will occur unless called for by a change in input command signal. Therefore, the ability of a closed-loop system to precisely control the position or velocity of a load is not affected by internal leakages due to changes in pressure (load changes) and oil viscosity (temperature changes).

In recent years, programmable logic controllers (PLCs) have increasingly been used in lieu of electromechanical relays to control fluid power systems. A PLC is a user-friendly electronic computer designed to perform logic functions such as AND, OR, and NOT for controlling the operation of industrial equipment and processes. A PLC consists of solid-state digital logic elements for making logic decisions and providing corresponding outputs. Unlike general-purpose computers, a PLC is designed to operate in industrial environments where high ambient temperature and humidity levels may exist. PLCs offer a number of advantages over electromechanical relay control systems. Unlike electromechanical relays, PLCs are not hard-wired to perform specific functions. Thus, when system operation requirements change, a software program is readily changed instead of having to physically rewire relays. In addition, PLCs are more reliable, faster in operation, smaller in size, and can be readily expanded.

12.2 ELECTRICAL COMPONENTS

There are five basic types of electric switches used in electrically controlled fluid power circuits: push-button, limit, pressure, temperature, and relay switches.

1. Push-button switches: Figure 12-4 shows the four common types of push-button switches. Figures 12-4(a) and (b) show the single-pole, single-throw type. These single-circuit switches can be wired either normally open or closed. Figure 12-4(c) depicts the double-pole, single-throw type. This double-contact type has one normally open and one normally closed pair of contacts. Depressing the push button opens the normally closed pair and closes the normally open pair of contacts.

Figure 12-4(d) illustrates the double-pole, double-throw arrangement. This switch has two pairs of normally open and two pairs of normally closed contacts to allow the inverting of two circuits with one input.

2. Limit switches: In Fig. 12-5 we see the various types of limit switches. Basically, limit switches perform the same functions as push-button switches. The difference is that they are mechanically actuated rather than manually. Figure 12-5(a) shows a normally open limit switch, which is abbreviated LS—NO. Figure 12-5(b) shows a normally open switch that is held closed. In Fig. 12-5(c) we see the normally closed type, whereas Fig. 12-5(d) depicts a normally closed type that is held open. There are a large number of different operators available for limit switches. Among these are cams, levers, rollers, and plungers. However, the symbols used for limit switches do not indicate the type of operator used.

3. Pressure switches: The symbols used for pressure switches are given in Fig. 12-6. Figure 12-6(a) gives the normally open type, whereas Fig. 12-6(b) depicts the normally closed symbol.

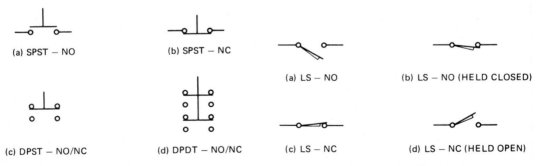

(a) SPST — NO   (b) SPST — NC

(a) LS — NO   (b) LS — NO (HELD CLOSED)

(c) DPST — NO/NC   (d) DPDT — NO/NC

(c) LS — NC   (d) LS — NC (HELD OPEN)

Figure 12-4. Push-button switch symbols. (a) SPST-N.O., (b) SPST-N.C., (c) DPST-N.O./N.C., (d) DPDT-N.O./N.C.

Figure 12-5. Limit switch symbols. (a) LS-N.O., (b) SL-N.O. (held closed), (c) LS-N.C., (d) LS-N.C. (held open).

SPST = single pole single throw.

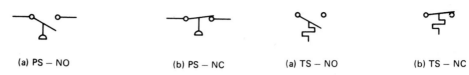

(a) PS – NO   (b) PS – NC   (a) TS – NO   (b) TS – NC

Figure 12-6. Pressure switch symbols.   **Figure 12-7.** Temperature switch
(a) PS-N.O., (b) PS-N.C.   symbols. (a) TS-N.O., (b) TS-N.C.

4. Temperature switches: This type of switch is depicted symbolically in Fig. 12-7. Figure 12-7(a) gives the symbol for a normally closed type, whereas Fig. 12-7(b) provides the normally open symbol.

5. Electrical relays: A relay is an electrically actuated switch. As shown schematically in Fig. 12-8(a), when switch 1-SW is closed, the coil (electromagnet) is energized. This pulls on the spring-loaded relay arm to open the upper set of normally closed contacts and close the lower set of normally open contacts. Figure 12-8(b) shows the symbol for the relay coil and the symbols for the normally open and closed contacts.

Timers are used in electrical control circuits when a time delay from the instant of actuation to the closing of contacts is required. Figure 12-9 gives the symbol used for timers. Figure 12-9(a) shows the symbol for the normally open switch that is time closed when energized. This type is one that is normally open but that when energized closes after a predetermined time interval. Figure 12-9(b) gives the normally closed switch that is time opened when energized. Figure 12-9(c) depicts the normally open type that is timed open when deenergized. Thus, it is normally open, and when the signal to close is removed (deenergized), it reopens after a predetermined time interval. Figure 12-9(d) gives the symbol for the normally closed type that is time closed when deenergized.

The symbol used to represent a solenoid, which is used to actuate valves, is shown in Fig. 12-10(a). Figure 12-10(b) gives the symbol used to represent indicator lamps. An indicator lamp is often used to indicate the state of a specific circuit component. For example, indicator lamps are used to determine which solenoid operator of a directional control valve is energized. They are also used to indicate whether a hydraulic cylinder is extending or retracting. An indicator lamp wired across each valve solenoid provides the troubleshooter with a quick means of pinpointing trouble in case of an electrical malfunction. If they are mounted on an operator's display panel, they should be mounted in the same order as they are actuated. Since indicator lamps are not a functional part of the electrical system, their inclusion in the ladder diagram is left to the discretion of the designer.

The remaining portion of this chapter is devoted to discussing a number of basic electrically controlled fluid power systems. In these systems, the standard electrical symbols are combined with ANSI fluid power symbols to indicate the operation of the total system. Fluid-power-operated electrical devices such as

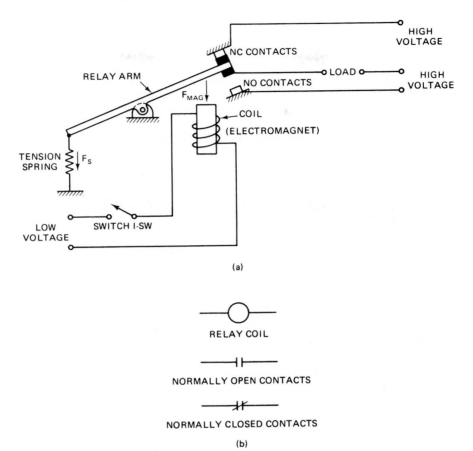

(a)

RELAY COIL

NORMALLY OPEN CONTACTS

NORMALLY CLOSED CONTACTS

(b)

Figure 12-8.   Electrical relay.

(a) NO (TIMED CLOSED WHEN
ENERGIZED)

(b) NC (TIMED OPEN WHEN
ENERGIZED)

(c) NO (TIMED OPEN WHEN
DE-ENERGIZED)

(d) NC (TIMED CLOSED WHEN
DE-ENERGIZED)

Figure 12-9.   Electrical timer symbols.
(a) N.O. (timed closed when energized), (b) N.C. (timed open when
energized), (c) N.O. (timed open when
deenergized), (d) N.C. (timed closed
when deenergized).

(a) SOLENOID   (b) INDICATOR LAMP

Figure 12-10. Solenoid and indicator lamp symbols. (a) Solenoid, (b) indicator lamp.

pressure switches and limit switches are shown on the fluid power circuits to correspond to symbols used in the electrical diagrams.

12.3 CONTROL OF A CYLINDER USING A SINGLE LIMIT SWITCH

Figure 12-11 shows a system that uses a single solenoid valve and a single limit switch to control a double-acting hydraulic cylinder. Figure 12-11(a) gives the hydraulic circuit in which the limit switch is labeled 1-LS and the solenoid is labeled SOL A. This method of labeling is required since many systems require more than one limit switch or solenoid.

In Fig. 12-11(b) we see the electrical diagram that shows the use of one relay with a coil designated as 1-CR and two separate, normally open sets of contacts labeled 1-CR (N.O.). The limit switch is labeled 1-LS (N.C.), and also included are one normally closed and one normally open push-button switch labeled STOP and START, respectively. This electrical diagram is called a *ladder diagram* because of its resemblance to a ladder. The two vertical electric power supply

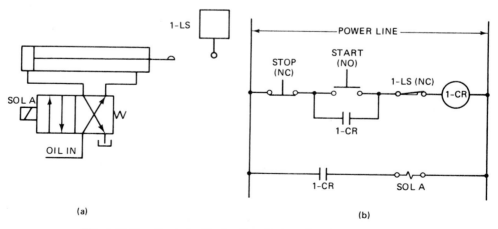

(a)   (b)

Figure 12-11. Control of hydraulic cylinder using single limit switch.

lines are called *legs* and the horizontal lines containing electrical components are called *rungs*.

When the START button is pressed momentarily, the cylinder extends because coil 1-CR is energized, which closes both sets of contacts of 1-CR. Thus, the upper 1-CR set of contacts serves to keep coil 1-CR energized even though the START button is released. The lower set of contacts closes to energize solenoid A to extend the cylinder. When 1-LS is actuated by the piston rod cam, it opens to deenergize coil 1-CR. This reopens the contacts of 1-CR to deenergize solenoid A. Thus, the valve returns to its spring offset mode and the cylinder retracts. This closes the contacts of 1-LS, but coil 1-CR is not energized because the contacts of 1-CR and the START button have returned to their normally open position. The cylinder stops at the end of the retraction stroke, but the cycle is repeated each time the START button is momentarily pressed. The STOP button is actually a panic button. When it is momentarily pressed, it will immediately stop the extension stroke and fully retract the cylinder.

12.4 RECIPROCATION OF A CYLINDER USING PRESSURE OR LIMIT SWITCHES

In Fig. 12-12 we see how pressure switches can be substituted for limit switches to control the operation of a double-acting hydraulic cylinder. Each of the two pressure switches has a set of normally open contacts. When switch 1-SW is closed, the cylinder reciprocates continuously until 1-SW is opened. The sequence of operation is as follows, assuming solenoid A was last energized: The pump is turned on, and oil flows through the valve and into the blank end of the cylinder. When the cylinder has fully extended, the pressure builds up to actuate pressure

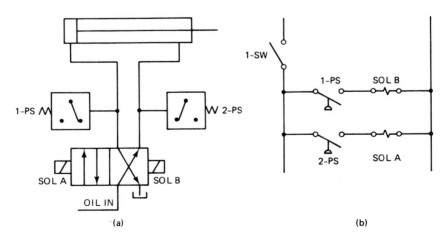

Figure 12-12.   Reciprocation of cylinder using pressure switches.

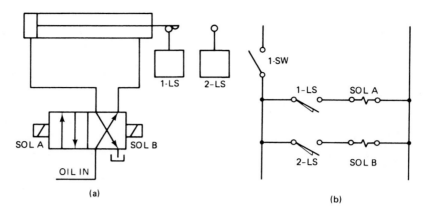

Figure 12-13. Reciprocation of cylinder using limit switches.

switch 1-PS. This energizes SOL B to switch the valve. Oil then flows to the rod end of the cylinder. Upon full retraction, the pressure builds up to actuate 2-PS. In the meantime, 1-PS has been deactuated to deenergize SOL B. The closing of the contacts of 2-PS energizes SOL A to begin once again the extending stroke of the cylinder.

Figure 12-13 gives the exact same control capability except each pressure switch is replaced by a normally open limit switch as illustrated. Observe that switches are always shown in their unactuated mode in the electrical circuits.

12.5 DUAL CYLINDER SEQUENCE CIRCUITS

Figure 12-14 shows a circuit that provides a cycle sequence of two pneumatic cylinders. When the start button is momentarily pressed, SOL A is momentarily energized to shift valve $V1$, which extends cylinder 1. When 1-LS is actuated, SOL C is energized, which shifts valve $V2$ into its left flow path mode. This extends cylinder 2 until it actuates 2-LS. As a result, SOL B is energized to shift valve $V1$ into its right flow path mode. As cylinder 1 begins to retract, it deactuates 1-LS, which deenergizes SOL C. This puts valve $V2$ into its spring offset mode, and cylinders 1 and 2 retract together. The complete cycle sequence established by the momentary pressing of the start button is as follows:

1. Cylinder 1 extends.
2. Cylinder 2 extends.
3. Both cylinders retract.
4. Cycle is ended.

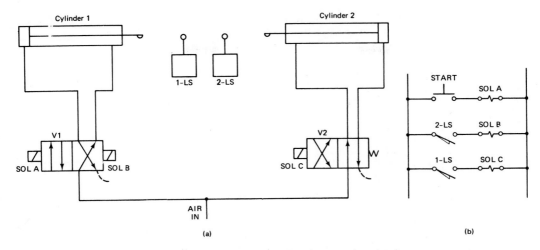

Figure 12-14.   Dual cylinder sequencing circuit.

A second dual cylinder sequencing circuit is depicted in Fig. 12-15. The operation is as follows: When the START button is depressed momentarily, SOL A is energized to allow flow through valve $V1$ to extend cylinder 1. Actuation of 1-LS deenergizes SOL A and energizes SOL B. Notice that limit switch 1-LS is a double-pole, single-throw type. Its actuation opens the holding circuit for relay 1-CR and simultaneously closes the holding circuit for relay 2-CR. This returns valve $V1$ to its spring offset mode and switches valve $V2$ into its solenoid-actuated mode. As a result, cylinder 1 retracts while cylinder 2 extends. When 2-LS is actuated, SOL B is deenergized to return valve $V2$ back to its spring offset mode to retract cylinder 2. The STOP button is used to retract both cylinders instantly. The complete cycle initiated by the START button is as follows:

1. Cylinder 1 extends.
2. Cylinder 2 extends while cylinder 1 retracts.
3. Cylinder 2 retracts.
4. Cycle is ended.

12.6 BOX-SORTING SYSTEM

An electropneumatic system for sorting two different-sized boxes moving on a conveyor is presented in Fig. 12-16. Low boxes are allowed to continue on the same conveyor, but high boxes are pushed on to a second conveyor by a pneumatic cylinder. The operation is as follows: When the START button is momentarily depressed, coil 2-CR is energized to close its two normally open sets of contacts. This turns on the compressor and conveyor motors. When a high box

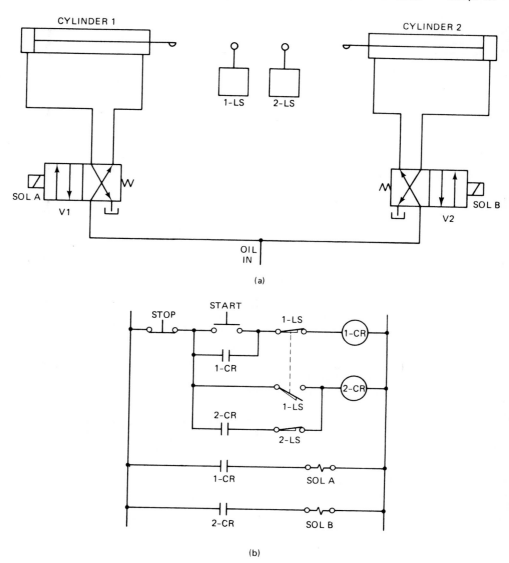

Figure 12-15. Second dual cylinder sequencing circuit.

actuates 1-LS, coil 1-CR is energized. This closes the 1-CR (N.O.) contacts and opens the 1-CR (N.C.) contacts. Thus, the conveyor motor stops, and SOL A is energized. Air then flows through the solenoid-actuated valve to extend the sorting cylinder to the right to begin pushing the high box onto the second conveyor. As 1-LS becomes deactuated, it does not deenergize coil 1-CR because contact set 1-CR (N.O.) is in its closed position. After the high box has been completely

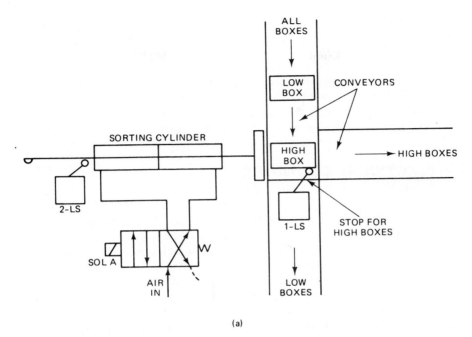

(a)

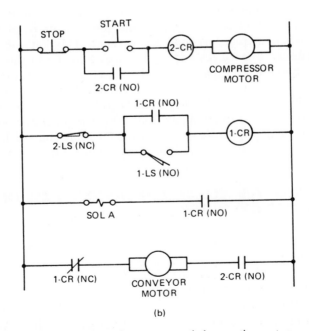

(b)

Figure 12-16.   Electro-pneumatic box-sorting system.

positioned onto the second conveyor, 2-LS is actuated. This deenergizes coil 1-CR and SOL A. The valve returns to its spring offset mode, which retracts the cylinder to the left. It also returns contact set 1-CR (N.C.) to its normally closed position to turn the conveyor motor back on to continue the flow of boxes.

When the next high box actuates 1-LS, the cycle is repeated. Depressing the STOP button momentarily turns off the compressor and conveyor motors because this causes coil 2-CR to become deenergized. The production line can be put back into operation by the use of the START button.

12.7 ELECTRICAL CONTROL OF REGENERATIVE CIRCUIT

Figure 12-17 shows a circuit that provides for the electrical control of a regenerative cylinder. The operation is as follows: Switch 1-SW is manually placed into the *extend* position. This energizes SOL A, which causes the cylinder to extend. Oil from the rod end passes through check valve V3 to join the incoming oil from the pump to provide rapid cylinder extension. When the cylinder starts to pick up its intended load, oil pressure builds up to actuate normally open pressure switch 1-PS. As a result, coil 1-CR and SOL C become energized. This vents rod oil directly back to the oil tank through valve V2. Thus, the cylinder extends slowly as it drives its load. Relay contacts 1-CR hold SOL C energized during the slow extension movement of the cylinder to prevent any *fluttering* of the pressure switch. This would occur because fluid pressure drops at the blank end of the cylinder when the regeneration cycle is ended. This can cause the pressure switch to oscillate as it energizes and deenergizes SOL C. In this design, the pressure switch is bypassed during the cylinder slow speed extension cycle. When switch 1-SW is placed into the *retract* position, SOL B becomes energized while the relay coil and SOL C become deenergized. Therefore, the cylinder retracts in a normal fashion to its fully retracted position. When the operator puts switch 1-SW into the *unload* position, all the solenoids and the relay coil are deenergized. This puts valve V1 in its spring-centered position to unload the pump.

12.8 COUNTING, TIMING, AND RECIPROCATION OF HYDRAULIC CYLINDER

Figure 12-18 shows an electrohydraulic system that possesses the following operating features:

1. A momentary push button starts a cycle in which a hydraulic cylinder is continuously reciprocated.
2. A second momentary push button stops the cylinder motion immediately, regardless of the direction of motion. It also unloads the pump.

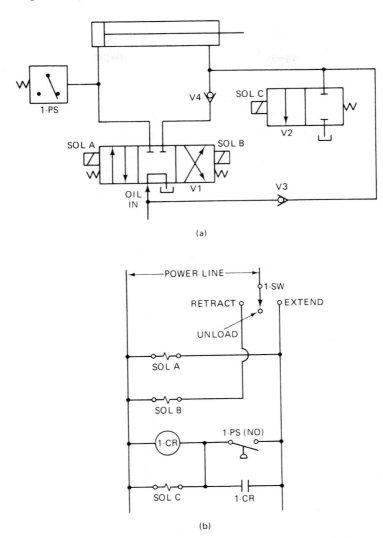

(a)

(b)

Figure 12-17. Electrical control of regenerative circuit.

3. If the START button is depressed after the operation has been terminated by the STOP button, the cylinder will continue to move in the same direction.

4. An electrical counter is used to count the number of cylinder strokes delivered from the time the START button is depressed until the operation is halted via the STOP button. The counter registers an integer increase in value each time an electrical pulse is received and removed.

5. An electrical timer is included in the system to time how long the system has

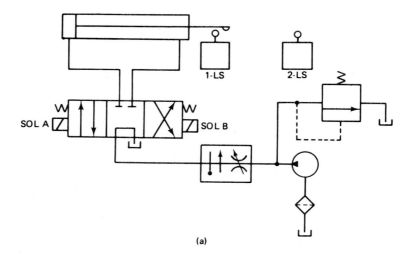

(a)

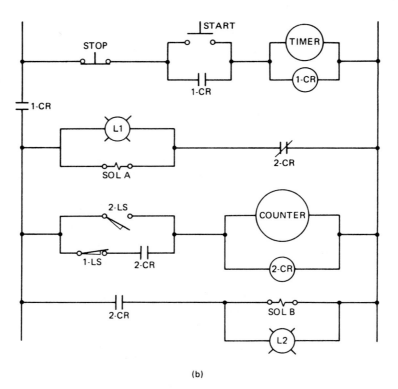

(b)

Figure 12-18.   Counting, timing, and reciprocation of hydraulic cylinders.

been operating since the START button was depressed. The timer runs as long as a voltage exists across its terminals. The timer runs only while the cylinder is reciprocating.

6. Two lamps ($L1$ and $L2$) are wired into the electric circuit to indicate whether the cylinder is extending or retracting. When $L1$ is ON, the cylinder is extending, and when $L2$ is ON, the cylinder is retracting.

7. The cylinder speed is controlled by the pressure- and temperature-compensated flow control valve.

Notice that the resistive components (lamps, solenoids, coils, timer, and counter) are connected in parallel in the same branch to allow the full line voltage to be impressed across each resistive component. It should be noted that switches (including relay contacts) are essentially zero resistance components. Therefore, a line that contains only switches will result in a short and thus should be avoided.

12.9 *ELECTROHYDRAULIC SERVO SYSTEM*

The primary purpose of a fluid power circuit is to control the position or velocity of an actuator. All the circuits discussed so far in this chapter accomplished this objective using open-loop controls. In these systems, precise control of speed is not possible. The speed will decrease when the load increases. This is due to the higher pressures that increase internal leakage inside pumps, actuators, and valves. Temperature changes that affect fluid viscosity and thus leakage also affect the accuracy of open-loop systems.

In Fig. 12-19 we see a closed-loop (servo) electrohydraulic control system. It is similar to the open-loop hydraulic system of Fig. 12-18 except the flow control and directional control valves have been replaced by a servo valve such as that shown in Fig. 12-20. A servo valve is a device containing an electrical torque motor that positions a spool valve. The spool valve position is proportional to the electrical signal to the torque motor coils (see Fig. 12-21 for a schematic drawing of a servo valve).

The system of Fig. 12-19 is a servo system because the loop is closed by a feedback device attached to the actuator. This feedback device, either linear or rotary, senses the actuator position or speed and transmits a corresponding electrical signal. This signal is compared electronically with the electrical input signal. If the actuator position or speed is not that intended, an error signal is generated by the electronic summer. This error signal is amplified and fed to the torque motor to correct the difference. Therefore, a closed-loop system is not affected by internal leakages due to pressure and temperature changes.

Feedback devices are called *transducers* because they perform the function of converting one source of energy into another, such as mechanical to electrical. A velocity transducer is one that senses the linear or angular velocity of the

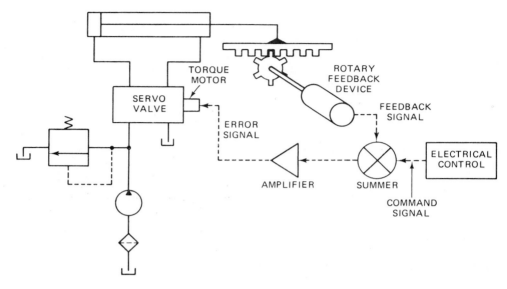

Figure 12-19. Electrohydraulic servo system.

Figure 12-20. Servo valve. (*Courtesy of Moog Inc., Industrial Division, East Aurora, New York.*)

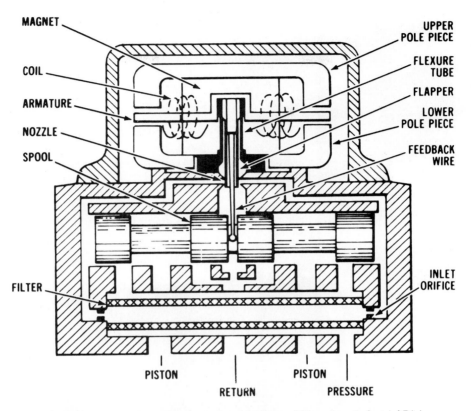

Figure 12-21. Schematic of servo valve. (*Courtesy of Moog Inc., Industrial Division, East Aurora, New York.*)

system output and generates a signal proportional to the measured velocity. The type most commonly used is the tachometer/generator, which produces a voltage (ac or dc) that is proportional to its rotational speed. Figure 12-22 shows a dc tachometer/generator. A positional transducer senses the linear or angular position of the system output and generates a signal proportional to the measured position. The most commonly used type of electrical positional transducer is the potentiometer, which can be of the linear or rotary motion type. Figure 12-23 shows a rotary potentiometer in which the wiper is attached to the moving member of the machine and the body to the stationary member. The positional signal is taken from the wiper and one end.

The electrical control box, which generates the command signal (see Fig. 12-19), can be a manual control unit such as that shown in Fig. 12-24. The scale graduations can represent either the desired position or velocity of the hydraulic

Figure 12-22. DC tachometer/generator. (*Courtesy of Oil Gear Co., Milwaukee, Wisconsin.*)

Figure 12-23. Rotary feedback potentiometer. (*Courtesy of Bourns Inc., Riverside, California.*)

Figure 12-24.   Remote manual control unit. (*Courtesy of Oil Gear Co., Milwaukee, Wisconsin.*)

actuator. An operator would position the hand lever to the position location or velocity level desired.

12.10 *ANALYSIS OF ELECTROHYDRAULIC SERVO SYSTEMS*

The analysis of servo system performance is accomplished using block diagrams in which each component is represented by a rectangle (block). The block diagram representation of a single component is a single rectangle shown as follows:

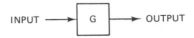

The gain G or transfer function of the component is defined as the output divided by the input.

$$G = \text{Gain} = \text{Transfer Function} = \frac{\text{output}}{\text{input}} \qquad (12\text{-}1)$$

For example, the block diagram for a pump is as follows where the input to the pump is shaft speed N(rpm) and the pump output is fluid flow Q(gpm). If a pump delivers 10 gpm when running at 2000 rpm, the pump gain is:

$$N(\text{rpm}) \longrightarrow \boxed{G_P} \longrightarrow Q(\text{gpm})$$

$$G_P = \text{Pump Gain} = \text{PumpTransfer Function}$$

$$= \frac{Q(\text{gpm})}{N(\text{rpm})} = \frac{10}{2000} = 0.005 \text{ gpm/rpm}$$

A block diagram of a complete closed-loop servo system (or simply servo system) is shown in Fig. 12-25. Block G represents the total gain of all the system components between the error signal and the output. This total gain (transfer function) is commonly called the forward transfer function because it is in the forward path. Block H represents the gain of the feedback component between the output and the summer (error detector). This transfer function is commonly called the feedback transfer function because it is in the feedback path. The sum of the input signal and negative feedback signal represents the error signal.

Another important parameter of a servo system is the open-loop gain (transfer function), which is defined as the gain from the error signal to the feedback signal. Thus it is the product of the forward path gain and feedback path gain.

$$\text{Open-loop gain} = GH \qquad (12\text{-}2)$$

The closed-loop gain (transfer function) is defined as the system output divided by the system input.

$$\text{Closed-loop transfer function} = \frac{\text{System Output}}{\text{System Input}} \qquad (12\text{-}3)$$

It can be shown that the closed-loop transfer function (which shows how the system output compares to the system input) equals the forward gain divided by

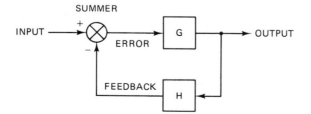

Figure 12-25. Block diagram of a closed-loop system.

the expression "one plus the product of the forward gain and feedback gain."

$$\text{Closed-loop transfer function} = \frac{G}{1 + GH} \tag{12-4}$$

Thus we have:

$$\text{Closed-loop transfer function} = \frac{\text{forward gain}}{1 + \text{open loop gain}} \tag{12-5}$$

It should be noted that each component of a servo system generates a phase lag as well as an amplitude gain. For a sinusoidal input command (which can be used to determine the frequency response of a servo system), if the open-loop gain is greater than unity when the phase shift from input signal to feedback signal is 180 degrees, the system will be unstable. This is because the feedback sine wave is in phase with the input sine wave, causing the output oscillations to become larger and larger until something breaks or corrective action is taken.

Figure 12-26 (b and c) shows the transient response of a stable and unstable servo system responding to a step input command. Figure 12-26(a) shows the step input command, which is a constant input voltage applied at time $t = 0$ to produce a desired output. Observe that even though the stable system exhibits overshoot, the output oscillation quickly dampens as it approaches the steady-state desired value. The unstable system is one that depicts repeated undamped oscillations, and thus the desired steady-state output value is not achieved.

Figure 12-27 shows a more detailed block diagram of an electrohydraulic positional closed-loop system. The components in the forward path include the amplifier, servo valve, and hydraulic cylinder. The gain of each component is specified as follows:

$$G_A = \text{amplifier gain (milliamps per volt or ma/v)}$$

$$G_{SV} = \text{servo valve gain} \left(\text{in.}^3/\text{sec per milliamp or } \frac{\text{in.}^3/\text{sec}}{\text{ma}} \right)$$

$$G_{CYL} = \text{cylinder gain (in./in.}^3)$$

$$H = \text{feedback transducer gain (volts per in. or V/in.)}$$

The open-loop gain can be determined as a product of the individual component gains.

$$\text{Open-loop gain} = GH = G_A \times G_{SV} \times G_{CYL} \times H \tag{12-6}$$

$$= \frac{\text{ma}}{\text{V}} \times \frac{\text{in.}^3/\text{sec}}{\text{ma}} \times \frac{\text{in.}}{\text{in.}^3} \times \frac{\text{V}}{\text{in.}} = \frac{1}{\text{sec}}$$

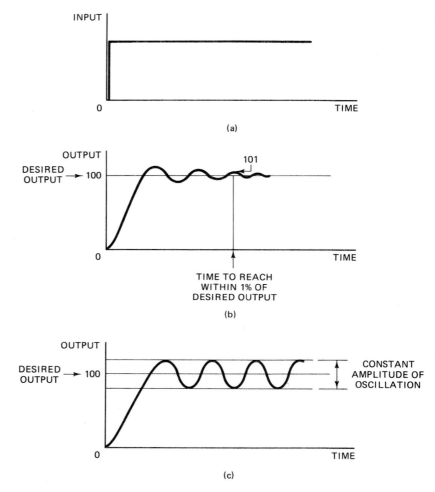

Figure 12-26. Stable and unstable transient responses to a step input command. (a) Step input command, (b) stable (output dampens out), (c) unstable (output oscillations repeat).

Thus, the units of open-loop gain are the reciprocal of time.

The accuracy of an electrohydraulic servo system depends on the system deadband and open-loop gain. System deadband is the composite deadband of all the components in the system. Deadband is defined as that region or band of no response where an input signal will not cause an output. For example, with respect to a servo valve, no flow will occur until the electric current to the torque motor reaches a threshold minimum value. This is shown in Fig. 12-28, which is obtained by cycling a servo valve through its rated input current range and record-

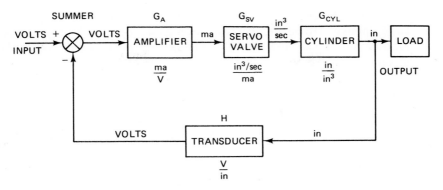

Figure 12-27.   Detailed block diagram of electrohydraulic closed-loop system.

ing the output flow for one cycle of input current. The resulting flow curve shows the deadband region and hysteresis of the servo valve. Hysteresis is defined as the difference between the response of the valve to an increasing signal and the response to a decreasing signal.

The repeatable error (discrepancy from the programmed output position) can be determined from Equation 12-7.

$$RE = \text{repeatable error (in.)} = \frac{\text{system deadband (ma)}}{G_A \text{ (ma/V)} \times H(\text{V/in.})} \qquad (12\text{-}7)$$

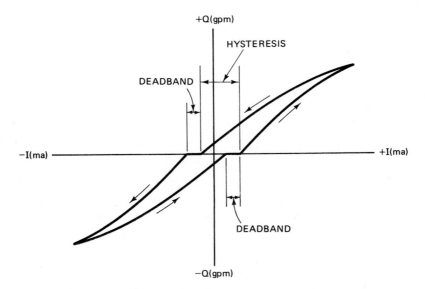

Figure 12-28.   Flow curve showing deadband and hysteresis for a servo valve.

To regulate system output accurately requires that the open-loop gain be large enough. However, the open-loop gain is limited by the component of the loop having the lowest natural frequency. Since hydraulic oil (although highly incompressible) is more compressible than steel, the oil under compression is the lowest natural frequency component. The following equation allows for the calculation of the natural frequency of the oil behaving as a spring-mass system.

$$\omega_H = A \sqrt{\frac{2\beta}{VM}} \tag{12-8}$$

where ω_H = natural frequency (radians/sec)
  A = area of cylinder (in.^{2})
  β = bulk modulus of oil (lb/in.^{2})
  V = volume of oil under compression (in.^{3})
  M = mass of load (lb · sec^2/in.)

The volume of oil under compression equals the volume of oil between the servo valve and cylinder in the line that is pressurized. For a typical servo system, the approximate value of the open-loop gain is one-third of the natural frequency of the oil under compression.

$$\text{Open-loop gain} = \frac{\omega_H \text{ (radians/sec)}}{3} \tag{12-9}$$

It should be noted that the amount of open-loop gain that can be used is limited. This is because the system would be unstable if the feedback signal has too large a value and adds directly to the input command signal.

Examples 12-1 and 12-2 show how to determine the accuracy of an electro-hydraulic servo system.

EXAMPLE 12-1

An electrohydraulic servo system contains the following characteristics:

1. G_{SV} = (0.15 in.^{3}/sec)/ma. The servo valve saturates at 300 ma to provide a maximum fluid flow of 45 in.^{3}/sec.
2. G_{CYL} = 0.20 in./in.^{3} The cylinder piston area equals 5 in.^{2} allowing for a maximum velocity of 9 in./sec. Cylinder stroke is 6 inches.
3. H = 4 V/in. Thus the maximum feedback voltage is 24.
4. Volume of oil under compression = 50 in.^{3}
5. Weight of load = 1000 lb.

$$\text{Mass} = \frac{1000 \text{ lb}}{386 \text{ (in./sec}^2)} = 2.59 \text{ lb} \cdot \text{sec}^2/\text{in.}$$

6. System deadband = 4 ma
7. Bulk modulus of oil = 175,000 lb/in.^{2}

Determine the system accuracy.

Solution   First calculate the natural frequency of the oil. This is the resonant frequency, which is the frequency at which oil would freely vibrate as would a spring-mass system.

$$\omega_H = 5 \sqrt{\frac{(2)(175,000)}{(50)(2.59)}} = 260 \text{ rad/sec}$$

The value of the open-loop gain is

$$\omega_H/3 = 86.7/\text{sec}$$

Solving for the amplifier gain from Equation 12-6, we have:

$$G_A = \frac{\text{open loop gain}}{G_{SV} \times G_{CYL} \times H} = \frac{86.7}{0.15 \times 0.20 \times 4} = 723 \text{ ma/volt}$$

The repeatable error (RE) can now be calculated using Equation 12-7.

$$RE = \frac{4}{723 \times 4} = 0.00138 \text{ in.}$$

EXAMPLE 12-2

Determine the system accuracy for a servo system containing the following characteristics (note that this system is identical to that given in Example 12-1 except that the units are metric rather than English).

1. G_{SV} = (2.46 cm^3/s)/ma
2. G_{CYL} = 0.031 cm/cm^3, cylinder area = 32.3 cm^2
3. H = 1.57 V/cm
4. V_{OIL} = 819 cm^3
5. Mass of load = 450 kg
6. System deadband = 4 ma
7. Bulk modulus of oil = 1200 MPa

Solution

$$\omega_H = 32.3 \times 10^{-4} \sqrt{\frac{(2)(1200 \times 10^6)}{(819 \times 10^{-6})(450)}} = 260 \text{ rad/sec}$$

$$\text{Open-loop gain} = \frac{260}{3} = 86.7/\text{s}$$

$$G_A = \frac{86.7}{2.46 \times 0.031 \times 1.57} = 732 \text{ ma/volt}$$

$$RE = \frac{4}{(723)(1.57)} = 0.00352 \text{ cm} = 0.00138 \text{ in.}$$

Another parameter that identifies the performance of a servo system is called tracking error, which is the distance by which the output lags the input command signal while the load is moving. The maximum tracking error (TE) is mathematically defined as:

$$TE = \frac{\text{servo valve maximum current (ma)}}{\text{amplifier gain (ma/V)} \times \text{transducer gain (Vin.)}} \quad (12\text{-}10)$$

Although the analysis presented here is applicable for a cylinder positional servo, the technique can be applied to hydraulic motor drive systems. In addition, if velocity (angular or linear) is to be controlled, then velocity units would be used instead of distance units. Although this analysis method is approximate, it provides a useful technique for determining design values for various system parameters.

EXAMPLE 12-3

For the Examples 12-1 and 12-2, find the maximum tracking error. Solution:

a. For English units:

$$TE = \frac{300 \text{ ma}}{723 \text{ ma/V} \times 4 \text{ V/in.}} = 0.104 \text{ inches}$$

Thus, the hydraulic cylinder position will lag its desired position by 0.104 inches based on the command signal when the cylinder is moving at the maximum velocity of 9 in./sec.

b. For metric units:

$$TE = \frac{300 \text{ ma}}{723 \text{ ma/V} \times 1.57 \text{ V/cm}} = 0.264 \text{ cm} = 0.104 \text{ in.}$$

12.11 PROGRAMMABLE LOGIC CONTROLLERS (PLCs)

A programmable logic controller (PLC) is a user-friendly electronic computer designed to perform logic functions such as AND, OR, and NOT for controlling the operation of industrial equipment and processes. PLCs, which are used in lieu of electromechanical relays (which are described in Sections 12.2 through 12.8), consist of solid-state digital logic elements for making logic decisions and providing corresponding outputs. Unlike general-purpose computers, a PLC is designed to operate in industrial environments where high ambient temperature and humidity levels may exist. In addition, PLCs are designed to not be affected by electrical noise commonly found in industrial plants. Figure 12-29 shows three PLCs designed for small- to medium-sized automation applications. These PLCs were designed to provide user-friendly service, from installation to troubleshooting and maintenance. The compact design (dimensions of 9.5 in. by 6.1 in. by 2.4 in. and 2.6 lb weight) permits the unit to be mounted directly onto an installation panel. The integrated keypad allows commands to be entered and supports displays to continuously monitor the status of programs.

PLCs provide the following advantages over electromechanical relay control systems:

1. Electromechanical relays (as shown in Sections 12.2 through 12.8) have to be hard-wired to perform specific functions. Thus when system operation requirements change, the relays have to be rewired.
2. They are more reliable and faster in operation.
3. They are smaller in size and can be more readily expanded.
4. They require less electrical power and are less expensive for the same number of control functions.

Figure 12-29. Programmable logic controllers (PLCs). (*Courtesy of Festo Corp., Hauppauge, N.Y.*)

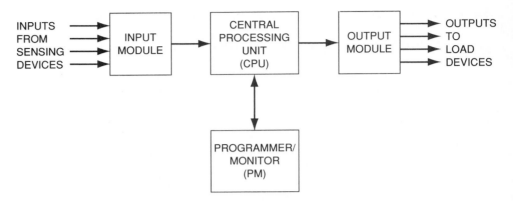

Figure 12-30. Block Diagram of a PLC.

As shown in Fig. 12-30, a PLC consists of the following three major units:

1. Central processing unit (CPU). This unit represents the "brains" of the PLC. It contains a microprocessor with a fixed memory and an alterable memory. The fixed memory contains the program set by the manufacturer. It is set into integrated circuit (IC) chips called read only memory (ROM) and this memory can not be changed during operation or lost when electrical power to the CPU is turned off. The alterable memory is stored on IC chips that can be programmed and altered by the user. This memory is stored on random access memory (RAM) chips and information stored on RAM chips is lost (volatile memory) when electrical power is removed. The PLC of Fig. 12-29 contains a battery backup that prevents the loss of any program in RAM if electrical power is lost. In general, the CPU receives input data from various sensing devices such as switches, executes the stored program, and delivers corresponding output signals to various load control devices such as relay coils and solenoids.

2. Programmer/monitor (PM). This unit allows the user to enter the desired program into the RAM memory of the CPU. The program, which is entered in relay ladder logic (similar to the relay ladder logic diagrams in Sections 12.2 through 12.8), determines the sequence of operation of the fluid power system being controlled. The PM needs to be connected to the CPU only when entering or monitoring the program. Programming is accomplished by pressing keys on the PM's keypad. The programmer/monitor may be either a hand-held device with a light emmiting diode (LED) or a desktop device with a cathode-ray tube (CRT) display. Figure 12-31 shows the details of the built-in programming/command keypad and LED display and status indicator of the PLC of Fig. 12-29. The keypad allows the operator to run programs continuously or in single-step mode, monitor all real-time functions, edit or monitor programs, and print programs and output code. Figure 12-32 shows a test application of this PLC.

Figure 12-31. Details of keypad and LED display of PLC. (*Courtesy of Festo Corp., Hauppauge, N.Y.*)

Figure 12-32. Test application of PLC. (*Courtesy of Festo Corp., Hauppauge, N.Y.*)

 3. Input/output module (I/O). This module is the interface between the fluid power system input sensing and output load devices and the CPU. The purpose of the I/O module is to transform the various signals received from or sent to the fluid power interface devices such as push-button switches, pressure switches, limit switches, motor relay coils, solenoid coils, and indicator lights. These interface devices are hard-wired to terminals of the I/O module. The PLC shown in Figs. 12-29, 12-31, and 12-32 contains 32 I/Os expandable to 128 I/Os, a memory

Figure 12-33. Off-line programming of PLC using a microcomputer. (*Courtesy of Festo Corp., Hauppauge, N.Y.*)

of 32K, and a serial port for a microcomputer interface. This PLC can be programmed off-line in a ladder logic diagram using a microcomputer as shown in Fig. 12-33. Capabilities include project creation, program editing, loading, and documentation. The system allows the user to monitor the controlled process during operation and provides immediate information regarding the status of timers, counters, inputs, and outputs.

To show how a PLC operates, let's look at the system of Fig. 12-11 (repeated in Fig. 12-34), which shows the control of a hydraulic cylinder using a single limit switch.

For this system, the wiring connections for the input and output modules are shown in Figs. 12-35(a) and (b), respectively. Note that there are three sensing input devices to be connected to the input module and one output control/load device to be connected to the output module. The electrical relay is not included in the I/O connection diagram since its function is replaced by an internal PLC control relay.

The PLC ladder logic diagram that would be constructed and programmed into the memory of the CPU is shown in Fig. 12-36(a). Note that the layout of the PLC ladder diagram [Fig. 12-36(a)] is similar to the layout of the hard-wired relay ladder diagram (Fig. 12-34). The two rungs of the relay ladder diagram are converted to two rungs of the PLC ladder logic diagram. The terminal numbers used on the I/O connection diagram are the same numbers used to identify the electrical devices on the PLC ladder logic diagram. The symbol ─╫─ represents a normally closed set of contacts and the symbol ─┤ ├─ represents a normally open set of contacts. The symbol ─( )─ with number 030 represents the relay coil that controls the two sets of contacts with number 030, which is the address in memory

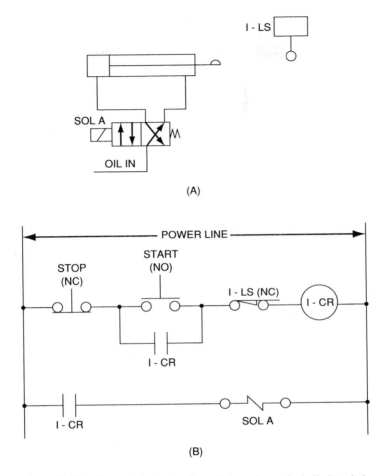

Figure 12-34.   Control of a hydraulic cylinder using a single limit switch.

for this internal relay. The relay coil and its two sets of contacts are programmed as internal relay equivalents. The symbol —( )— with number 010 represents the solenoid.

A PLC is a digital solid-state device and thus performs operations based on the three fundamental logic functions: AND, OR, and NOT. Each rung of a ladder diagram can be represented by a Boolean equation. The hard-wired ladder diagram logic is fixed and can be changed only by modifying the way the electrical components are wired. However, the PLC ladder diagram contains logic functions that are programmable and thus easily changed. Figure 12-36(b) shows the PLC ladder logic diagram of Fig. 12-36(a) with capital letters used to represent each electrical component. With the letter designation, Boolean equations can be writ-

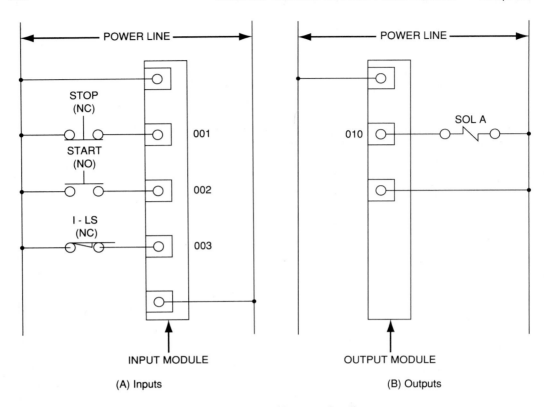

Figure 12-35. I/O connection diagram.

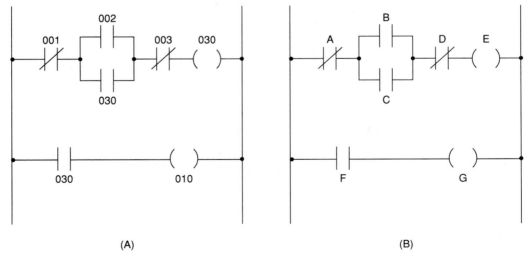

Figure 12-36. PLC ladder logic diagrams.

ten for each rung as follows:

Top rung:

$$\overline{A} \cdot (B + C) \cdot \overline{D} = E$$

This equation means: NOT A AND (B OR C) AND NOT D EQUALS E. It can also be stated as follows noting that 0 is the OFF state and 1 is the ON state: E is energized when A is NOT actuated AND B OR C is actuated AND D is NOT actuated.

Bottom rung:

$$F = G$$

G is energized when F is actuated.

Since PLCs use logic ladder diagrams, the conversion from existing electrical relay logic to programmed logic is easy to accomplish. Each rung contains devices connected from left to right, with the device to the far right being the output. The devices are connected in series or parallel to produce the desired logical result. PLC programming can thus be readily implemented to provide the desired control capability for a particular fluid power system.

Figure 12-37 shows a PLC-controlled electrohydraulic linear servo actuator designed to move and position loads quickly, smoothly, and accurately. As shown in the cutaway view of Fig. 12-38, the linear actuator includes, in its basic configu-

Figure 12-37.   PLC-controlled electrohydraulic linear servo actuator. (*Courtesy of Atos Systems, Inc., East Brunswick, New Jersey.*)

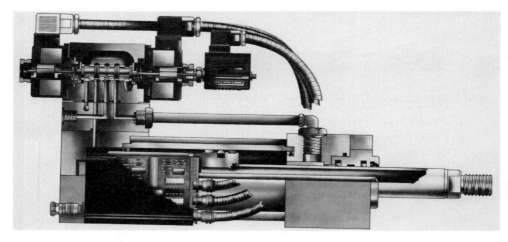

Figure 12-38. Cutaway view of PLC-controlled electrohydraulic linear servo actuator. (*Courtesy of Atos Systems, Inc., East Brunswick, New Jersey.*)

ration, a hydraulic servo cylinder (with built-in electronic transducer) complete with electrohydraulic valves and control electronics so that it forms one integrated machine element. As shown in the block diagram of Fig. 12-39, the control electronics is of the ''closed loop'' type and programmable cycles are capable of

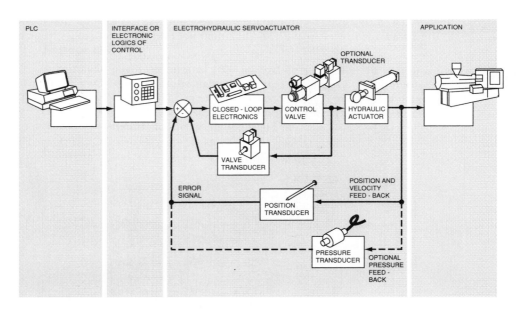

Figure 12-39. Block diagram of PLC control for electrohydraulic linear servo actuator. (*Courtesy of Atos Systems, Inc., East Brunswick, New Jersey.*)

generating the reference command signal. Typical applications for this system include sheet metal punching and bending, lifting and mechanical handling, radar and communications control systems, blow molding machines, machine tools, and mobile machinery.

EXERCISES

12-1. In recent years, the trend has been toward electrical control of fluid power systems and away from manual controls. Give one reason for this trend.

12-2. What is the difference between a pressure switch and a temperature switch?

12-3. How does a limit switch differ from a push-button switch?

12-4. What is an electrical relay? How does it work?

12-5. What is the purpose of an electrical timer?

12-6. How much resistance do electrical switches possess?

12-7. Give one reason for having an indicator lamp in an electrical circuit for a fluid power system.

12-8. Name two reasons an open-loop system does not provide a perfectly accurate control of its output actuator.

12-9. Name two types of feedback devices used in closed-loop systems. What does each device accomplish?

12-10. What is the definition of a feedback transducer?

12-11. What two basic components of an open-loop system does a servo valve replace?

12-12. What happens to the cylinder of Fig. 12-40 when the push button is momentarily depressed?

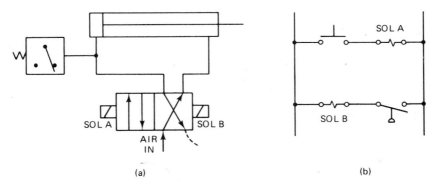

(a)

(b)

Figure 12-40.   Circuit for Exercise 12-12.

12-13. What happens to cylinders 1 and 2 of Fig. 12-41 when the switch 1-SW is closed? What happens when 1-SW is opened?

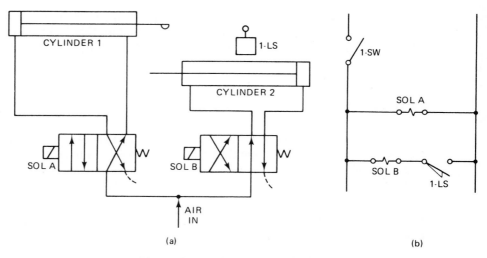

Figure 12-41.   Circuit for Exercise 12-13.

12-14. For the system of Fig. 12-42, what happens to the two cylinders if

    (a) Push button 1-PB is momentarily depressed?

    (b) Push button 2-PB is momentarily depressed?

        Cylinder 2 does not actuate 1-LS at the end of its extension stroke.

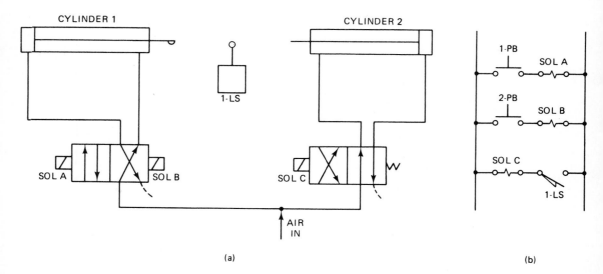

Figure 12-42.   Circuit for Exercise 12-14.

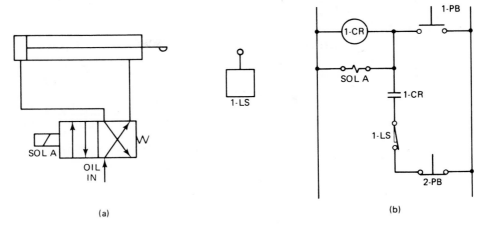

(a)   (b)

Figure 12-43.   System for Exercise 12-15.

12-15. Explain the complete operation of the system shown in Fig. 12-43.

12-16. What happens to cylinders 1 and 2 of Fig. 12-44 when
   (a) 1-PB is momentarily depressed?
   (b) 2-PB is momentarily depressed?

12-17. What is a transfer function?

12-18. What is the difference between deadband and hysteresis?

12-19. Define the term *open-loop gain.*

12-20. Define the term *closed-loop transfer function.*

12-21. What is meant by the term *repeatable error?*

12-22. What is meant by the term *tracking error?*

12-23. What is the difference between the forward and feedback paths of a closed-loop system?

12-24. An electrohydraulic servo system contains the following characteristics:
   1. $G_{CYL} = (0.15 \text{ in./in.}^3)$, cylinder piston area $= 4 \text{ in.}^2$
   2. $H = 3.5 \text{ V/in.}$
   3. $V_{OIL} = 40 \text{ in.}^3$
   4. Weight of load $= 750 \text{ lb}$
   5. System deadband $= 3.5 \text{ ma}$
   6. Bulk modulus of oil $= 200,000 \text{ lb/in.}^2$
   7. System accuracy $= 0.002 \text{ in.}$
   Determine the gain of the servo valve in units of $(\text{in.}^3/\text{sec})/\text{ma}$.

12-25. What is the closed-loop transfer function for the system of Exercise 12-24?

12-26. An electrohydraulic servo system contains the following characteristics:
   1. $G_{CYL} = 0.04 \text{ cm/cm}^3$, cylinder piston area $= 25 \text{ cm}^2$
   2. $H = 1.75 \text{ V/cm}$

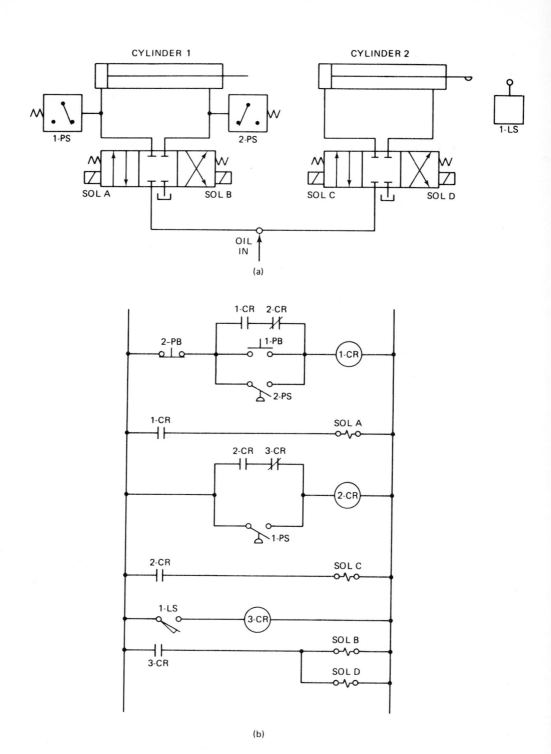

Figure 12-44. System for Exercise 12-16.

    3. $V_{OIL} = 750 \text{ cm}^3$

    4. Mass of load $= 300 \text{ kg}$

    5. System deadband $= 3.5 \text{ ma}$

    6. Bulk modulus of oil $= 1400 \text{ MPa}$

    7. System accuracy $= 0.004 \text{ cm}$
        Determine the gain of the servo valve in units of $(\text{cm}^3/\text{s})/\text{ma}$.

12-27. What is the closed-loop transfer function for the system of Exercise 12-26?

12-28. For the system of Exercise 12-24, if the maximum amplifier current is 250 ma, find the maximum tracking error.

12-29. For the system of Exercise 12-26, if the maximum amplifier current is 250 ma, find the maximum tracking error.

12-30. What is a programmable logic controller?

12-31. How does a PLC differ from a general-purpose computer?

12-32. What is the difference between a programmable logic controller and an electromechanical relay control?

12-33. Name three advantages that PLCs provide over electromechanical relay control systems.

12-34. State the main function of each of the following elements of a PLC:

    (a) CPU

    (b) Programmer/monitor

    (c) I/O module

12-35. What is the difference between read only memory (ROM) and random access memory (RAM)?

12-36. Draw the PLC logic ladder diagram and write the Boolean statements and equations for the hard-wired relay ladder diagram shown in Fig. 12-12.

12-37. Draw the PLC logic ladder diagram and write the Boolean statements and equations for the hard-wired relay ladder diagram shown in Fig. 12-14.

12-38. Draw the PLC logic ladder diagram and write the Boolean statements and equations for the hard-wired relay ladder diagram shown in Fig. 12-41.

12-39. Draw the I/O connection diagram and PLC logic ladder diagram that will replace the hard-wired relay logic diagram shown in Fig. 12-42.

12-40. Draw the I/O connection diagram and PLC logic ladder diagram that will replace the hard-wired relay logic diagram shown in Fig. 12-44.

12-41. Draw the PLC logic ladder diagram rung for each of the following Boolean algebra equations:

    (a) $Z = A + B$.

    (b) $Z = A \cdot B$.

    (c) $Z = A \cdot (B + C)$.

    (d) $Z = (A + B) \cdot C \cdot D$.

    (e) $Z = A \cdot \overline{B} \cdot C + \overline{D} + E$.

13

Fluid Power Maintenance and Safety

13.1 INTRODUCTION

In the early years of fluid power systems, maintenance was frequently performed on a hit or miss basis. The prevailing attitude was to fix the problem when the system broke down. However, with today's highly sophisticated machinery and the advent of mass production, industry can no longer afford to operate on this basis. The cost of downtime is prohibitive.

The following is a list of the most common causes of hydraulic system breakdown:

1. Clogged or dirty oil filters
2. Inadequate supply of oil in the reservoir
3. Leaking seals
4. Loose inlet lines that cause the pump to take in air
5. Incorrect type of oil
6. Excessive oil temperature
7. Excessive oil pressure

Most of these kinds of problems can be eliminated if a planned preventative maintenance program is undertaken. This starts with the fluid power designer in the selection of quality and properly sized components. The next step is the proper assembly of the various components. This includes applying the correct amount of torque to the various tube fittings to prevent leaks and, at the same time, not distort the fitting. Parts should be cleaned as they are assembled, and the system should be completely flushed with clean oil prior to putting it into service. It is important for the total system to provide easy access to components requiring periodic inspection such as filters, strainers, sight gages, drain and fill plugs, flowmeters, and pressure and temperature gages.

Over half of all hydraulic system problems have been traced directly to the oil. This is why the sampling and testing of the fluid is one of the most important preventative maintenance measures that can be undertaken. Figure 13-1 shows a hydraulic fluid test kit, which provides a quick, easy method to test for hydraulic system contamination. Even small hydraulic systems may be checked. The test kit may be used on the spot to determine whether fluid quality permits continued use. Tests that can be performed include the determination of viscosity, water content, and particulate contamination level. Viscosity is measured using a Visgage viscosity comparator. Water content is determined by the hot plate method. Contamination is evaluated by filtering a measured amount of hydraulic fluid, examining the particles caught on the filter under a microscope, and comparing what is seen with a series of photos indicating contamination levels. The complete test requires only approximately 10 min of time.

It is vitally important for maintenance personnel and machine operators to be trained to recognize early symptoms of potential hydraulic problems. For example, a noisy pump may be due to cavitation caused by a clogged inlet filter.

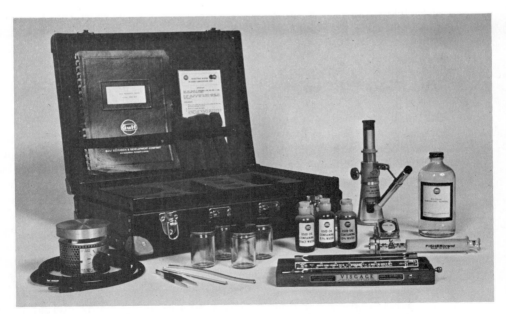

Figure 13-1. Hydraulic fluid test kit. (*Courtesy of Gulf Oil Corp., Houston, Texas.*)

This may also be due to a loose intake fitting, which allows air to be taken into the pump. If the cavitation noise is due to such an air leak, the oil in the reservoir will be covered with foam. When air becomes entrained in the oil, it causes spongy operation of hydraulic actuators. A sluggish actuator may be due to fluid having too high a viscosity. However, it can also be due to excessive internal leakage through the actuator or one of its control valves.

For preventative maintenance techniques to be truly effective, it is necessary to have a good report and records system. These reports should include the

they were detected, and the date. ·rformed. This should include the time, and the date. lded, or changed. Dates of filter

ect to external oil leaks are also the floor and around machinery external oil leakage is prohibitive because of contamination or the ~~~ ... oose mounting bolts or brackets should be tightened as soon as they are detected. The reason is that they can cause misalignment of the shafts of actuators and pumps, which can result in shaft seal or packing damage. A premature external oil leak can occur that will require costly downtime for repair.

13.2 SEALING DEVICES

Oil leakage, located anywhere in a hydraulic system, reduces efficiency and increases power losses. Internal leakage does not result in loss of fluid from the system because the fluid returns to the reservoir. Most hydraulic components possess clearances that permit a small amount of internal leakage. This leakage increases as component clearances between mating parts increase due to wear. If the entire system leakage becomes large enough, most of the pump's output is bypassed, and the actuators will not operate properly. External leakage represents a loss of fluid from the system. In addition, it is unsightly and represents a safety hazard. Improperly assembled pipe fittings is the most common cause of external leakage. Overtightened fittings may become damaged, or vibration can cause properly tightened fittings to become loose. Shaft seals on pumps and cylinders may become damaged due to misalignment or excessive pressure.

Seals are used in hydraulic systems to prevent excessive internal and external leakage and to keep out contamination. Seals can be of the positive or nonpositive type and can be designed for static or dynamic applications. Positive seals do

<u>BASIC FLANGE JOINTS</u>

GASKET

<u>METAL-TO-METAL JOINTS</u>

Figure 13-2.   Static seal flange joint applications. (*Courtesy of Sperry Vickers, Sperry Rand Corp., Troy, Michigan.*)

not allow any leakage whatsoever (external or internal). Nonpositive seals (such as the clearance used to provide a lubricating film between a valve spool and its housing bore) permit a small amount of internal leakage.

Static seals are used between mating parts that do not move relative to each other. Figure 13-2 shows some typical examples which include flange gaskets and seals. Notice that these seals are compressed between two rigidly connected parts. They represent a relatively simple and nonwearing joint, which should be trouble-free if properly assembled. Figure 13-3 shows a number of die cut gaskets used for flange-type joints.

Dynamic seals are assembled between mating parts that move relative to each other. Hence, dynamic seals are subject to wear because one of the mating parts rubs against the seal. The following represent the most widely used types of seal configurations:

1. O-rings
2. Compression packings (V and U shapes)

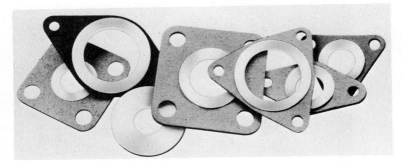

Figure 13-3.   Die cut gaskets used for flanged joints. (*Courtesy of Crane Packing Co., Morton Grove, Illinois.*)

3. Piston cup packings

4. Piston rings

5. Wiper rings

The O-ring is one of the most widely used seals for hydraulic systems. It is a molded, synthetic rubber seal that has a round cross section in its free state. See Fig. 13-4 for several different-sized O-rings, which can be used for most static and dynamic conditions. These O-ring seals give effective sealing through a wide range of pressures, temperatures, and movements with the added advantages of sealing pressure in both directions and low running friction on moving parts.

As illustrated in Fig. 13-5, an O-ring is installed in an annular groove machined into one of the mating parts. When it is initially installed, it is compressed at both its inside and outside diameters. When pressure is applied, the O-ring is forced against a third surface to create a positive seal. The applied pressure also forces the O-ring to push even harder against the surfaces in contact with its inside and outside diameters. As a result, the O-ring is capable of sealing against high pressures. However, O-rings are not generally suited for sealing rotating shafts or where vibration is a problem.

At very high pressures, the O-ring may extrude into the clearance space between mating parts, as illustrated in Fig. 13-6. This is unacceptable in a dynamic application because of the rapid resulting seal wear. This extrusion is prevented by installing a backup ring, as shown in Fig. 13-6. If the pressure is applied in both directions, a backup ring must be installed on both sides of the O-ring.

V-ring packings are compression-type seals that are used in virtually all types of reciprocating motion applications. These include rod and piston seals in hydraulic and pneumatic cylinders, press rams, jacks, and seals on plungers and pistons in reciprocating pumps. They are also readily suited to certain slow rotary applications such as valve stems. These packings (which can be molded into U

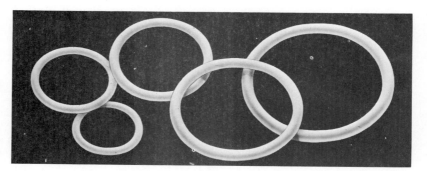

Figure 13-4.   Several different-sized O-rings. (*Courtesy of Crane Packing Co., Morton Grove, Illinois.*)

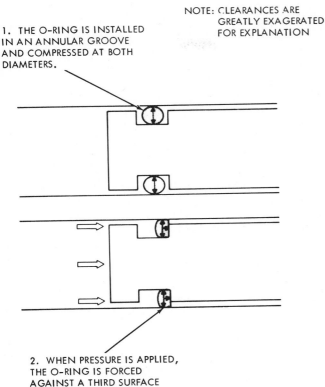

1. THE O-RING IS INSTALLED
IN AN ANNULAR GROOVE
AND COMPRESSED AT BOTH
DIAMETERS.

NOTE: CLEARANCES ARE
GREATLY EXAGERATED
FOR EXPLANATION

2. WHEN PRESSURE IS APPLIED,
THE O-RING IS FORCED
AGAINST A THIRD SURFACE
CREATING A POSITIVE SEAL.

Figure 13-5.   Operation of O-ring.
(*Courtesy of Sperry Vickers, Sperry Rand Corp., Troy, Michigan.*)

shapes as well as V shapes) are frequently installed in multiple quantities for more effective sealing. As illustrated in Fig. 13-7, these packings are compressed by tightening a flanged follower ring against them. Proper adjustment is essential since excessive tightening will hasten wear.

In many applications these packings are spring-loaded to control the correct force as wear takes place. However, springs are not recommended for high speed or quick reverse motion on reciprocating applications. Figure 13-8(a) shows several different-sized V-ring packings, whereas Fig. 13-8(b) shows two different-sized sets of V-ring packings stacked together.

Piston cup packings are designed specifically for pistons in reciprocating pumps and pneumatic and hydraulic cylinders. They offer the best service life for this type of application, require a minimum recess space and minimum recess machining, and are simply and quickly installed. Figure 13-9 shows the typical installation for single-acting and double-acting operations. Sealing is accomplished when pressure pushes the cup lip outward against the cylinder barrel. The backing plate and retainers clamp the cup packing tightly in place, allowing it to

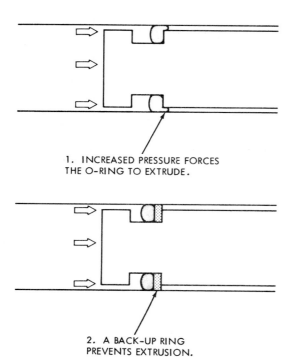

NOTE: CLEARANCES ARE GREATLY
EXAGERATED FOR
EXPLANATION.

1. INCREASED PRESSURE FORCES
THE O-RING TO EXTRUDE.

2. A BACK-UP RING
PREVENTS EXTRUSION.

Figure 13-6.   Backup ring prevents extrusion of O-ring. (*Courtesy of Sperry Vickers, Sperry Rand Corp., Troy, Michigan.*)

handle very high pressures. Figure 13-10 shows several different-sized piston cup packings.

Piston rings are seals that are universally used for cylinder pistons, as shown in Fig. 13-11. Metallic piston rings are made of cast iron or steel and are usually plated or given an outer coating of materials such as zinc phosphate or manganese phosphate to prevent rusting and corrosion. Piston rings offer substantially less opposition to motion than do synthetic rubber (elastomer) seals. Sealing against high pressures is readily handled if several rings are used, as illustrated in Fig. 13-11.

Figure 13-12 shows a number of nonmetallic piston rings made out of tetra-fluoroethylene (TFE), a chemically inert, tough, waxy solid. Their extremely low coefficient of friction (0.04) permits them to be run completely dry and at the same time prevents scoring of the cylinder walls. This type of piston ring is an ideal solution to many applications where the presence of lubrication can be detrimen-

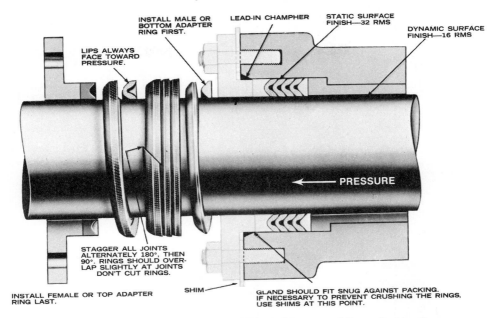

Figure 13-7. Application of V-ring packings. (*Courtesy of Crane Packing Co., Morton Grove, Illinois.*)

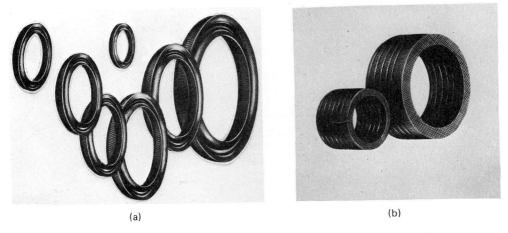

(a)   (b)

Figure 13-8. V-ring packings. (a) Several different sizes, (b) two sizes in stacked arrangement. (*Courtesy of Crane Packing Co., Morton Grove, Illinois.*)

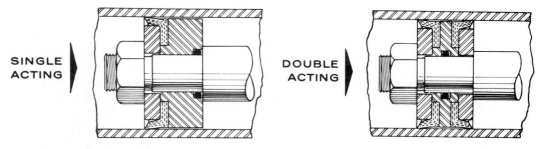

SINGLE ACTING

DOUBLE ACTING

Figure 13-9. Typical applications of piston cup packings. (*Courtesy of Crane Packing Co., Morton Grove, Illinois.*)

Figure 13-10. Several different-sized piston cup packings. (*Courtesy of Crane Packing Co., Morton Grove, Illinois.*)

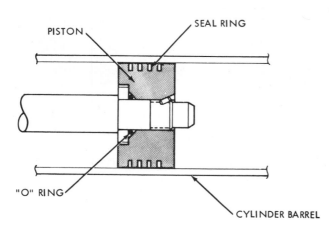

PISTON

SEAL RING

"O" RING

CYLINDER BARREL

Figure 13-11. Use of piston rings for cylinder pistons. (*Courtesy of Sperry Vickers, Sperry Rand Corp., Troy, Michigan.*)

Figure 13-12.   TFE nonmetallic piston rings. (*Courtesy of Crane Packing Co., Morton Grove, Illinois.*)

tal or even dangerous. For instance, in an oxygen compressor, just a trace of oil is a fire or explosion hazard.

Figure 13-13(a) shows the four standard types of TFE piston ring configurations and their applications: endless, butt-cut, level-cut, and step-cut. In Fig. 13-13(b) we see a new flexicut expandable ring design. They install as easily as split rings yet seal completely as endless rings. They have been developed especially for applications where difficulties may be encountered in the installation of standard endless rings. In addition, the flexicut rings can be easily installed in cylinder grooves, which is not possible with standard endless rings. The plain-cut ring is recommended for use without an expander; the lock-cut is for use with an expander.

Wiper rings are seals designed to prevent foreign abrasive or corrosive material from entering a cylinder. They are not designed to seal against pressure. They provide insurance against rod scoring and add materially to packing life. Figure 13-14(a) shows several different-sized wiper rings, whereas Fig. 13-14(b) shows a typical installation arrangement. The wiper ring is molded from a synthetic rubber, which is stiff enough to wipe all dust or dirt from the rod yet pliable enough to maintain a snug fit. The rings are easily installed with a snap fit into a machined groove in the gland. This eliminates the need and expense of a separate retainer ring.

Natural rubber is rarely used as a seal material because it swells and deterio-

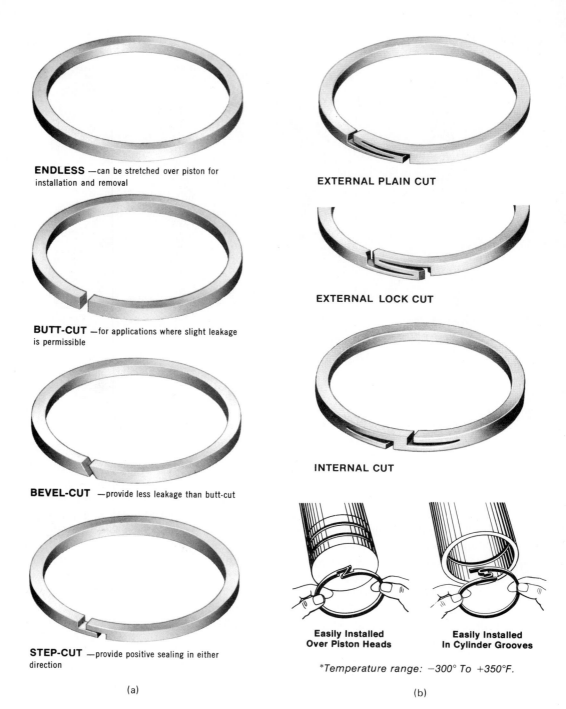

ENDLESS —can be stretched over piston for installation and removal

EXTERNAL PLAIN CUT

BUTT-CUT —for applications where slight leakage is permissible

EXTERNAL LOCK CUT

BEVEL-CUT —provide less leakage than butt-cut

INTERNAL CUT

STEP-CUT —provide positive sealing in either direction

Easily Installed Over Piston Heads

Easily Installed In Cylinder Grooves

*Temperature range: −300° To +350°F.

(a)

(b)

Figure 13-13. TFE piston ring configurations. (a) Four standing types of TFE piston rings, (b) new flexicut design. (*Courtesy of Crane Packing Co., Morton Grove, Illinois.*)

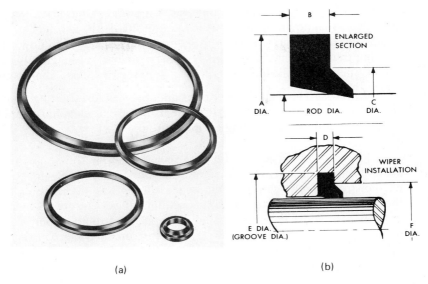

Figure 13-14. Wiper rings. (a) Various sizes, (b) installation arrangement. (*Courtesy of Crane Packing Co., Morton Grove, Illinois.*)

rates with time in the presence of oil. In contrast, synthetic rubber materials are compatible with most oils. The most common types of materials used for seals are leather, Buna-N, silicone, neoprene, tetrafluoroethylene, viton, and, of course, metals.

1. *Leather:* This material is rugged and inexpensive. However, it tends to squeal when dry and cannot operate above 200°F, which is inadequate for many hydraulic systems. Leather does operate well at cold temperatures to about −60°F.

2. *Buna-N:* This material is rugged and inexpensive and wears well. It has a rather wide operating temperature range (−50°F to 230°F) during which it maintains its good sealing characteristics.

3. *Silicone:* This elastomer has an extremely wide temperature range (−90°F to 450°F). Hence, it is widely used for rotating shaft seals and static seals where a wide operating temperature is expected. Silicone is not used for reciprocating seal applications because it has low tear resistance.

4. *Neoprene:* This material has a temperature range of −65°F to 250°F. It is unsuitable above 250°F because it has a tendency to vulcanize.

5. *Tetrafluoroethylene:* This material is the most widely used plastic for seals of hydraulic systems. It is a tough, chemically inert, waxy solid, which can be processed only by compacting and sintering. It has excellent resistance to chemical breakdown up to temperatures of 700°F. It also has an extremely low coefficient of friction. One major drawback is its tendency to flow under pressure, forming thin, feathery films. This tendency to flow can be greatly reduced by the use of filler materials such as graphite, metal wires, glass fibers, and asbestos.

6. *Viton:* This material contains about 65% fluorine. It has become almost a standard material for elastomer-type seals for use at elevated temperatures up to 500°F. Its minimum operating temperature is −20°F.

Physical properties frequently used to describe the behavior of elastomers are as follows: hardness, coefficient of friction, volume change, compression set, tensile strength, elongation modulus, tear strength, squeeze stretch, coefficient of thermal expansion, and permeability. Among these physical properties, hardness is among the most important since it has a direct relationship to service performance. A durometer (see Fig. 13-15) is an instrument used to measure the indentation hardness of rubber and rubberlike materials. As shown, the hardness scale has a range from 0 to 100. The durometer measures 100 when pressed firmly on flat

Figure 13-15. Durometer. (*Courtesy of Shore Instruments & Mfg. Co., Jamaica, New York.*)

glass. High durometer readings indicate a great resistance to denting and thus a hard material. A durometer hardness of 70 is the most common value for seal materials. A hardness of 80 is usually specified for rotating motion to eliminate the tendency toward side motion and bunching in the groove. Values between 50 and 60 are used for static seals on rough surfaces. Hard seal materials (values between 80 and 90) have less breakaway friction than softer materials, which have a greater tendency to deform and flow into surface irregularities. As a result, harder materials are used for dynamic seals.

13.3 THE RESERVOIR SYSTEM

The proper design of a suitable reservoir for a hydraulic system is essential to the overall performance and life of the individual components. The reservoir not only serves as a storage space for the hydraulic fluid used by the system but also as the principal location where the fluid is conditioned. The reservoir is where sludge, water, and metal chips settle and where entrained air picked up by the oil is allowed to escape. The dissipation of heat is also accomplished by a properly designed reservoir.

Figure 13-16 illustrates the construction features of a reservoir that satisfies industry's standards. This reservoir is constructed of welded steel plates. The inside surfaces are painted with a sealer to prevent rust, which can occur due to condensed moisture. Observe that this reservoir allows for easy fluid maintenance. The bottom plate is dished and contains a drain plug at its lowest point to allow complete drainage of the tank when required. Removable covers are included to provide easy access for cleaning. A sight glass is also included to permit a visual check of the fluid level. A vented breather cap is also included and contains an air filtering screen. This allows the tank to breathe as the oil level changes due to system demand requirements. In this way, the tank is always vented to the atmosphere.

As shown in Fig. 13-17, a baffle plate extends lengthwise across the center of the tank. Its height is about 70% of the height of the oil level. The purpose of the baffle plate is to separate the pump inlet line from the return line to prevent the same fluid from recirculating continuously within the tank. In this way all the fluid is uniformly used by the system.

Essentially the baffle serves the following functions:

1. Permits foreign substances to settle to the bottom
2. Allows entrained air to escape from oil
3. Prevents localized turbulence in reservoir
4. Promotes heat dissipation through reservoir walls

As illustrated in Fig. 13-16, the reservoir is constructed so that the pump and driving motor can be installed on its top surface. A smooth machined surface is

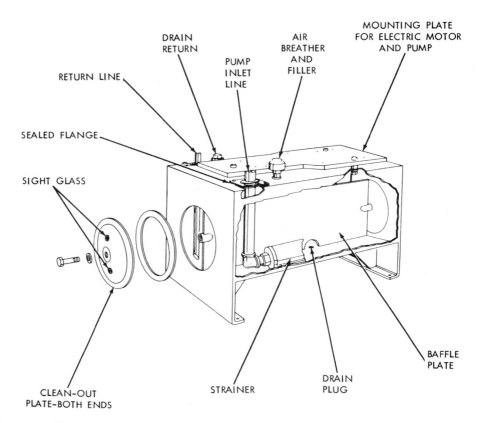

Figure 13-16. Reservoir construction. (*Courtesy of Sperry Vickers, Sperry Rand Corp., Troy, Michigan.*)

provided that has adequate strength to support and hold the alignment of the two units.

The return line should enter the reservoir on the side of the baffle plate that is opposite from the pump suction line. To prevent foaming of the oil, the return line should be installed within two pipe diameters from the bottom of the tank. The pump suction strainer should be well below the normal oil level in the reservoir and at least 1 in. from the bottom. If the strainer is located too high, it will cause a vortex or crater to form, which will permit air to enter the suction line. In addition, suction line connections above the oil level must be tightly sealed to prevent the entrance of air into the system.

The sizing of a reservoir is based on the following criteria:

1. It must make allowance for dirt and chips to settle and for air to escape.

2. It must be able to hold all the oil that might drain into the reservoir from the system.

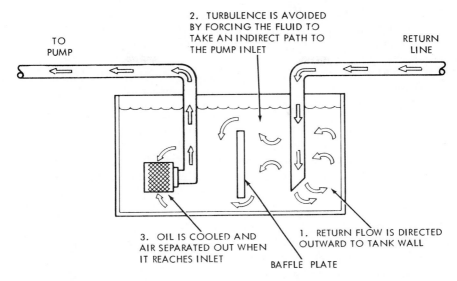

Figure 13-17.   Baffle plate controls direction of flow in reservoir. (*Courtesy of Sperry Vickers, Sperry Rand Corp., Troy, Michigan.*)

3. It must maintain the oil level high enough to prevent a *whirlpool* effect at the pump inlet line opening. Otherwise air will be drawn into the pump.

4. It should have surface area large enough to dissipate most of the heat generated by the system.

5. It should have adequate air space to allow for thermal expansion of the oil.

A reservoir having a capacity of three times the volumetric flow rate of the pump has been found to be adequate for most hydraulic systems where average demands are expected. This relationship is given by

$$\text{reservoir size (gal)} = 3 \times \text{pump flow rate (gpm)} \tag{13-1}$$

Thus, a hydraulic system using a 10-gpm pump would require a 30-gal reservoir. However, the benefits of a large reservoir are usually sacrificed for mobile and aerospace applications due to weight and space limitations.

13.4 *FILTERS AND STRAINERS*

Modern hydraulic systems must be dependable and provide high accuracy. This requires highly precision-machined components. The worst enemy of a precision-made hydraulic component is contamination of the fluid. Essentially, contamination is any foreign material in the fluid that results in detrimental operation of any component of the system. Contamination may be in the form of a liquid, gas, or solid and can be caused by any of the following:

1. *Built into system during component maintenance and assembly:* Contaminants here include metal chips, bits of pipe threads, tubing burrs, pipe dope, shreds of plastic tape, bits of seal material, welding beads, bits of hose, and dirt.

2. *Generated within system during operation:* During the operation of a hydraulic system, many sources of contamination exist. They include moisture due to water condensation inside the reservoir, entrained gases, scale caused by rust, bits of worn seal materials, particles of metal due to wear, and sludges and varnishes due to oxidation of the oil.

3. *Introduced into system from external environment:* The main source of contamination here is due to the use of dirty maintenance equipment such as funnels, rags, and tools. Disassembled components should be washed using clean hydraulic fluid before assembly. Any oil added to the system should be free of contaminants and poured from clean containers.

As indicated in Sec. 13-3, reservoirs help to keep the hydraulic fluid clean. In fact, some reservoirs contain magnetic plugs at their bottom to trap iron and steel particles carried by the fluid (see Fig. 13-18). However, this is not adequate, and in reality the main job of keeping the fluid clean is performed by filters and strainers.

Filters and strainers are devices for trapping contaminants. Specifically, a filter is a device whose primary function is to retain, by some porous medium,

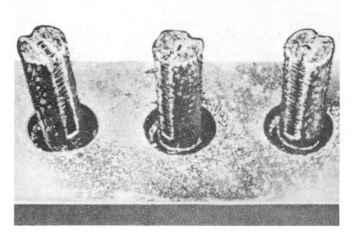

Figure 13-18.   Magnetic plugs trap iron and steel particles. (*Courtesy of Sperry Vickers, Sperry Rand Corp., Troy, Michigan.*)

Figure 13-19. Inlet strainer. (*Courtesy of Sperry Vickers, Sperry Rand Corp., Troy, Michigan.*)

insoluble contaminants from a fluid. Basically, a strainer (see Fig. 13-19) is a coarse filter. Strainers are constructed of a wire screen that rarely contains openings less than 100 mesh (U.S. Sieve No.). The screen is wrapped around a metal frame. As shown in Fig. 13-20, a 100-mesh screen has openings of 0.0059 in., and thus a strainer removes only the larger particles. Observe that the lower the mesh number, the coarser the screen. Because strainers have low-pressure drops, they are usually installed in the pump suction line to remove contaminants large enough to damage the pump. A pressure gage is normally installed in the suction line between the pump and strainer to indicate the condition of the strainer. A drop in pressure indicates that the strainer is becoming clogged. This can starve the pump, resulting in cavitation and increased pump noise.

A filter can consist of materials in addition to a screen. Particle sizes removed by filters are measured in microns. As illustrated in Fig. 13-20, 1 micron is one-millionth of a meter or 0.000039 in. The smallest-sized particle that can normally be removed by a strainer is 0.0059 in. or approximately 150 microns (M). On the other hand, filters can remove particles as small as 1 M. Studies have shown that particles as small as 1 M can have a damaging effect on hydraulic systems (especially servo systems) and can also accelerate oil deterioration. Figure 13-20 also gives the relative sizes of micronic particles magnified 500 times. Another way to visualize the size of a micron is to note the following comparisons:

A grain of salt has a diameter of about 100 M.

A human hair has a diameter of about 70 M.

The lower limit of visibility is about 40 M.

One-thousandth of an inch equals about 25 M.

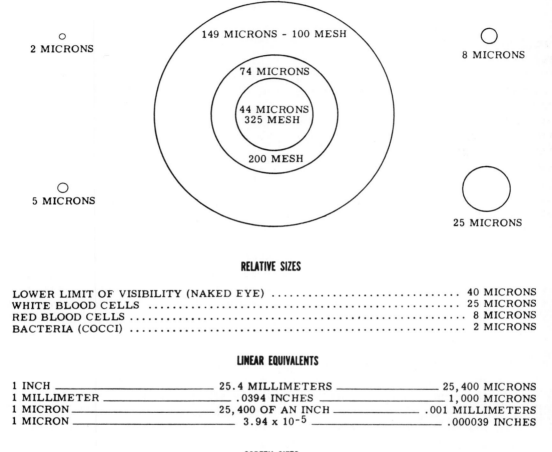

RELATIVE SIZE OF MICRONIC PARTICLES

MAGNIFICATION 500 TIMES

RELATIVE SIZES

LOWER LIMIT OF VISIBILITY (NAKED EYE) .... 40 MICRONS
WHITE BLOOD CELLS .... 25 MICRONS
RED BLOOD CELLS .... 8 MICRONS
BACTERIA (COCCI) .... 2 MICRONS

LINEAR EQUIVALENTS

1 INCH	25.4 MILLIMETERS		25,400 MICRONS
1 MILLIMETER	.0394 INCHES		1,000 MICRONS
1 MICRON	25,400 OF AN INCH		.001 MILLIMETERS
1 MICRON	3.94×10^{-5}		.000039 INCHES

SCREEN SIZES

MESHES PER LINEAR INCH	U.S. SIEVE NO.	OPENING IN INCHES	OPENING IN MICRONS
52.36	50	.0117	297
72.45	70	.0083	210
101.01	100	.0059	149
142.86	140	.0041	105
200.00	200	.0029	74
270.26	270	.0021	53
323.00	325	.0017	44
		.00039	10
		.000019	.5

Figure 13-20. Relative and absolute sizes of micronic particles. (*Courtesy of Sperry Vickers, Sperry Rand Corp., Troy, Michigan.*)

There are three basic types of filtering methods used in hydraulic systems: mechanical, absorbent, and adsorbent.

1. *Mechanical:* This type normally contains a metal or cloth screen or a series of metal disks separated by thin spacers. Mechanical-type filters are capable of removing only relatively coarse particles from the fluid.

2. *Absorbent:* These filters are porous and permeable materials such as paper, wood pulp, diatomaceous earth, cloth, cellulose, and asbestos. Paper filters are normally impregnated with a resin to provide added strength. In this type of filter, the particles are actually absorbed as the fluid permeates the material. As a result, these filters are used for extremely small particle filtration.

3. *Adsorbent:* Adsorption is a surface phenomenon and refers to the tendency of particles to cling to the surface of the filter. Thus, the capacity of such a filter depends on the amount of surface area available. Adsorbent materials used include activated clay and chemically treated paper. Charcoal and Fuller's earth should not be used because they remove some of the essential additives from the hydraulic fluid.

Some filters are designed to be installed in the pressure line and normally are used in systems where high-pressure components such as valves are more dirt sensitive than the pump. Pressure line filters are accordingly designed to sustain system operating pressures. Return line filters are used in systems that do not have a very large reservoir to permit contaminants to settle to the bottom. A return line filter is needed in systems containing close tolerance, high-performance pumps, because inlet line filters, which have limited pressure drop allowance, cannot filter out extremely small particles in the 1- to 5-M range.

Figure 13-21(a) shows a versatile filter that can be directly welded into reservoirs for suction or return line installations or mounted into the piping for pressure line applications to 150 psi. This filter can remove particles as small as 3 M. It also contains an indicating element that signals the operator when cleaning is required. The operation of this Tell Tale filter is dependent on fluid passing through a porous filter media, which traps contamination. The Tell Tale indicator monitors the pressure differential buildup due to dirt, reporting the condition of the filter element. The filter works as follows [refer to Fig. 13-21(b)].

Fluid enters the inlet at the bottom of the tube, passes from inside to outside through the filter element, and exits at the side outlet. If the magnet option has been selected, iron and steel particles will be attracted. With a new or recently serviced filter element in the housing, the Tell Tale indicator shows "clean."

As dirt is deposited on the element's surface, pressure differential from inside to outside the element rises. This difference in pressure is sensed at the bypass piston. The piston is held seated by a spring. But when the element requires cleaning, pressure differential is high enough to compress the spring,

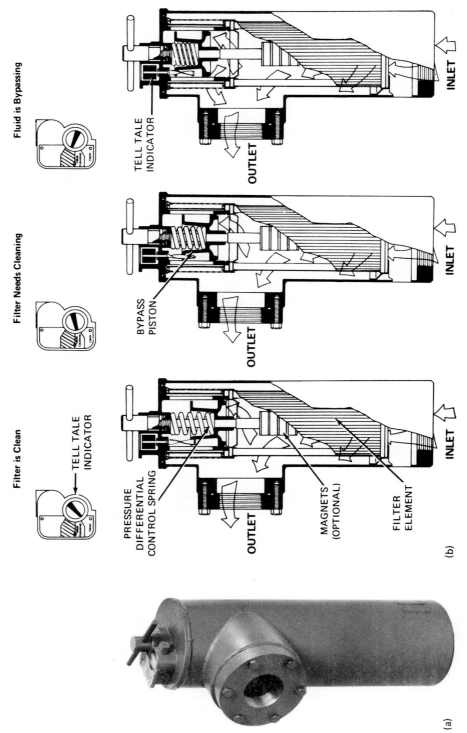

Filter is Clean

TELL TALE INDICATOR

PRESSURE DIFFERENTIAL CONTROL SPRING

MAGNETS (OPTIONAL)

FILTER ELEMENT

OUTLET

INLET

Filter Needs Cleaning

BYPASS PISTON

OUTLET

INLET

Fluid is Bypassing

TELL TALE INDICATOR

OUTLET

INLET

(b)

(a)

Figure 13-21. Tell Tale filter. (*Courtesy of Parker Hannifin Corp., Hazel Park, Michigan.*)

forcing the piston off its seat. This piston movement causes the Tell Tale indicator to point to the ''needs cleaning'' position.

If the filter element is not serviced when indicated, pressure differential will continue to increase, causing the piston to uncover a bypass passage in the cover. This action limits the rise in pressure differential to a value equal to the rating of the spring. During this condition the indicator will be pointing to the ''red'' region. If the no-bypass feature is ordered, the indicator still functions as before.

In Fig. 13-22, we see a cutaway view of a Tell Tale filter that filters particles from 0.062 in. to 3 M. It can handle flow rates to 700 gpm and pressures from suction to 300 psi. This filter is available with or without bypass relief valves. The optional bypass assures continued flow no matter how dirt-clogged the filter might become. The bypass valves can be set to provide the maximum allowable pressure drop to match system requirements.

Figure 13-23 shows the four typical locations where filters are installed in hydraulic circuits. Figure 13-23(a) shows the location for a proportional flow filter. As the name implies, proportional flow filters are exposed to only a percentage of the total system flow during operation. It is assumed that on a recirculating basis the probability of mixture of the fluid within the system will force all the fluid through the proportional filter. The primary disadvantages of proportional flow filtration are that there is no positive protection of any specific components within the system, and there is no way to know when the filter is dirty. Figures 13-23(b), (c), and (d) show the full flow filtration filters, which accept all the flow of the pumps. Figure 13-23(b) shows the location on the suction side of the pump, whereas Fig. 13-23(c) and (d) show the filter on the pressure side of the pump and

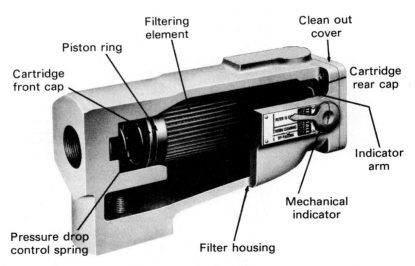

Figure 13-22. Cutaway view of a Tell-Tale filter. (*Courtesy of Parker Hannifin Corp., Hazel Park, Michigan.*)

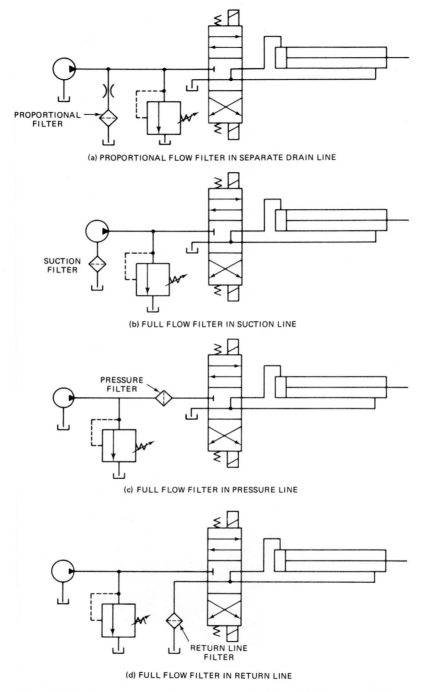

(a) PROPORTIONAL FLOW FILTER IN SEPARATE DRAIN LINE

PROPORTIONAL FILTER

(b) FULL FLOW FILTER IN SUCTION LINE

SUCTION FILTER

(c) FULL FLOW FILTER IN PRESSURE LINE

PRESSURE FILTER

(d) FULL FLOW FILTER IN RETURN LINE

RETURN LINE FILTER

Figure 13-23. Four common circuit locations for filters. (a) Proportional flow filter in separate drain line, (b) full flow filter in suction line, (c) full flow filter in pressure line, (d) full flow filter in return line.

in the return line, respectively. In general, there is no best single place to put a filter. The basic rule of thumb is the following: Consider where the dirt enters the system, and put the filter/filters where they do the most good.

13.5 BETA RATIO OF FILTERS

Filters are rated according to the smallest size of particles that they can trap. Filter ratings used to be identified by nominal and absolute values in microns. A filter with a nominal rating of 10 microns is supposed to trap 95% of the entering particles greater than 10 microns in size. The absolute rating represents the size of the largest opening or pore in the filter and thus indicates the largest size particle that could pass through the filter. Hence the absolute rating of a 10 micron nominal size filter would be larger than 10 microns.

A better parameter for establishing how well a filter traps particles is called the Beta ratio or Beta rating. The Beta ratio is determined during laboratory testing of a filter receiving a specified steady state flow containing a fine dust of selected particle size. The test begins with a clean filter and ends when the pressure drop across the filter reaches a specified value indicating that the filter has reached the saturation point. This is when the contaminant capacity has been reached, which is a measure of the service life or acceptable time interval between filter element changes in an actual operating system.

By mathematical definition, the Beta ratio equals the number of upstream particles of greater size than N microns divided by the number of downstream particles greater in size than N microns (as counted during the test) where N is the selected particle size for the given filter. This ratio is represented by the following equation:

$$\text{Beta ratio} = \frac{\text{no. upstream particles of size greater than } N \text{ microns}}{\text{no. downstream particles of size greater than } N \text{ microns}} \qquad (13\text{-}2)$$

A Beta ratio of one would mean that no particles above the specified size N are trapped by the filter. A Beta ratio of 50 means that 50 particles are trapped for every one that gets through the filter. Most filters have Beta ratings greater than 75 when N equals the absolute rating. A filter efficiency value can be calculated using the following equation:

$$\text{Beta efficiency} = \frac{\text{no. upstream particles} - \text{no. downstream particles}}{\text{no. upstream particles}} \qquad (13\text{-}3)$$

where the particle size is greater than a specified value of N microns.

Thus we have the following relationship between Beta efficiency and Beta ratio:

$$\text{Beta efficiency} = 1 - \frac{1}{\text{Beta ratio}} \qquad (13\text{-}4)$$

Hence a filter with a Beta ratio of 50 would have an efficiency of $1 - 1/50$ or 98%. Note from Eq. (13-4), that the higher the Beta ratio the higher the Beta efficiency. The designation $B_{20} = 50$ identifies a particle size of 20 microns and a Beta ratio of 50 for a particular filter. Thus a designation of $B_{20} = 50$ means that 98% of the particles larger than 20 microns would be trapped by the filter during the time a clean filter becomes saturated.

13.6 *TEMPERATURE CONTROL*

Heat is generated in hydraulic systems because no component can operate at 100% efficiency. Significant sources of heat include the pump, pressure relief valves, and flow control valves. This can cause the hydraulic fluid temperature to exceed its normal operating range of 110°F to 150°F. Excessive temperature hastens oxidation of the hydraulic oil and causes it to become too thin. This promotes deterioration of seals and packings and accelerates wear between closely fitting parts of hydraulic components of valves, pumps, and actuators.

The steady-state temperature of the fluid of a hydraulic system depends on the heat generation rate and the heat dissipation rate of the system. If the fluid operating temperature in a hydraulic system becomes excessive, this means that the heat generation rate is too large relative to the heat dissipation rate. Assuming that the system is reasonably efficient, the solution is to increase the heat dissipation rate. This is accomplished by the use of "*coolers*," which are commonly called *heat exchangers*. In some applications, the fluid must be heated to produce a satisfactory value of viscosity. This is typical when, for example, mobile hydraulic equipment is to operate in below-0°F temperatures. In these cases, the heat exchangers are called *heaters*. However, for most hydraulic systems, the natural heat generation rate is sufficient to produce high enough temperatures after an initial warm-up period. Hence, the problem usually becomes one of using a heat exchanger to provide adequate cooling.

There are two main types of heat dissipation heat exchangers: air coolers and water coolers. Figure 13-24 shows an air cooler in which the hydraulic fluid is pumped through tubes banded to fins. It can handle oil flow rates up to 200 gpm and employs a fan to increase the heat transfer rate.

The air cooler shown uses tubes that contain special devices—turbulators to mix the warmer and cooler oil for better heat transfer—because the oil near the center of the tube is warmer than that near the wall. Light, hollow, metal spheres are randomly inserted inside the tubes. This causes the oil to tumble over itself to provide thorough mixing to produce a lighter and better cooler.

In Fig. 13-25 we see a water cooler. In this type heat exchanger water is circulated through the unit by flowing around the tubes, which contain the hydraulic fluid. The design shown has a tough ductile, red-brass shell; unique flanged, yellow-brass baffles; seamless nonferrous tubes; and cast-iron bonnets. It provides heat transfer surface areas up to 124 ft^2.

Figure 13-24.   Air-cooled heat exchanger. (*Courtesy of American Standard, Heat Transfer Division, Buffalo,
New York.*)

The following equations permit the calculation of the fluid temperature rise
as it flows through a restriction such as a pressure relief valve:

$$\text{temperature increase (°F)} = \frac{\text{heat generation rate (Btu/min)}}{\text{oil specific heat (Btu/lb/°F)} \times \text{oil flow rate (lb/min)}} \quad (13\text{-}5)$$

$$1 \text{ hp} = 42.4 \text{ Btu/min} = 2544 \text{ Btu/hr} \quad (13\text{-}6)$$

$$\text{specific heat of oil} = 0.42 \text{ Btu/lb/°F} \quad (13\text{-}7)$$

$$\text{oil flow rate (lb/min)} = 7.42 \times \text{oil flow rate (gpm)} \quad (13\text{-}8)$$

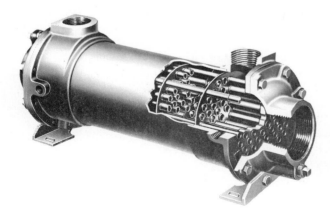

Figure 13-25.   Water-cooled shell and
tube heat exchangers. (*Courtesy of
American Standard, Heat Transfer
Division, Buffalo, New York.*)

EXAMPLE 13-1

Oil at 120°F and 1000 psi is flowing through a pressure relief valve at 10 gpm. What is the downstream oil temperature?

Solution   First calculate the horsepower lost and convert to the heat generation rate in units of Btu/min:

$$HP = \frac{P \text{ (psi)} \cdot Q \text{ (gpm)}}{1714} = \frac{(1000)(10)}{1714} = 5.83$$

$$\text{Btu/min} = 5.83 \times 42.4 = 247$$

Next calculate the oil flow rate in units of lb/min and the temperature increase:

$$\text{oil flow rate} = 7.42(10) = 74.2 \text{ lb/min}$$

$$\text{temperature increase} = \frac{247}{0.42 \times 74.2} = 7.9°F$$

$$\text{downstream oil temperature} = 120 + 7.9 = 127.9°F$$

When sizing heat exchangers, a heat load value is calculated for the entire system in units of Btu/hr. The calculation of the system heat load can be readily calculated by noting that 1 hp equals 2544 Btu/hr.

EXAMPLE 13-2

A hydraulic pump operates at 1000 psi and delivers oil at 20 gpm to a hydraulic actuator. Oil discharges through the pressure relief valve (PRV) during 50% of the cycle time. The pump has an overall efficiency of 85%, and 10% of the power is lost due to frictional pressure losses in the hydraulic lines. What rating heat exchanger is required to dissipate all the generated heat?

Solution

$$\text{pump } HP \text{ loss} = 0.15 \times \frac{1000 \times 20}{1714} = 1.75$$

$$\text{PRV average } HP \text{ loss} = 0.50 \times \frac{1000 \times 20}{1714} = 5.83$$

$$\text{line average } HP \text{ loss} = 0.50 \times 0.10 \times \frac{1000 \times 20}{1714} = 0.58$$

$$\text{total average } HP \text{ loss} = 8.16$$

$$\text{heat exchanger rating} = 8.16 \times 2544 = 20,759 \text{ Btu/hr}$$

The following metric units equations permit the calculation of the fluid temperature rise as it flows through a restriction such as a pressure relief valve.

$$\text{temperature increase (°C)} = \frac{\text{heat generation rate (kW)}}{\text{oil specific heat (kJ/kg°C)} \times \text{oil flow rate (kg/s)}} \tag{13-9}$$

$$\text{specific heat of oil} = 1.8 \text{ kJ/kg°C} \tag{13-10}$$

$$\text{oil flow rate (kg/s)} = 895 \times \text{oil flow rate (m}^3\text{/s)} \tag{13-11}$$

EXAMPLE 13-3

Oil at 50°C and 70 bars is flowing through a pressure relief valve at 0.000632 m³/s. What is the downstream oil temperature?

Solution   First, calculate the heat generation rate in units of kW.

$$\text{kW} = \frac{P(Pa)\ Q(\text{m}^3\text{/s})}{1000} = \frac{(7 \times 10^6)(632 \times 10^{-6})}{1000} = 4.42 \text{ kW}$$

Next, calculate the oil flow rate in units of kg/s and the temperature in units of °C.

$$\text{oil flow rate} = (895)(0.000632) = 0.566 \text{ kg/s}$$

$$\text{temperature increase} = \frac{4.42}{(1.8)(0.566)} = 4.3°C$$

$$\text{downstream oil temperature} = 50 + 4.3 = 54.3°C$$

EXAMPLE 13-4

A hydraulic pump operates at 70 bars and delivers oil at 0.00126 m³/s to a hydraulic actuator. Oil discharges through the pressure relief valve (PRV) during 50% of the cycle time. The pump has an overall efficiency of 85% and 10% of the power is lost due to frictional pressure losses in the hydraulic lines. What rating heat exchanger is required to dissipate all the generated heat?

Solution

$$\text{pump kW loss} = 0.15 \times \frac{7 \times 10^6 \times 1260 \times 10^{-6}}{1000} = 1.32$$

$$\text{PRV average kW loss} = 0.50 \times \frac{7 \times 10^6 \times 1260 \times 10^{-6}}{1000} = 4.41$$

$$\text{line average kW loss} = 0.50 \times 0.10 \times \frac{7 \times 10^6 \times 1260 \times 10^{-6}}{1000} = 0.44$$

$$\text{total kW loss} = 6.17$$

$$\text{heat exchanger rating} = 6.17 \text{ kW}$$

13.7 TROUBLESHOOTING FLUID POWER CIRCUITS

Hydraulic systems depend on proper flow and pressure from the pump to provide the necessary actuator motion for producing useful work. Therefore, flow and pressure measurements are two important means of troubleshooting faulty operating hydraulic circuits. Temperature is a third parameter, which is frequently monitored when troubleshooting hydraulic systems because it affects the viscosity of the oil. Viscosity, in turn, affects leakage, pressure drops, and lubrication.

The use of flowmeters can tell whether or not the pump is producing proper flow. Flowmeters can also indicate whether or not a particular actuator is receiving the expected flow rate. Figure 13-26 shows a flowmeter that can be installed permanently or used for hydraulic system checkout or troubleshooting in pressure lines to 3000 psi.

This direct-reading flowmeter monitors fluid flow rates to determine pump performance, flow regulator settings, or hydraulic system performance. It is intended for use in mobile and industrial oil hydraulic circuits. It can be applied to pressure lines, return lines, or drain lines. A moving indicator on the meter provides direct readings and eliminates any need for electrical connections on readout devices. These flowmeters are available for capacities up to 100 gpm and also can be calibrated to read pump rpm or fluid velocity in ft/sec when connected in a pipe of known diameter.

Pressure measurements can provide a good indication of leakage problems and faulty components such as pumps, flow control valves, pressure relief valves, strainers, and actuators. Excessive pressure drops in pipelines can also be detected by the use of pressure measurements. The Bourdon gage is the most com-

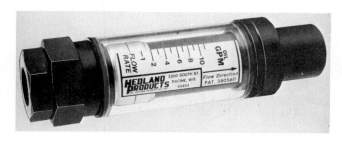

Figure 13-26. In-line flowmeter. (*Courtesy of Heland Products, Racine, Wisconsin.*)

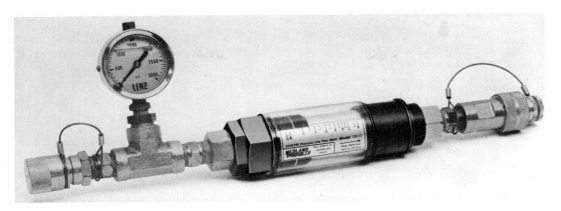

Figure 13-27.   Flow-pressure test kit. (*Courtesy of Heland Products, Racine, Wisconsin.*)

monly used type of pressure-measuring device. This type of gage can be used to measure vacuum pressures in suction lines as well as absolute pressures anywhere in a hydraulic circuit.

Figure 13-27 shows a combination flow-pressure test kit. This unit measures both pressure (using a Bourdon gage) as well as flow rate and can be quickly installed in a hydraulic line because it uses quick couplers at each end.

A portable hydraulic circuit tester (all components are built into a convenient-to-carry container) is shown in Fig. 13-28. This unit not only measures

Figure 13-28.   Portable hydraulic circuit tester. (*Courtesy of Schroeder Brothers Corp., McKees Rocks, Pennsylvania.*)

pressure and flow rate but also temperature. By connecting this tester to the hydraulic circuit as shown in Fig. 13-29, a visual means is provided to determine the efficiency of the system and to determine which component in the system, if any, is not working properly. Testing a hydraulic system with this tester consists of the following:

1. Measure pump flow at no-load conditions.
2. Apply desired pressure with the tester load valve on each component to find out how much of the fluid is not available for power because it may be
   a. Flowing at a lower rate because of slippage inside the pump due to worn parts.
   b. Flowing over pressure relief valves due to worn seats or weak or improperly set springs.
   c. Leaking past valve spools and seats back into the fluid supply reservoir without having reached the working cylinder or motor.
   d. Leaking past the cylinder packing or motor parts directly into the return line without having produced any useful work.

When troubleshooting hydraulic circuits, it should be kept in mind that a pump produces the flow of a fluid. However, there must be resistance to flow in order to have pressure. The following is a list of hydraulic system operating problems and the corresponding probable causes which should be investigated during troubleshooting:

1. *Noisy pump*
   a. Air entering pump inlet
   b. Misalignment of pump and drive unit
   c. Excessive oil viscosity
   d. Dirty inlet strainer
   e. Chattering relief valve
   f. Damaged pump
   g. Excessive pump speed
   h. Loose or damaged inlet line
2. *Low or erratic pressure*
   a. Air in the fluid
   b. Pressure relief valve set too low
   c. Pressure relief valve not properly seated
   d. Leak in hydraulic line
   e. Defective or worn pump
   f. Defective or worn actuator

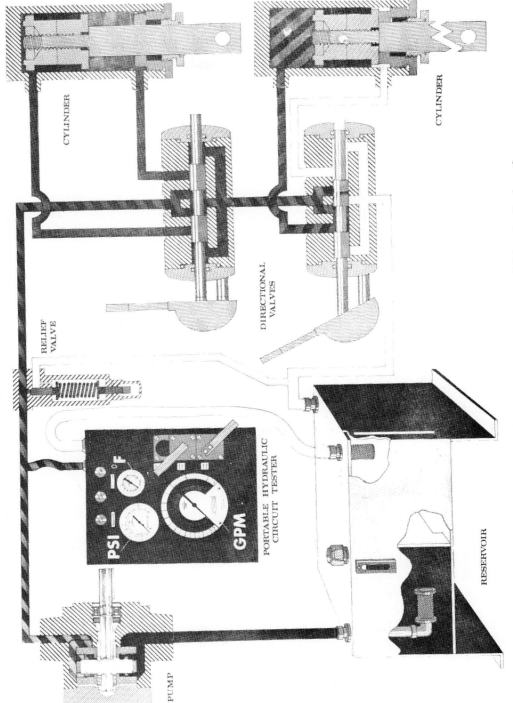

Figure 13-29. Hydraulic circuit "tee test." (*Courtesy of Schroeder Brothers Corp., McKees Rocks, Pennsylvania.*)

CYLINDER

CYLINDER

RELIEF VALVE

DIRECTIONAL VALVES

PORTABLE HYDRAULIC CIRCUIT TESTER

PSI

°F

GPM

PUMP

RESERVOIR

3. *No pressure*
  a. Pump turning in wrong direction
  b. Ruptured hydraulic line
  c. Low oil level in reservoir
  d. Pressure relief valve stuck open
  e. Broken pump shaft
  f. Full pump flow bypassed to tank due to faulty valve or actuator

4. *Actuator fails to move*
  a. Faulty pump
  b. Directional control valve fails to shift
  c. System pressure too low
  d. Defective actuator
  e. Pressure relief valve stuck open
  f. Actuator load is excessive
  g. Check valve in backwards

5. *Slow or erratic motion of actuator*
  a. Air in system
  b. Viscosity of fluid too high
  c. Worn or damaged pump
  d. Pump speed too low
  e. Excessive leakage through actuators or valves
  f. Faulty or dirty flow control valves
  g. Blocked air breather in reservoir
  h. Low fluid level in reservoir
  i. Faulty check valve
  j. Defective pressure relief valve

6. *Overheating of hydraulic fluid*
  a. Heat exchanger turned off or faulty
  b. Undersized components or piping
  c. Incorrect fluid
  d. Continuous operation of pressure relief valve
  e. Overloaded system
  f. Dirty fluid
  g. Reservoir too small
  h. Inadequate supply of oil in reservoir
  i. Excessive pump speed
  j. Clogged or inadequate-sized air breather

13.8 *SAFETY CONSIDERATIONS*

There should be no compromise in safety when hydraulic circuits are designed, operated, and maintained. However, human errors are unavoidable, and accidents can occur, resulting in injury to operating and maintenance personnel. This can be greatly reduced by eliminating all unsafe conditions dealing with the operation and maintenance of the system. Many safe practices have been proven effective in preventing safety hazards, which can be harmful to the health and safety of personnel.

The Occupational Safety and Health Administration (OSHA) of the Department of Labor describes and enforces safety standards at the industry location where hydraulic equipment is operated. For detailed information on OSHA standards and requirements, the reader should request a copy of OSHA publication 2072, General Industry Guide for Applying Safety and Health Standards, 29 CFR 1910. These standards and requirements deal with the following categories:

1. *Work-place standards:* In this category are included the safety of floors, entrance and exit areas, sanitation, and fire protection.

2. *Machines and equipment standards:* Important items are machine guards, inspection and maintenance techniques, safety devices, and the mounting, anchoring, and grounding of fluid power equipment. Of big concern are noise levels produced by operating equipment.

3. *Materials standards:* These standards cover items such as toxic fumes, explosive dust particles, and excessive atmospheric contamination.

4. *Employee standards:* Concerns here include employee training, personnel protective equipment, and medical and first-aid services.

5. *Power source standards:* Standards are applied to power sources such as electrohydraulic, pneumatic, and steam supply systems.

6. *Process standards:* Many industrial processes are included such as welding, spraying, abrasive blasting, part dipping, and machining.

7. *Administrative regulations:* Industry has many administrative responsibilities which it must meet. These include the displaying of OSHA posters stating the rights and responsibilities of both the employer and employees. Industry is also required to keep safety records on accidents, illnesses, and other exposure-type occurrences. An annual summary must also be posted.

It is important that safety be incorporated into hydraulic systems to ensure compliance with OSHA regulations. The basic rule to follow is that there should be no compromise when it comes to the health and safety of people at the place of their employment.

EXERCISES

13-1. Name five of the most common causes of hydraulic system breakdown.

13-2. To what source has over half of all hydraulic system problems been traced?

13-3. Name two items that should be included in reports dealing with maintenance procedures.

13-4. What is the difference between a positive and nonpositive seal?

13-5. Explain, by examples, the difference between internal and external leaks.

13-6. What is the difference between a static and dynamic seal?

13-7. Why are backup rings sometimes used with O-rings?

13-8. Name three types of seals in addition to an O-ring.

13-9. Are wiper seals designed to seal against pressure? Explain your answer.

13-10. Name four types of materials used for seals.

13-11. What is the purpose of a durometer?

13-12. Name four criteria by which the size of a reservoir is determined.

13-13. What size reservoir should be used for a hydraulic system using a 15-gpm pump?

13-14. What is the difference between a filter and strainer?

13-15. Name the three ways in which the hydraulic fluid becomes contaminated.

13-16. What is a 10-micron filter?

13-17. Name the three basic types of filtering methods.

13-18. What is the purpose of an indicating filter?

13-19. Name the four locations where filters are typically installed in hydraulic circuits.

13-20. What is the purpose of a heat exchanger?

13-21. Oil at 130°F and 2000 psi is flowing through a pressure relief valve at 15 gpm. What is the downstream oil temperature?

13-22. A hydraulic pump operates at 2000 psi and delivers oil at 15 gpm to a hydraulic actuator. Oil discharges through the pressure relief valve (PRV) during 60% of the cycle time. The pump has an overall efficiency of 82%, and 15% of the power is lost due to frictional pressure losses in the hydraulic lines. What rating heat exchanger is required to dissipate all the generated heat?

13-23. What three devices are commonly used in the troubleshooting of hydraulic circuits?

13-24. What single most important concept should be understood when troubleshooting hydraulic systems?

13-25. Name five things that can cause a noisy pump.

13-26. Name four causes of low or erratic pressure.

13-27. Name four causes of no pressure.

13-28. If an actuator fails to move, name five possible causes.

13-29. If an actuator has slow or erratic motion, name five possible causes.

13-30. Give six reasons for the overheating of the fluid in a hydraulic system.

13-31. What does OSHA stand for? What is OSHA attempting to accomplish?

13-32. Name and give a brief explanation of the seven categories of safety for which OSHA has established standards.

13-33. Oil at 60°C and 140 bars is flowing through a pressure relief valve at 0.001 m³/s. What is the downstream temperature?

13-34. A hydraulic pump operates at 140 bars and delivers oil at 0.001 m³/s to a hydraulic actuator. Oil discharges through the pressure relief valve (PRV) during 60% of the cycle time. The pump has an overall efficiency of 82%, and 15% of the power is lost due to frictional pressure losses in the hydraulic lines. What rating heat exchanger is required to dissipate all the generated heat?

13-35. Why is loss of pressure in a hydraulic system not a symptom of pump malfunction?

13-36. What are the three most common reservoir designs?

13-37. What is the purpose of the reservoir breather?

13-38. What is a reservoir baffle plate used for?

13-39. What would be an adequate size reservoir for a hydraulic system using a 0.001 m³/s pump?

13-40. What happens when a filter becomes filled with contaminants?

13-41. What factors influence cylinder friction?

13-42. A BTU is the amount of heat required to raise one pound of water one degree Fahrenheit. Derive the conversion factor between BTUs and Joules.

13-43. A hydrostatic transmission that is driven by a 12 HP electric motor delivers 10 HP to the output shaft. Assuming that 75% of the power loss is due to heat loss, calculate the total BTU heat loss over a 5-hour period.

13-44. A hydraulic press with a 12 gpm pump, cycles every 6 minutes. During the 2-minute high-pressure portion of the cycle, oil is dumped over a high-pressure relief valve at 3000 psi. During the 4-minute, low-pressure portion of the cycle, oil is dumped over a low-pressure relief valve at 600 psi. If the travel time of the cylinder is negligibly small, calculate the total BTU heat loss per hour generated by dumping oil through both relief valves.

13-45. A pump delivers oil to a hydraulic motor at 20 liters/minute at a pressure of 15 MPa. If the motor delivers 4 kW and 80% of the power loss is due to internal leakage which heats the oil, calculate the heat generation rate in kJ/min.

13-46. Name the important factors to consider when selecting a heat exchanger.

13-47. Determine the Beta ratio of a filter when during test operation, 30,000 particles greater than 20 microns enter the filter and 1050 of these particles pass through the filter.

13-48. For the filter in Ex. 13-47, what is the Beta efficiency?

13-49. What is the relationship between Beta ratio and Beta efficiency?

13-50. What is meant by the Beta ratio designation of $B_{10} = 75$?

13-51. A Beta ratio of 75 means that 75 particles are trapped for every _____ that gets through the filter.

13-52. What is the difference between nominal and absolute ratings of filters?

Appendixes

A

Properties of Common Liquids (60°F)

Liquid	S_g	$\gamma \ (lb/ft^3)$	$\mu \ (cP)$	$v \ (cS)$
Hydraulic Liquids				
Mineral oil	0.89	55.6	133.4	150
Water-oil emulsion	0.90	56.2	149	166
Water-glycol solution	1.10	68.6	110	100
Phosphate ester	1.10	68.6	220	200
Silicone oil	1.04	64.8	41.6	40
MIL-5606	0.86	53.6	19.1	22
Miscellaneous Liquids				
Castor oil	0.97	60.5	986	1016
Ethyl alcohol	0.79	49.4	1.20	1.51
Ethylene glycol	1.12	69.9	19.9	17.8
Gasoline	0.68	42.5	2.64	3.88
Glycerol	1.26	78.6	1490	1180
Linseed oil	0.94	58.8	65	68.9
Mercury	13.6	849	1.55	0.114
Mineral oil, SAE 10	0.91	56.7	114	125.3
Olive oil	0.92	57.1	84	91.8
Turpentine	0.87	54.3	1.49	1.71
Water, fresh	1.00	62.4	1.00	1.00
Water, salt	1.03	64.0	1.06	1.03

B

Properties of Common Gases (60°F)

Gas	Gas Constant R ($ft \cdot lb/lb \cdot °R$)	γ (lb/ft^3)	$\mu(cP)$
Air	53.3	0.0752	0.0180
Ammonia	89.5	0.0448	0.0108
Carbon dioxide	34.9	0.1146	0.0160
Methane	96.3	0.0416	0.0120
Nitrogen	55.1	0.0726	0.0184
Oxygen	48.3	0.0830	0.0206
Sulfur dioxide	23.6	0.1695	0.0125

C

Weight Density of Air (lb/ft³)

psi °F	0	20	40	60	80	100	120	140
30	0.0811	0.1915	0.302	0.412	0.522	0.633	0.743	0.853
40	0.0795	0.1876	0.295	0.404	0.512	0.620	0.728	0.836
50	0.0782	0.1846	0.291	0.397	0.504	0.610	0.717	0.823
60	0.0764	0.1804	0.248	0.388	0.492	0.596	0.700	0.804
70	0.0750	0.1770	0.279	0.381	0.483	0.585	0.687	0.789
80	0.0736	0.1737	0.274	0.374	0.474	0.574	0.674	0.774
90	0.0722	0.1705	0.269	0.367	0.465	0.564	0.662	0.760
100	0.0709	0.1675	0.264	0.361	0.457	0.554	0.650	0.747
110	0.0697	0.1645	0.259	0.354	0.449	0.544	0.639	0.734
120	0.0685	0.1617	0.255	0.348	0.441	0.535	0.628	0.721
130	0.0673	0.1590	0.251	0.342	0.434	0.525	0.617	0.709
140	0.0662	0.1563	0.246	0.337	0.427	0.517	0.607	0.697
150	0.0651	0.1537	0.242	0.331	0.420	0.508	0.597	0.686
175	0.0626	0.1477	0.233	0.318	0.403	0.488	0.573	0.659
200	0.0602	0.1412	0.224	0.306	0.388	0.470	0.552	0.634
250	0.0559	0.1321	0.208	0.284	0.361	0.437	0.513	0.589
300	0.0523	0.1234	0.195	0.266	0.337	0.408	0.479	0.550

D

Sizes of Steel Pipe

Nominal Pipe Size (*in.*)	Outside Diameter (*in.*)	Inside Diameter (*in.*)	Wall Thickness (*in.*)	Internal Area (*in.*^{2})
		Schedule 40		
$\frac{1}{8}$	0.405	0.269	0.068	0.0568
$\frac{1}{4}$	0.540	0.364	0.088	0.1041
$\frac{3}{8}$	0.675	0.493	0.091	0.1910
$\frac{1}{2}$	0.840	0.622	0.109	0.304
$\frac{3}{4}$	1.050	0.824	0.113	0.533
1	1.315	1.049	0.133	0.864
$1\frac{1}{4}$	1.660	1.380	0.140	1.496
$1\frac{1}{2}$	1.900	1.610	0.145	2.036
2	2.375	2.067	0.154	3.36
$2\frac{1}{2}$	2.875	2.469	0.203	4.79
3	3.500	3.068	0.216	7.39
$3\frac{1}{2}$	4.000	3.548	0.226	9.89
4	4.500	4.026	0.237	12.73
5	5.563	5.047	0.258	20.01
6	6.625	6.065	0.280	28.89
8	8.625	7.981	0.322	50.0
10	10.750	10.020	0.365	78.9
12	12.750	11.938	0.406	111.9

Nominal Pipe Size (in.)	Outside Diameter (in.)	Inside Diameter (in.)	Wall Thickness (in.)	Internal Area (in.²)
Schedule 80				
$\frac{1}{8}$	0.405	0.215	0.095	0.0364
$\frac{1}{4}$	0.540	0.302	0.119	0.0716
$\frac{3}{8}$	0.675	0.423	0.126	0.1405
$\frac{1}{2}$	0.840	0.546	0.147	0.2340
$\frac{3}{4}$	1.050	0.742	0.154	0.432
1	1.315	0.957	0.179	0.719
$1\frac{1}{4}$	1.660	1.278	0.191	1.283
$1\frac{1}{2}$	1.900	1.500	0.200	1.767
2	2.375	1.939	0.218	2.953
$2\frac{1}{2}$	2.875	2.323	0.276	4.24
3	3.500	2.900	0.300	6.61
$3\frac{1}{2}$	4.000	3.364	0.318	8.89
4	4.500	3.826	0.337	11.50
5	5.563	4.813	0.375	18.19
6	6.625	5.761	0.432	26.07
8	8.625	7.625	0.500	45.7
10	10.750	9.564	0.593	71.8
12	12.750	11.376	0.687	101.6

E

Sizes of Steel Tubing

Outside Diameter (in.)	Wall Thickness (in.)	Inside Diameter (in.)	Inside Area (in.²)	Outside Diameter (in.)	Wall Thickness (in.)	Inside Diameter (in.)	Inside Area (in.²)
$\frac{1}{8}$	0.028	0.069	0.00374	$\frac{1}{2}$	0.035	0.430	0.1452
	0.032	0.061	0.00292		0.049	0.402	0.1269
	0.035	0.055	0.00238		0.065	0.370	0.1075
					0.083	0.334	0.0876
$\frac{3}{16}$	0.032	0.1235	0.01198				
	0.035	0.1175	0.01084	$\frac{5}{8}$	0.035	0.555	0.2419
					0.049	0.527	0.2181
$\frac{1}{4}$	0.035	0.180	0.02544		0.065	0.495	0.1924
	0.049	0.152	0.01815		0.083	0.459	0.1655
	0.065	0.120	0.01131				
				$\frac{3}{4}$	0.049	0.652	0.3339
$\frac{5}{16}$	0.035	0.2425	0.04619		0.065	0.620	0.3019
	0.049	0.2145	0.03614		0.083	0.584	0.2679
	0.065	0.1825	0.02616		0.109	0.532	0.2223
				$\frac{7}{8}$	0.049	0.777	0.4742
$\frac{3}{8}$	0.035	0.305	0.07306		0.065	0.745	0.4359
	0.049	0.277	0.06026		0.083	0.709	0.3948
	0.065	0.245	0.04714		0.109	0.657	0.3390

Outside Diameter (in.)	Wall Thickness (in.)	Inside Diameter (in.)	Inside Area (in.^{2})	Outside Diameter (in.)	Wall Thickness (in.)	Inside Diameter (in.)	Inside Area (in.^{2})
1	0.049	0.902	0.6390	$1\frac{3}{4}$	0.065	1.620	2.061
	0.065	0.870	0.5945		0.083	1.584	1.971
	0.083	0.834	0.5463		0.109	1.532	1.843
	0.109	0.782	0.4803		0.134	1.482	1.725
$1\frac{1}{4}$	0.049	1.152	1.0423	2	0.065	1.870	2.746
	0.065	1.120	0.9852		0.083	1.834	2.642
	0.085	1.084	0.9229		0.109	1.782	2.494
	0.109	1.032	0.8365		0.134	1.732	2.356
$1\frac{1}{2}$	0.065	1.370	1.474				
	0.083	1.334	1.398				
	0.109	1.282	1.291				

F

Unit Conversion Factors

Quantity	English Unit	SI Unit	Metric Symbol	Equivalent Unit
Length	1 foot (ft)	0.3048 meter	m	—
Mass	1 slug	14.59 kilograms	kg	—
Time	1 second (sec)	1.0 second	s	—
Force	1 pound (lb)	4.448 newtons	N	kg·m/s
Pressure	1 lb/in.^{2} (psi)	6895 pascals	Pa	N/m^2 or kg/m·s^2
Temperature	Fahrenheit (°F) $°F = 1.8°C + 32$	Celsius (°C) $°C = (°F - 32)/1.8$	°C	—
Absolute Temperature	Rankine (°R) $°R = °F + 460$ $°R = 1.8 °K$	Kelvin (°K) $°K = °C + 273$ $°K = °R/1.8$	°K	—
Energy	1 ft·lb	1.356 joules	J	kg·m^2/s or N·m
Power	1 ft·lb/sec	1.356 watts	W	J/s or N·m/s

ADDITIONAL CONVENIENT CONVERSION FACTORS

Parameter	English Unit	SI Unit	SI Symbol
Length	1 inch (in.)	2.540 centimeters	cm
	1 foot (ft)	0.3048 meter	m
	1 yard (yd)	0.9144 meter	m
	1 mile (mi)	1.609 kilometers	km
Area	1 in.^{2}	6.452 square centimeters	cm^2
	1 ft^2	0.0929 square meter	m^2
Volume	1 in.^{3}	16.39 cubic centimeters	cm^3
	1 ft^3	0.0283 cubic meter	m^3
Velocity	1 ft/sec	30.48 centimeters/second	cm/s
	1 ft/sec	0.3048 meter/second	m/s
	1 mi/hr	1.609 kilometers/hour	km/h
Volumetric flow rate	1 in.^{3}/sec	16.39 cubic centimeters/second	cm^3/s
	1 in.^{3}/min	983.2 cubic centimeters/second	cm^3/s
	1 ft^3/sec	28.32 liters/second	l/s
	1 ft^3/min	1699 liters/second	l/s
	1 gal/sec	3.785 liters/second	l/s
	1 gal/min (gpm)	0.0631 liter/second	l/s
Note: 1000 liters = 1 m^3			
Mass	1 lb	453.6 grams	g
	1 lb	0.4536 kilogram	kg
Pressure	1 standard atmosphere (14.7 psi)	101.3 kilopascals	kPa
	1 in. of water (at 39.2°F)	249.1 pascals	kPa
	1 lb force/square inch (psi)	6.894 kilopascals	kPa
Energy	1 foot pound (ft·lb)	1.356 joules	J
Force	1 lb (lb$_f$)	4.448 newtons	N
Torque	1 lb force-inch (lb·in.)	0.1130 newton-meter	N·m
	1 lb force-foot (lb·ft)	1.356 newton-meters	N·m
Power	1 ft·lb/sec	1.356 watts	W
	1 ft·lb/min	0.0226 watt	W
	1 hp	745.7 watts	W
Absolute viscosity	1 pound force-sec/foot squared (lb·sec/ft^2)	47.88 pascal-seconds *Note:* 1 pascal-second = 10 poise	Pa·s or N·s/m^2
Kinematic viscosity	1 ft squared/second (ft^2/sec)	0.0929 square meter/second *Note:* 1 square meter/second = 10,000 stokes	m^2/s

G

Nomenclature

Symbol	Definition	Units
a	Acceleration	in./sec^2, ft/sec^2, cm/s^2, m/s^2
A	Area	in.^{2}, ft^2, cm^2, m^2
BP	Burst pressure	psi (lb/in.^{2})
CF	Coefficient of friction	None
CR	Compression ratio	None
C_v	Capacity constant	None
d	Diameter	in., ft, cm, m
D	Diameter	in., ft, cm, m
D_i	Inside diameter	in., ft, cm, m
D_o	Outside diameter	in., ft, cm, m
e	Eccentricity	in., ft, cm, m
E	Energy	lb·in, lb·ft, N·cm, N·m
F	Force	lb, N
FS	Factor of safety	None
f	Friction factor	None
g	Acceleration of gravity	in./sec^2, ft/sec^2, cm/s^2, m/s^2
G	Gain (Transfer Function)	in./v, cm/v, mm/v
G_A	Amplifier gain	ma/v
G_{sv}	Servo valve gain	(in.^{3}/sec)/ma, (cm^3/s)/ma
G_{CYL}	Cylinder gain	in./in.^{3}, cm/cm^3

ROMAN ALPHABETIC SYMBOLS (CONT'D)

Symbol	Definition	Units
H	Feedback transducer gain	V/in., V/cm, V/mm
H	Head	in., ft, cm, m
HP	Horsepower	hp
H_L	Head loss	in., ft, cm, m
H_m	Motor head	in., ft, cm, m
H_p	Pump head	in., ft, cm, m
ID	Inside diameter	in., ft, cm, m
J	Joule	N·m
K	K factor	None
KE	Kinetic energy	lb·in., lb·ft, N·cm, N·m
L	Length	in., ft, cm, m
L_e	Equivalent length	in., ft, cm, m
m	Mass	Slugs, kg
M	Micron	in., ft, cm, m
n	Rotational (speed)	rpm (rev/min)
N	Force in newtons	kg·m/s^2
N_R	Reynolds number	None
OD	Outside diameter	in., ft, cm, m
P	Power	ft·lb/sec, J/s
P	Pressure	lb/in.^{2}, lb/ft^2, N/m^2, bars
Pa	Pascal	N/m^2
Q	Volumetric flow rate	in.^{3}/sec, ft^3/sec, cm^3/s, m^3/s, gpm (gal/min), l/s
Q_A	Actual flow rate	in.^{3}/sec, ft^3/sec, cm^3/s, m^3/s, gpm (gal/min), l/s
Q_T	Theoretical flow rate	in.^{3}/sec, ft^3/sec, cm^3/s, m^3/s, gpm (gal/min), l/s
r	Moment arm	in., ft, cm, m
RE	Repeatable error	in., cm, mm
s	Tensile strength	lb/in.^{2} (psi), N/m^2 (Pa)
S	Distance	in., ft, cm, m
S_g	Specific gravity	None
t	Wall thickness	in., cm
t	Time	sec, s
T	Temperature	°F, °R, °C, °K
T	Torque	lb·in., lb·ft, N·cm, N·m
T_A	Actual torque	lb·in., lb·ft, N·cm, N·m
T_T	Theoretical torque	lb·in., lb·ft, N·cm, N·m
TE	Tracking error	in., cm, mm
v	Velocity	in./sec, ft/sec, cm/s, m/s
V	Volume	in.^{3}, ft^3, cm^3, m^3, l(liter)

Roman Alphabetic Symbols (cont'd)

Symbol	Definition	Units
V_D	Volumetric displacement	in.^{3}, ft^3, cm^3, m^3, l(liter)
VI	Viscosity index	None
w	Weight flow rate	lb/sec, lb/min, N/s, N/min
W	Weight	lb, N
W	Work	lb·in., lb·ft, N·cm, N·m
W	Watt	J/s
WP	Working pressure	lb/in.^{2} (psi), N/m^2 (Pa)
y	Oil film thickness	ft
Z	Elevation	in., ft, cm, m
ΔP	Pressure drop	lb/in.^{2}, lb/ft^2, N/cm^2, N/m^2

Greek Alphabetic Symbols

Symbol	Phonetic	Definition	Units
β	Beta	Bulk modulus	lb/in.^{2} (psi), Pa
γ	Gamma	Weight density	lb/in.^{3}, lb/ft^3, N/cm^3, N/m^3
η	Eta	Efficiency	None
η_m	—	Mechanical efficiency	None
η_o	—	Overall efficiency	None
η_v	—	Volumetric efficiency	None
θ	Theta	Angle	Degrees
ω	Omega	Natural frequency	radians/s
μ	Mu	Absolute viscosity	lb·sec/ft^2, poise, cP
ν	Nu	Kinematic viscosity	ft^2/sec, cm^2/s, stoke, cS
π	Pi	A constant	None
ρ	Rho	Mass density	Slubs/ft^3, kg/m^3
τ	Tau	Shear stress	lb/in.^{2}, lb/ft^2

Metric System Prefixes

Symbol	Name	Multiplication Factor
G	giga	10^9
M	mega	10^6
k	kilo	10^3
c	centi	10^{-2}
m	milli	10^{-3}
μ	micro	10^{-6}
n	nano	10^{-9}

H

Fluid Power Symbols

THE SYMBOLS SHOWN CONFORM TO THE AMERICAN NATIONAL STANDARDS INSTITUTE (ANSI) SPECIFICATIONS. BASIC SYMBOLS CAN BE COMBINED IN ANY COMBINATION. NO ATTEMPT IS MADE TO SHOW ALL COMBINATIONS.			
LINES AND LINE FUNCTIONS		**LINES AND LINE FUNCTIONS (CONT.)**	
LINE, WORKING	——————	DIRECTION OF FLOW, HYDRAULIC PNEUMATIC	
LINE, PILOT (L>20W)	– – – – –	LINE TO RESERVOIR ABOVE FLUID LEVEL BELOW FLUID LEVEL	
LINE, DRAIN (L<5W)	– – – – – – –		
CONNECTOR	•		
LINE, FLEXIBLE		LINE TO VENTED MANIFOLD	
LINE, JOINING		PLUG OR PLUGGED CONNECTION	✕
LINE, PASSING		RESTRICTION, FIXED	

PUMPS	
PUMP, SINGLE FIXED DISPLACEMENT	
PUMP, SINGLE VARIABLE DISPLACEMENT	

MOTOR AND CYLINDERS	
MOTOR, ROTARY FIXED DISPLACEMENT	
MOTOR, ROTARY VARIABLE DISPLACEMENT	
MOTOR, OSCILLATING	
CYLINDER, SINGLE ACTING	
CYLINDER, DOUBLE ACTING	
CYLINDER, DIFFERENTIAL ROD	
CYLINDER, DOUBLE END ROD	
CYLINDER, CUSHIONS BOTH ENDS	

MISCELLANEOUS UNITS	
DIRECTION OF ROTATION (ARROW IN FRONT OF SHAFT)	
COMPONENT ENCLOSURE	
RESERVOIR, VENTED	
RESERVOIR, PRESSURIZED	
PRESSURE GAGE	

MISCELLANEOUS UNITS (CONT.)	
TEMPERATURE GAGE	
FLOW METER (FLOW RATE)	
ELECTRIC MOTOR	
ACCUMULATOR, SPRING LOADED	
ACCUMULATOR, GAS CHARGED	
FILTER OR STRAINER	
HEATER	
COOLER	
TEMPERATURE CONTROLLER	
INTENSIFIER	
PRESSURE SWITCH	

BASIC VALVE SYMBOLS	
CHECK VALVE	
MANUAL SHUT OFF VALVE	
BASIC VALVE ENVELOPE	
VALVE, SINGLE FLOW PATH, NORMALLY CLOSED	

BASIC VALVE SYMBOLS (CONT.)	
VALVE, SINGLE FLOW PATH, NORMALLY OPEN	
VALVE, MAXIMUM PRESSURE (RELIEF)	
BASIC VALVE SYMBOL, MULTIPLE FLOW PATHS	
FLOW PATHS BLOCKED IN CENTER POSITION	
MULTIPLE FLOW PATHS (ARROW SHOWS FLOW DIRECTION)	

VALVE EXAMPLES		METHODS OF OPERATION	
UNLOADING VALVE, INTERNAL DRAIN, REMOTELY OPERATED		PRESSURE COMPENSATOR	
DECELERATION VALVE, NORMALLY OPEN		DETENT	
SEQUENCE VALVE, DIRECTLY OPERATED, EXTERNALLY DRAINED		MANUAL	
PRESSURE REDUCING VALVE		MECHANICAL	
COUNTER BALANCE VALVE WITH INTEGRAL CHECK		PEDAL OR TREADLE	
TEMPERATURE AND PRESSURE COMPENSATED FLOW CONTROL WITH INTEGRAL CHECK		PUSH BUTTON	
DIRECTIONAL VALVE, TWO POSITION, THREE CONNECTION		LEVER	
		PILOT PRESSURE	
DIRECTIONAL VALVE, THREE POSITION, FOUR CONNECTION		SOLENOID	
		SOLENOID CONTROLLED, PILOT PRESSURE OPERATED	
VALVE, INFINITE POSITIONING (INDICATED BY HORIZONTAL BARS)		SPRING	
		SERVO	

/

Answers to Odd-Numbered Exercises

Chapter 1

1-1. 1. Steers and brakes automobiles. 2. Moves earth. 3. Harvests crops. 4. Mines coal. 5. Drives machine tools.

1-3. Advantage of fluid power system: 1. Not hindered by geometry of machine. 2. Provides remote control. 3. Complex mechanical linkages are eliminated. 4. Instantly reversible motion. 5. Automatic protection against overloads. 6. Infinitely variable speed control. Advantages of mechanical system: 1. No mess due to oil leakage problems. 2. No danger of bursting of hydraulic lines. 3. No fire hazard due to oil leak.

1-5. Hydraulic fluid power uses liquids which provide a very rigid medium for transmitting power. Thus, huge forces can be provided to move loads with utmost accuracy and precision. Pneumatic systems exhibit spongy characteristics due to the compressibility of air. However, pneumatic systems are less expensive to build and operate.

1-7. Hydraulic cylinder.

1-9. 1. Liquids are a very rigid medium. 2. Power capacity of fluid systems is limited only by the strength capacity of the component material.

1-11. An electric motor or other power source to drive the pump or compressor.

1-13. 1. Compressed air tank. 2. Compressor. 3. Prime mover. 4. Valves. 5. Actuators. 6. Piping.

1-15. Research project.

1-17. Research project.

1-19. Electrical components such as pressure switches, limit switches, and relays can be used to operate electrical solenoids to control the operation of valves that direct fluid to the hydraulic actuators. This permits the design of a very versatile fluid power circuit.

1-21. A closed-loop system is one which uses feedback, whereas an open-loop system does not use feedback.

1-23. Dr. Robert Jarvik made medical history with the design of an artificial, pneumatically actuated heart which sustained the life of Dr. Barney Clark for over 100 days. Two examples are artificial kidneys and valve-assisted bladders.

1-25. MPL devices are miniature valve-type devices which by the action of internal moving parts, perform switching operations in fluid logic systems.

1-27. Fluidic devices, which have no moving parts, utilize fluid flow phenomena to perform a wide variety of sensory and control functions.

1-29. Air has entered the hydraulic oil line and has greatly reduced the bulk modulus (measure of stiffness or incompressibility) of the oil-air combination fluid.

1-31. The terms *fluid power* and *hydraulics and pneumatics* are synonymous.

1-33. Changes in oil leakage past seals due to changes in viscosity resulting from temperature variations.

1-35. Hydraulic brakes. Components are oil lines, oil pump, valve, and cylinders.

1-37. Research project.

1-39. A PLC is a user-friendly electronic computer designed to perform logic functions such as AND, OR, and NOT for controlling the operation of industrial equipment and processes.

1-41. Unlike electromechanical relays, PLCs are not hard-wired to perform specific functions.

Chapter 2

2-1. 1. Transmit power. 2. Lubricate moving parts. 3. Seal clearances between mating parts. 4. Dissipate heat.

2-3. Generally speaking, a fluid should be changed when its viscosity and acidity increase due to fluid breakdown or contamination.

2-5. Advantages of air: 1. Fire resistant. 2. Not messy. Disadvantages of air: 1. Due to its compressibility, it cannot be used in an application where accurate positioning or rigid holding is required. 2. Because it is compressible, it tends to be sluggish.

2-7. $S_g = 0.881$, $\rho_{fluid} = 1.71$ slugs/ft^3.

2-9. Pressure is force per unit area.

2-11. Bulk modulus is a measure of the incompressibility of a hydraulic fluid.

2-13. Viscosity is a measure of the sluggishness with which a fluid flows. Viscosity index is a relative measure of an oil's viscosity change with respect to temperature change.

2-15. 1. Increased leakage losses past seals. 2. Excessive wear due to breakdown of the oil film between moving parts.

2-17. ν (cS) = 43.3 cS, μ (cP) = 39.0 cP.

2-19. Pour point is the lowest temperature at which a fluid will flow.

2-21. In applications where human safety is of concern.

2-23. 1. Water-glycol solutions. 2. Water-in-oil emulsions. 3. Straight synthetics. 4. High water content fluids.

2-25. Air can become dissolved or entrained in hydraulic fluids. This can cause pump cavitation and also greatly reduce the bulk modulus of the hydraulic fluid. Foam-resistant fluids contain chemical additives which break out entrained air to separate quickly the air from the oil while it is in the reservoir.

2-27. Coefficient of friction is the proportionality constant between a normal force and the frictional force it creates between two mating surfaces sliding relative to each other. $CF = F/N$.

2-29. The neutralization number is a measure of the relative acidity or alkalinity of a hydraulic fluid and is specified by a pH factor.

2-31. It may cause cavitation problems in the pump due to excessive vacuum pressure in the pump inlet line unless proper design steps are implemented.

2-33. $1 \text{ (lb-sec/ft}^2) = 47.88 \text{ N} \cdot \text{s/m}^2$

2-35. $\rho = 896 \text{ kg/m}^3$

2-37. 99 kPa

2-39. $\gamma = 55.7 \text{ lb/ft}^3$

2-41. The constant 0.433 is the pressure in psi that is produced at the base of a one-foot column of water. Multiplying this constant by the head of the liquid in feet times the specific gravity of the liquid gives the pressure at the base in units of psi.

2-43. By atmospheric pressure at the base of the mercury column.

2-45. When the fluid power system operates in an environment undergoing large temperature variations such as in outdoor machines like automobiles.

2-47. High viscosity and excessive contamination.

2-49. $\beta = 507,000 \text{ psi}$.

2-51. $F = 1000 \text{ N}$.

2-53. P = 19.5 psi.

2-55. A high *VI* should be specified, indicating small changes in viscosity with respect to changes in temperature.

Chapter 3

3-1. The total energy at station 1 in a pipeline plus the energy added by a pump minus the energy removed by a motor minus the energy loss due to friction equals the total energy at station 2. If a section of pipe contains no pump or motor, the pressure at a small-diameter location will be less than the pressure at a large-diameter location. Kinetic energy is substituted for pressure energy.

3-3. $Q = 24.5 \text{ gpm}$.

3-5. $P_2 = -1 \text{ atmosphere} = -14.7 \text{ psi}$ (perfect vacuum).

3-7. As shown in Fig. 3-19, in order for a siphon to work, two conditions must be met: 1. The elevation of the free end must be lower than the elevation of the liquid surface inside the container. 2. The fluid must be initially forced to flow up from the container into the center portion of the U-tube. This is normally done by temporarily providing a suction pressure at the free end of the siphon.

3-9. $WZ = 834,167 \text{ ft} \cdot \text{lb}$.

3-11. $Q = 98.3 \text{ gpm}$.

3-13. Per Fig. 3-16, the volume of air flow is determined by the opening position of the butterfly valve. As the air flows through the venturi, it speeds up and loses some of its pressure. This produces a differential pressure between the fuel bowl and the venturi throat. This causes gasoline to flow into the air stream.

3-15. 1. A force is required to change the motion of a body. 2. If a body is acted upon by a force, the body will have an acceleration proportional to the magnitude of the force and inversely to the mass of the body. 3. If one body exerts a force on a second body, the second body must exert an equal but opposite force on the first body.

3-17. Torque equals the product of a force and moment arm which is measured from the center of a shaft (center of rotation) perpendicularly to the line of action of the force.

3-19. (a) F_{load} = 320.4 1b, (b) $S_{load\ piston}$ = 15 in., (c) $HP_{90\%\ efficiency}$ = 0.0437 hp.

3-21. Mechanical power = force × velocity, electrical power = volts × amps, hydraulic power = pressure × flow rate.

3-23. P_2 = 248 psi.

3-25. P_2 = 1714 kPa.

3-27. v = 3.74 m/s

3-29. v = 1.1 ft/sec

3-31. Q = 0.0236 m³/s

3-33. (a) $A = F/P$ = 0.004 m², (b) $Q = Av$ = 0.0015 m³/s, (c) Power = PQ = 15 kW

3-35. (a) W = 24,500 ft · lb. (b) P = 69.7 psi. (c) Power = 4.45 HP. (d) v = 1.02 in./sec; Q = 82.2 gpm.

3-37. t = 3.27 sec.

3-39. H = 12.8 ft.

3-41. Q = 20.4 gpm.

3-43. N = 86 cycles/min.

3-45. F = 13.9 lb, S = 0.0347 in.

3-47. Elevation head is potential energy per unit weight of fluid. Pressure head is pressure energy per unit weight of fluid. Velocity head is kinetic energy per unit weight of fluid.

Chapter 4

4-1. To carry the fluid from the reservoir through operating components and back to the reservoir.

4-3. 4 ft/sec.

4-5. Zinc, magnesium, and cadmium.

4-7. 1. Tensile strength of conductor material. 2. Conductor outside diameter. 3. Operating pressure levels.

4-9. D = 1.428-in. inside diameter.

4-11. 1. When a joint is taken apart, the pipe must be tightened further to reseal. 2. Pipes cannot be bent around obstacles.

4-13. Average fluid velocity is defined as the volumetric flow rate divided by the pipe cross-sectional area.

4-15. Malleable iron can be used for hydraulic fittings for low-pressure lines such as inlet, return, and drain lines.

4-17. Plastic tubing is relatively inexpensive. Also, it can readily be bent to fit around obstacles, it is easy to handle, and it can be stored on reels.

4-19. The quick disconnect coupling is used mainly where a conductor must be frequently disconnected from a component.

4-21. Figure 4-8 shows the flared-type fitting which was developed before the compression type for sealing against high pressures. Figure 4-7 shows a compression-type fitting which can be repeatedly taken apart and reassembled and remain perfectly sealed against leakage.

4-23. 1. Install so there is no kinking during operation of system. 2. There should always be some slack to relieve any strain and allow for the absorption of pressure surges. 3. If the hose is subject to rubbing, it should be encased in a protective sleeve.

4-25. $ID_{min} = 40.9$ mm, Select 42 mm ID

4-27. $WP = 129$ bars

4-29. $D_o = 1.17$ in.

4-31. Increases.

4-33. Four.

4-35. By nominal size and schedule number.

4-37. $1\frac{1}{2}$ in. OD, 1.310 in ID produces velocity of 7.13 ft/sec. Therefore, need a larger size than given in Figure 4-5 for pump inlet. $1\frac{1}{4}$ in. OD, 1.060 in ID produces velocity of 10.9 ft/sec. Therefore, this size is adequate for pump outlet.

Chapter 5

5-1. It is very important to keep all energy losses in a fluid power system to a minimum acceptable level.

5-3. 1. If N_R is less than 2000, the flow is laminar. 2. If N_R is greater than 4000, the flow is turbulent. 3. Reynolds number between 2000 and 4000 cover a critical zone between laminar and turbulent flow.

5-5. $\Delta P = 40.0$ psi.

5-7. (a) $f = 0.0735$, (b) $f = 0.046$.

5-9. $\Delta P = 0.378$ psi.

5-11. $L_e = 0.313$ ft.

5-13. $P_2 = 9.31$ psi.

5-15. 1. Bourdon gage. 2. Schrader gage.

5-17. $Q = 713$ gpm.

5-19. $P_2 = 1600$ kPa.

5-21. $N_R = 1800$, the flow is laminar.

5-23. (a) $N_R = 400$ laminar, $f = 0.16$, (b) $N_R = 2000$ laminar, $f = 0.032$

5-25. $L_e = 0.076$ m

5-27. $P_2 - P_1 = -4.93$ bars

5-29. $P_2 = 6.31$ bars

5-31. $\Delta P \propto K$, true

5-33. Easier to read since values are given in digits rather than by a needle pointing along a scale.

5-35. Square (provided flow is fully turbulent).

5-37. High velocity and large pipe roughness.

5-39. Flow coefficient and K factor values would be the same because these two parameters are dimensionless.

5-41. Heat generation rate = 14,000 BTU/hr.

Chapter 6

6-1. 1. Gear. 2. Vane 3. Piston.

6-3. All pumps operate on the principle whereby a partial vacuum is created at the pump inlet due to the internal operation of the pump. This allows atmospheric pressure to push the fluid out of the oil tank and into the pump intake. The pump then mechanically pushes the fluid out the discharge line, as shown by Fig. 6-1.

6-5. Volumetric efficiency equals actual flow rate produced by a pump divided by the theoretical flow rate based on volumetric displacement and pump speed. Actual flow rate is measured by a flowmeter, and theoretical flow rate is calculated from the equation $Q_T = V_D N/231$.

6-7. After the volumetric efficiency (η_v) and mechanical efficiency (η_m) have been found, the overall efficiency (η_o) is determined from the following equation: $\eta_o = \eta_v \eta_m/100$.

6-9. A fixed displacement pump is one in which the amount of fluid ejected per revolution (displacement) cannot be varied. In a variable displacement pump, the displacement can be varied by changing the physical relationships of various pump elements. This change in pump displacement produces a change in pump flow output even though pump speed remains constant.

6-11. 1. Lobe. 2. Gerotor.

6-13. 1. Flow-rate requirements (gpm). 2. Operating speed (rpm). 3. Pressure rating (psi). 4. Performance. 5. Reliability. 6. Maintenance. 7. Cost. 8. Noise.

6-15. Pump cavitation occurs when suction lift is excessive and the inlet pressure falls below the vapor pressure of the fluid (usually about 5-psi suction). As a result, air bubbles, which form in the low-pressure inlet region of the pump, are collapsed when they reach the high-pressure discharge region. This produces high velocity and explosive forces which severely erode the metallic components and shorten pump life.

6-17. $\eta_m = 95.7\%$.

6-19. Cavitation can occur due to entrained air bubbles. This occurs when suction lift is excessive and the inlet pressure falls below the vapor pressure of the fluid (usually about 5-psi suction). Cavitation produces explosive forces which severely erode the metallic components and shorten pump life.

6-21. The flow output of a centrifugal pump is reduced as circuit resistance is increased. Therefore centrifugal pumps are rarely used in hydraulic systems.

6-23. A balanced vane pump is one that has two intake and two outlet ports diametrically opposite each other. Thus, pressure ports are opposite each other, and a complete hydraulic balance is achieved. This eliminates the bearing side loads and thus permits higher operating pressures.

6-25. 1. Axial design. 2. Radial design.

6-27. The pressure rating is defined as the maximum pressure level at which the pump can operate safely and provide a good useful life.

6-29. Gear pumps are simple in design and compact in size. They are the least expensive. Vane pump efficiencies and costs fall between those for gear and piston pumps. Piston pumps are the most expensive and provide the highest level of overall performance.

6-31. $e = 0.371$ in.

6-33. (a) $\eta_o = 80.2\%$, (b) $T_T = 956$ in. · lb.

6-35. Hydraulic Power = 14.0 kW, Electric Power = 16.5 kW

6-37. $\eta = 86.3\%$

6-39. $\theta = 9.5°$

6-41. 0.000333 m³/s, 210 bars

6-43. Suction pressure because the oil tank is vented to the atmosphere and there is a pressure drop in the inlet line due to elevation, velocity, and friction.

6-45. Ejects a fixed amount of fluid per revolution of pump shaft rotation. Capable of overcoming the pressure resulting from the mechanical loads on the system as well as the resistance of flow due to friction.

6-47. Vane and piston pumps.

6-49. By varying the offset angle between the cylinder block centerline and the drive shaft centerline.

6-51. Change diameter of gear teeth and width of gear teeth.

6-53. $\eta_m = 96\%$. Frictional hp = 0.96

6-55. Prime mover hp = 11.7; Prime mover speed = 1540 rpm.

6-57. The eccentricity between the centerline of the rotor and the centerline of the cam ring can be changed by a handwheel or by a pressure compensator.

6-59. $Q(\text{m}^3/\text{min}) = V_D(\text{m}^3)N(\text{rpm})$

Chapter 7

7-1. A single-acting cylinder can exert a force in only the extending direction. Single-acting cylinders do not retract hydraulically. Retraction is accomplished by using gravity or by the inclusion of a compression spring in the rod end. Double-acting cylinders can be extended and retracted hydraulically.

7-3. (a) Pressure = 679 psi, (b) velocity = 4.54 ft/sec, (c) HP = 9.91 hp, (d) pressure = 902 psi, (e) velocity = 6.04 ft/sec, (f) HP = 13.18 hp.

7-5. $P_2 = 668$ psi.

7-7. Telescoping rod cylinders contain multiple cylinders which slide inside each other. They are used where long work strokes are required but the full retraction length must be minimized.

7-9. $P = 1423$ psi.

7-11. Simple design and subsequent low cost.

7-13. The vanes must have some means other than centrifugal force to hold them against the cam ring. Some designs use springs, while other types use pressure-loaded vanes.

7-15. (a) N = 577.5 rpm, (b) T = 1911 in. · 1b, (c) HP = 17.5 hp.

7-17. A motor uses more flow than it theoretically should because the motor inlet pressure is greater than the motor discharge pressure. Thus, leakage flow passes through a motor from the inlet port to the discharge port.

7-19. (a) η_v = 92.4%, (b) η_m = 94.2%, (c) η_o = 87.0%, (d) HP = 57.1 hp.

7-21. An electrohydraulic stepping motor is a device which uses a small electrical stepping motor to control the huge power available from a hydraulic motor. The electrohydraulic stepping motor consists of three components: electrical stepping motor, hydraulic servo valve, and hydraulic motor. The electric stepping motor (see Fig. 7-34) rotates a precise, fixed amount per each electrical pulse received. This motor is directly coupled to the rotary linear translator of the servo valve. The flow forces in the servo valve are directly proportional to the rate of flow through the valve. The speed and direction of rotation of the hydraulic motor reproduce the motion of the electric motor, and the servo valve provides feedback through mechanical linkage.

7-23. The effective cylinder area is not the same for the extension and retraction strokes. This is due to the effect of the piston rod.

7-25. (a) P = 3.98 MPa, (b) v = 1.27 m/s, (c) kW = 6.37 kW, (d) P = 5.31 MPa, (e) v = 1.70 m/s, (f) kW = 8.50 kW

7-27. (a) N = 600 rpm, (b) T = 222.9 $N \cdot m$, (c) Power = 14.0 kW

7-29. (a) V_{DM} = 133 cm³, (b) $T_{A(By\ Motor)}$ = 205 $N \cdot m$

7-31. True since $T_A = T_T \eta_m = (V_D P/6.28)\eta_m$

7-33. The speed and volumetric displacement requirements of the motoring unit (hydraulic cylinder or motor).

7-35. Displacement is the volume of oil required to produce one revolution of the motor. Torque rating is the torque delivered by the motor at rated pressure.

7-37. Pressure exerts a force on the pistons. The piston thrust is transmitted to the angled swash plate causing torque to be created in the drive shaft.

7-39. Piston motor.

7-41. Difference = $\frac{1}{4}$ (Pressure)(Piston Area).

7-43. Q = 8.16 gpm; HP = 4.76.

7-45. HP = 93.3.

7-47. T = 144 in. · lb.

7-49. For $\eta_m = \eta_v = \eta_o$ = 100%; V_D = 17.3 in.³, T = 8260 in.-lb.

Chapter 8

8-1. Directional control valves determine the path through which a fluid traverses within a given circuit.

8-3. A pilot check valve always permits free flow in one direction but permits flow in the normally blocked opposite direction only if pilot pressure is applied at the pilot pressure port of the valve.

8-5. This valve contains a spool which can be actuated into three different functioning positions. The center position is obtained by the action of the springs alone.

8-7. A solenoid is an electric coil. When the coil is energized, it creates a magnetic force that pulls the armature into the coil. This causes the armature to push on the push rod to move the spool of the valve.

8-9. A shuttle valve is another type of directional control valve. It permits a system to operate from either of two fluid power sources. One application is for safety in the event that the main pump can no longer provide hydraulic power to operate emergency devices.

8-11. A pressure-reducing valve is another type of pressure control valve. It is used to maintain reduced pressures in specified locations of hydraulic systems.

8-13. A sequence valve is a pressure control device. Its purpose is to cause a hydraulic system to operate in a pressure sequence.

8-15. Flow control valves are used to regulate the speed of hydraulic cylinders and motors by controlling the flow rate to these actuators.

8-17. A servo valve is a directional control valve which has infinitely variable positioning capability. Servo valves are coupled with feedback-sensing devices, which allows acceleration of an actuator.

8-19. See Fig. 8-46 for a block diagram of a closed-loop system.

8-21. A pressure switch is an instrument that automatically senses a change in pressure and opens or closes an electrical switching element when a predetermined pressure point is reached. Four types are diaphragm, Bourdon tube, sealed piston, and dia-seal piston.

8-23. A normally open switch is one in which no current can flow through the switching element until the switch is actuated. In a normally closed switch, current flows through the switching element until the switch is actuated.

8-25. $HP = 29.2$ hp.

8-27. In the design of Fig. 8-4, the check valve poppet has the pilot piston attached to the threaded poppet stem by a nut. The light spring holds the poppet seated in a no-flow condition by pushing against the pilot piston. The purpose of the separate drain port is to prevent oil from creating a pressure buildup on the bottom of the piston.

8-29. Flow can go through the valve in four unique ways depending on the spool position. (a) Spool position 1: Flow can go from P to A and B to T. (b) Spool position 2: Flow can go from A to T and P to B.

8-31. A compound pressure relief valve (see Fig. 8-30) is one which operates in two stages. Referring to Fig. 8-31, the operation is as follows: In normal operation, the balanced piston is in hydraulic balance. For pressures less than the valve setting, the piston is held on its seat by a light spring. As soon as pressure reaches the setting of the adjustable spring, the poppet is forced off its seat. This limits the pressure in the upper chamber. The restricted flow through the orifice and into the upper chamber results in an increase in pressure in the lower chamber. This causes an unbalance in hydraulic forces, which tends to raise the piston off its seat. When the pressure difference between the upper and lower chambers reaches 20 psi, the large piston lifts off its seat to permit flow directly to tank.

8-33. This design incorporates a hydrostat which maintains a constant 20-psi differential across the throttle, which is an orifice whose area can be adjusted by an external knob setting. The orifice area setting determines the flow rate to be controlled. The hydrostat is held normally open by a light spring. However, it starts to close as inlet pressure increases and overcomes the light spring force. This closes the opening through the hydrostat and thereby blocks off all flow in excess of the throttle setting. As a result, the only oil that will pass through the valve is the amount which 20 psi can force through the throttle. Flow exceeding this amount can be used by other parts of the circuit or return to the tank via the pressure relief valve.

8-35. These shock absorbers are filled completely with oil. Therefore, they may be mounted in any position or at any angle. The spring-return units are entirely self-contained, extremely compact types that require no external hoses, valves, or fittings. In this spring-returned type, a built-in cellular accumulator accommodates oil displaced by the piston rod as the rod moves inward. These shock absorbers are multiple-orifice hydraulic devices. When a moving load strikes the bumper of the shock absorber, it sets the rod and piston in motion. The moving piston pushes oil through the series of holes. The resistance to the oil flow caused by the holes creates a pressure that acts against the piston to oppose the moving load.

8-37. kW power = 0.32 kW

8-39. Three.

8-41. To shift the spool in directional control valves.

8-43. The pressure at which a pressure relief valve begins to open.

8-45. Control direction of flow. Control flow rate. Control pressure.

8-47. Position is the location of the spool inside the valve. Way is the flow path through the valve. Port is the opening in the valve body for the fluid to enter or exit.

8-49. Slip-in design cartridge valve uses a bolted cover while screw-type design uses threads for assembling into the manifold block.

8-51. Directional control, pressure relief, pressure reducing, unloading, and flow control functions.

Chapter 9

9-1. 1. Safety of operation. 2. Performance of desired function. 3. Efficiency of operation.

9-3. The load-carrying capacity for a regenerative cylinder equals the pressure times the piston rod area rather than the pressure times the piston area.

9-5. $P_1 = 1000$ psi.

9-7. A hydraulic motor may be driving a machine having a large inertia. This would create a flywheel effect on the motor, and stopping the flow of fluid to the motor would cause it to act as a pump. The circuit should be designed to provide fluid to the motor while it is pumping to prevent it from pulling in air.

9-9. An air-over-oil system is one using both air and oil to obtain the advantages of each medium.

9-11. 1. Piston type: the principal advantage is its ability to handle very high or low temperature system fluids through the utilization of compatible O-ring seals. 2. Diaphragm type: The primary advantage is its small weight-to-volume ratio, which makes it suitable almost exclusively for airborne applications. 3. Bladder type: The greatest advantage is the positive sealing between the gas and oil chambers.

9-13. A mechanical hydraulic servo system is a closed-loop system using a mechanical feedback. One application is an automotive power steering system.

9-15. Use the circuit of Fig. 9-11 entitled hydraulic cylinder sequence circuit. The left cylinder of Fig. 9-11 becomes the clamp cylinder of Fig. 9-45, and the right cylinder of Fig. 9-11 becomes the work cylinder of Fig. 9-45.

9-17. A check valve is needed in the hydraulic line just upstream of where the pilot line to the unloading valve is connected to the hydraulic line. Otherwise the unloading valve would behave like a pressure relief valve, and thus valuable energy would be wasted.

9-19. Both manually actuated directional control valves must be actuated in order to extend or retract the hydraulic cylinder.

9-21. (a) 0.25 m/s, 68,300N, (b) 0.25 m/s, 68,300N

9-23. Yes. Use a regenerative circuit with a cylinder having a rod area equal to one-half the piston area, or use a double rod cylinder having equal size rods at each end.

9-25. $F_1 = F_2 = 46,700$ lb.

9-27. Heat generation rate = 3790 BTU/hr.

9-29. Cylinder 2 will extend through its complete stroke receiving full pump flow while cylinder 1 does not move. As soon as cylinder 2 has extended through its complete stroke, cylinder 1 receives full pump flow and extends through its complete stroke. This is because system pressure builds up until load resistance is overcome to move cylinder 2 with the smaller load first. Then pressure continues to increase until the load on cylinder 1 is overcome. This causes cylinder 1 to then extend. In the retraction mode, the cylinders move in the same sequence.

9-31. $F = 51$ kN.

9-33. Heat generation rate = 3.57 kW.

9-35. Valve spool moves with the load. Valve sleeve moves with the input.

Chapter 10

10-1. 1. Liquids exhibit greater inertia than do gases. 2. Liquids exhibit greater viscosity than do gases. 3. Hydraulic systems require special reservoirs and no-leak design components.

10-3. $P_2 = 193.7$ psig.

10-5. $P_2 = 36.6$ psig.

10-7. 1. Piston type. 2. Screw type. 3. Sliding vane type.

10-9. $V_1 = 324$ cfm of free air.

10-11. The function of an air filter is to remove contaminants from the air before it reaches pneumatic components such as valves and actuators.

10-13. A lubricator ensures proper lubrication of internal moving parts of pneumatic components.

10-15. A pneumatic exhaust silencer (muffler) is used to control the noise caused by a rapidly exhausting airstream flowing into the atmosphere.

10-17. $P_f = 7.85$ psi.

10-19. $Q = 468$ cfm of free air.

10-21. Cylinder extends and retracts continuously.

10-23. (a) Cylinder 1 extends and then cylinder 2 extends, (b) cylinder 2 retracts and then cylinder 1 retracts.

10-25. $V_1 = 7.27$-gal accumulator.

10-27. $P_2 = 19$ bars gage

10-29. $P_2 = 2.45$ bars gage

10-31. $V_1 = 27.5$ liters

10-33. $V = 6.95$ ft³/min of free air.

10-35. 71.1 °C, 620 °R, 344.1 °K.

10-37. $t = 35.4$ min.

10-39. Determine pressure capacity requirement.
Establish number of stages required.
Determine cfm of free air required.
Size the air receiver and compressor.
Determine type of compressor (piston, vane, or screw).
Establish type of unloader control and pressure settings.

10-41. $V = 59.3$ ft³/min of free air.

10-43. 1. Actuate V_1 only: cylinder moves to the right.
2. Actuate V_2 only: cylinder moves to the left.
3. Actuate both V_1 and V_2: cylinder is pneumatically locked since both ends are exposed to system air pressure.
4. Unactuate both V_1 and V_2: cylinder is free to move since both ends are vented to the atmosphere.

10-45. Cylinder 2 extends through full stroke while cylinder 1 does not move. Then cylinder 1 extends through full stroke.

10-47. (a) $V = 1.57$ in.³, (b) $V = 1.57$ in.³, (c) $S = 0.89$ in., (d) $V = 101$ in.³, (e) $Q = 0.41$ gpm, (f) $Q = 25.4$ ft³/min of free air.

Chapter 11

11-1. Fluidics is the technology that utilizes fluid flow phenomena in components and circuits to perform a wide variety of control functions. These include sensing, logic, memory, timing, and interfacing to other control media.

11-3. Fluidic components operate at low power and pressure levels (normally below 15 psi). Yet these low-pressure fluidic signals can reliably control hydraulic systems operating with up to 10,000-psi oil to provide the muscle to do useful work at rates up to several hundred horsepower.

11-5. An OR function is one in which all control signals must be OFF in order for the output not to exist. Therefore, any one control signal will produce an output.

11-7. A flip-flop is a bistable digital control device. Thus, a flip-flop has two stable states when all control signals are OFF. However, each of the two stable states are predictable, as shown by the truth table in Fig. 11-18. For the operation of a flip-flop, refer to Fig. 11-17.

11-9. In some applications it is necessary to have a specific output when the power supply is first turned ON and all control signals are OFF. For these applications, a flip-flop with a start-up preference is used. Figure 11-19 shows such a flip-flop with its symbol (note the + sign) and truth table. As shown, when all control signals are OFF, the output is ON at 02 and OFF at 01. Otherwise it behaves exactly the same as a basic flip-flop.

11-11. An OR gate is a device which will have an output if any one or any combination of control signals is ON. An EXCLUSIVE OR gate is a device which will have an output only if one control signal (but not any combination of control signals) is ON.

11-13. The interruptible jet sensor uses a nozzle to transmit an airstream across a gap to a receiver. When the flow is unobstructed, adequate pressure is recovered by the receiver to hold the fluidic element in a particular mode [see Fig. 11-28(a)]. However, when an object enters the jet area, the output pressure in the receiver is reduced to cause the fluidic element to switch to the opposite mode [see Fig. 11-28(b)].

11-15. Contact sensing is accomplished by the sensing of objects by physical contact. It is usually done by a moving-part device called a limit valve or limit switch (see Fig. 11-29).

11-17. (a) Switch the outputs of the FLIP-FLOP to opposite sides of the interface valve. Also move the interruptible jets as close as possible to the ends of the hydraulic cylinder. (b) When the present system with a preferenced flip-flop is initially started, the cylinder rod will always begin by moving to the right. If the preferenced flip-flop is replaced by a regular flip-flop, the system may start up with the cylinder beginning its move to the left or to the right. The starting direction is indeterminant.

11-19. (a) The cylinder extends, retracts, and stops. Thus, we have one cycle of reciprocation. (b) The cylinder extends and retracts continuously. Thus, we have continuous reciprocation.

11-21. No. At the completion of the extension stroke, the cylinder will automatically reverse even though one or both buttons are still held. It retracts fully and another cycle cannot be started until both buttons are released and reactuated, because the cylinder cannot extend until all three control signals have been removed from the OR gate, which receives one of its three control signals from the output of the FLIP-FLOP. The bottom control signal to this OR gate is removed on the previous cycle when both buttons are released. The two top signals to this OR gate are removed when both buttons are pressed to start the next cycle.

11-23. Moving part logic devices are miniature valve-type devices which by the action of internal moving parts, perform switching operations in fluid logic systems.

11-25. Fluidic devices have no moving parts.

11-27. Since fluidic devices have no moving parts, they virtually do not wear out.

11-29. 1. It provides a means by which a logic circuit can be reduced to its simplest form. 2. It allows for the quick synthesis of a circuit which is to perform desired logic operations.

11-31. Logic inversion is the process that makes the output signal not equal to the input signal in terms of ON versus OFF.

11-33. DeMorgan's Theorem allows for the inversion of functions as follows:

$$\overline{A + B + C} = \overline{A} \cdot \overline{B} \cdot \overline{C} \text{ and } \overline{A \cdot B \cdot C} = \overline{A} + \overline{B} + \overline{C}$$

11-35. 3 variables produce $2^3 = 8$ possible combinations.

A	B	C	A + B	B + C	(A + B) + C	A + (B + C)
0	0	0	0	0	0	0
1	0	0	1	0	1	1
1	1	0	1	1	1	1
1	0	1	1	1	1	1
0	1	0	1	1	1	1
0	0	1	0	1	1	1
0	1	1	1	1	1	1
1	1	1	1	1	1	1

11-37.

A	B	C	B + C	A·(B + C)	A·B	A·C	(A·B) + (A·C)
0	0	0	0	0	0	0	0
1	0	0	0	0	0	0	0
1	1	0	1	1	1	0	1
1	0	1	1	1	0	1	1
0	1	0	1	0	0	0	0
0	0	1	1	0	0	0	0
0	1	1	1	0	0	0	0
1	1	1	1	1	1	1	1

11-39. From DeMorgan's Theorem we have:

$$\overline{A \cdot B \cdot C} = \overline{A} + \overline{B} + \overline{C}$$

hence

$$A \cdot B \cdot C = \overline{\overline{(A \cdot B \cdot C)}} = \overline{(\overline{A} + \overline{B} + \overline{C})} = \text{NOT } (\overline{A} + \overline{B} + \overline{C})$$

11-41. When the cylinder is fully retracted, the signals from A_1 and A_2 are both ON. The extension stroke begins when push button A is pressed since the output $P \cdot A_1$ of the AND gate produces an output Q from the Flip-Flop. The push button can be released because the Flip-Flop maintains its Q output even though $P \cdot A_1$ is OFF. When the cylinder is fully extended, A_2 is OFF, causing $\overline{A_2}$ to go ON switching the Flip-Flop to output $\overline{Q}$. This removes the signal to the DCV which retracts the cylinder. The push button must be pressed again to produce another cycle. If the push button is held depressed, the cycle repeats continuously.

11-43. $P = A \cdot (A + B)$
$P = A \cdot A + A \cdot B$
$P = A + A \cdot B$ (using Theorem 6)
Thus output P is ON when A is ON, or A and B are ON. Therefore control signal B (applied to valve 3) is not needed.

Chapter 12

12-1. One of the reasons for this trend is that more machines are being designed for automatic operation to be controlled with electrical signals from computers.

12-3. Limit switches open and close circuits when they are actuated at the end of the retraction or extension strokes of hydraulic or pneumatic cylinders. Push-button switches are actuated manually.

12-5. Timers are used in electrical control circuits when a time delay is required from the instant of actuation to the closing of contacts.

12-7. An indicator lamp is often used to indicate the state of a specific circuit component. For example, indicator lamps are used to determine which solenoid operator of a directional control valve is energized.

12-9. 1. Velocity transducer: senses the linear or angular velocity of the system output and generates a signal proportional to the measured velocity. 2. Positional transducer: senses the linear or angular position of the system output and generates a signal proportional to the measured position.

12-11. A servo valve replaces the flow control valve and directional control valve of an open-loop system.

12-13. Cylinder 1 extends. Cylinder 2 extends. Both cylinders remain extended until 1-SW opened. Then both cylinders retract together and stop.

12-15. When push-button switch 1-PB is actuated, coil 1-CR is energized. This closes normally open contact 1-CR, which energizes SOL A and holds. Thus, the cylinder extends until limit switch 1-LS is actuated. This opens the contacts of 1-LS, which deenergizes coil 1-CR. As a result, the contacts for 1-CR are returned back to their normally open mode, and SOL A is deenergized. This shifts the DCV back into its spring offset mode to retract the cylinder. If push button 2-PB is actuated while the cylinder is extending, the cylinder will immediately stop and then retract, because coil 1-CR is deenergized, which returns the contacts for 1-CR back to their normally open mode. This deenergizes SOL A, which shifts the DCV into its spring offset mode.

12-17. The transfer function of a component or a total system is defined as the output divided by the input.

12-19. Open loop gain is the gain (output divided by input) from the error signal to the feedback signal.

12-21. Repeatable error is the discrepancy between the actual output position and the programmed output position.

12-23. The forward path contains the amplifier, servo valve and cylinder. The feedback path contains the transducer.

12-25. 0.283 in./V

12-27. 0.565 cm/V

12-29. 0.286 cm

12-31. Unlike general purpose computers, a PLC is designed to operate in industrial environments where high ambient temperature and humidity levels may exist.

12-33. 1. Electromechanical relays have to be hard-wired to perform specific functions.
2. PLCs are more reliable and faster in operation.
3. PLCs are smaller in size and can be more readily expanded.

12-35. ROM memory can not be changed during operation or lost when electrical power to the CPU is turned off. RAM memory, which is lost when electrical power is removed, can be programmed and altered by the user.

12-37.

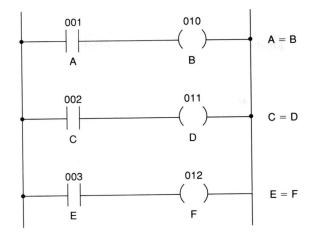

12-39.

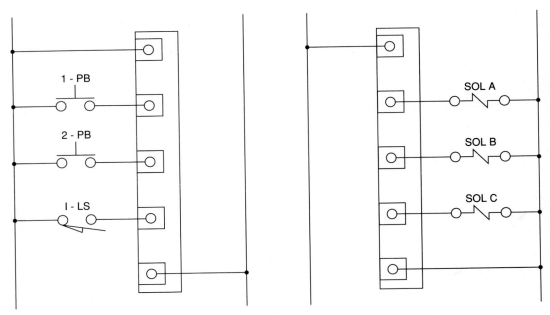

INPUT CONNECTION   OUTPUT CONNECTION

PLC LOGIC

12-41.

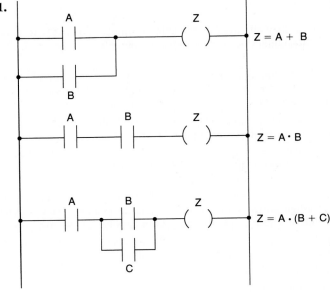

$Z = A + B$

$Z = A \cdot B$

$Z = A \cdot (B + C)$

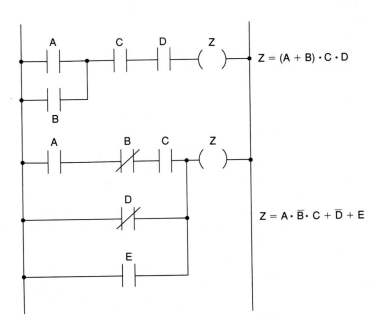

$Z = (A + B) \cdot C \cdot D$

$Z = A \cdot \overline{B} \cdot C + \overline{D} + E$

Chapter 13

13-1. 1. Clogged or dirty oil filters. 2. Inadequate supply of oil in the reservoir. 3. Leaking seals. 4. Loose inlet lines which cause the pump to take in air. 5. Incorrect type of oil.

13-3. 1. The type of symptoms encountered, how they were detected, and the date. 2. A description of the maintenance repairs performed. This should include the replacement of parts, the amount of downtime, and the date.

13-5. Internal leak: leakage past piston rings in hydraulic cylinders. External leak: leakage through pipe fittings which have become loose.

13-7. At very high pressures, O-rings may extrude into the clearance space between mating parts, as shown in Fig. 13-6. This extrusion is prevented by installing a backup ring, as illustrated in Fig. 13-6.

13-9. Wiper seals are not designed to seal against pressure. Instead they are designed to prevent foreign abrasive or corrosive material from entering a cylinder. As such they provide insurance against rod scoring and add materially to packing life.

13-11. A durometer (see Fig. 13-15) is an instrument used to measure the indentation hardness of rubber and rubberlike materials.

13-13. Reservoir size = 45 gal.

13-15. 1. Built into the system during component maintenance and assembly. 2. Generated within system during operation. 3. Introduced into system from external environment.

13-17. 1. Mechanical. 2. Absorbent. 3. Adsorbent.

13-19. 1. Proportional flow filter in separate drain line. 2. Full flow filter in suction line. 3. Full flow filter in pressure line. 4. Full flow filter in return line.

13-21. Downstream oil temperature = 145.9°F.

13-23. 1. Flowmeters. 2. Pressure gages. 3. Temperature gage.

13-25. 1. Air entering pump inlet. 2. Misalignment of pump and drive unit. 3. Excessive oil viscosity. 4. Dirty inlet strainer. 5. Chattering relief valve.

13-27. 1. Pump turning in wrong direction. 2. Ruptured hydraulic line. 3. Low oil level in reservoir. 4. Pressure relief valve stuck open.

13-29. 1. Air in system. 2. Viscosity of fluid too high. 3. Worn or damaged pump. 4. Pump speed too low. 5. Excessive leakage through actuators or valves.

13-31. OSHA stands for the Occupational Safety and Health Administration of the Department of Labor. OSHA is attempting to prevent safety hazards which can be harmful to the health and safety of personnel.

13-33. 68.7°C

13-35. Pumps do not pump pressure; instead, they produce fluid flow. The resistance to this flow, produced by the hydraulic system, is what determines pressure. Low oil level in the reservoir could be a cause of no pressure even though there is nothing wrong with the pump.

13-37. The purpose of a reservoir breather is to allow the reservoir to breathe as the oil level changes due to system demand requirements. In this way, the tank is always vented to the atmosphere.

13-39. Reservoir size $= 0.003$ m^3.

13-41. Cylinder friction is influenced by the type of materials in sliding contact, type of fluid lubricating sliding surfaces, and magnitude of the normal force between the mating surfaces.

13-43. Heat loss $= 19,000$ BTU.

13-45. Heat loss $= 48$ kJ/min.

13-47. Beta ratio $= 28.6$

13-49. Beta efficiency $= 1 - \dfrac{1}{\text{Beta ratio}}$

13-51. One.

Index

A

B

C